# Cereal Grains - Volume 2

*Edited by Aakash Kumar Goyal*

Published in London, United Kingdom

IntechOpen

*Supporting open minds since 2005*

Cereal Grains - Volume 2
http://dx.doi.org/10.5772/intechopen.100814
Edited by Aakash Kumar Goyal

Contributors
Muhammad Ikram, Haseeb-Ur-Rehman, Sanat Kumar Dwibedi, Mahendra Kumar Mohanty, Vimal Chandra Pandey, Donakonda Divyasree, Richard Olutayo Akinwale, Sripriya Das, Manoj Kumar Singh, Sneha Kumari, Manimala Mahato, M. Akhlasur Rahman, Hasina Khatun, Hosneara Hossain, M. Ruhul Amin Sarker, Khandakar M. Iftekharuddaula, M. Ruhul Quddus, M. Shahjahan Kabir, Ryusuke Oishi, Abu Mustapha Dadzie, Allen Oppong, Ebenezer Obeng-Bio, Marilyn L. Warburton, Jimmy Lamo, David Ocan, Desta Abebe, Zelalem Zewdu Ayalew, Anna Mlaki, Cyprien Ndikuryayo, Josefine Kant, Tobias Wojciechowski, Somchai Seviset, Songwut Egwutvongsa, Shyama R. Weerakoon, Veerendra Jaldhani, Durbha Sanjeeva Rao, Ponnuvel Senguttuvel, Bathula Srikanth, Puskur Raghuveer Rao, Desiraju Subrahmanyam, Damilola Omobowale Seyi-Amole, Abiodun A. Onilude, Yulin Jia, Melissa H. Jia



First published in London, United Kingdom, 2021 by IntechOpen
IntechOpen is the global imprint of INTECHOPEN LIMITED, registered in England and Wales, registration number: 11086078, 5 Princes Gate Court, London, SW7 2QJ, United Kingdom
Printed in Croatia

British Library Cataloguing-in-Publication Data
A catalogue record for this book is available from the British Library

Additional hard and PDF copies can be obtained from orders@intechopen.com

Cereal Grains - Volume 2
Edited by Aakash Kumar Goyal
p. cm.
Print ISBN 978-1-80355-615-4
Online ISBN 978-1-80355-616-1
eBook (PDF) ISBN 978-1-80355-617-8

For EU product safety concerns:
IN TECH d.o.o., Prolaz Marije Krucifikse Kozulić 3, 51000 Rijeka, Croatia, info@intechopen.com or visit our website at intechopen.com.

# Meet the editor

Aakash Goyal graduated with a degree in Biology from Maharshi Dayanand Saraswati University (MDSU), Ajmer, in 1999. He obtained a master's in Biotechnology with a specialization in Plant Biotechnology and Molecular Breeding from Guru Jambheshwar University of Science and Technology (GJUS&T), Hissar, in 2002, and a Ph.D. in Genetics and Plant Breeding with a specialization in Wheat Breeding from Chaudhary Charan Singh Universit (CCSU), Meerut, in 2007. After earning his Ph.D., Dr. Goyal completed a visiting fellowship with the Natural Sciences and Engineering Research Council of Canada (NSERC) and joined the wheat and triticale breeding program at Lethbridge Research Center, Agriculture and Agri Food Canada (AAFC), Lethbridge. In 2012, he received an honorable position of Wheat Breeder for Bayer Crop Science, Saskatoon, Canada. In 2014, he took a senior research scientist position with the International Center of Agriculture Research in Dry Areas (ICARDA). In 2017, he moved back to Canada and joined InnoTech Alberta as a Native Plant Research Scientist. Since November 2019 he has been working as an agriculture specialist with Palm Gardens, Inc. In July 2021, he joined RAYN Cultivation Inc., Edmonton, Alberta, Canada as Chief Executive Officer (CEO). Dr. Goyal has published ten books and fifty research papers, review articles, book chapters, and book reviews. He is also an elected fellow member of the International College of Nutrition (FICN) and Society of Applied Biotechnology (FSAB).

# Contents

# Preface

**Germplasm, Agronomy, Breeding and Bio-Waste Research in Cereal Grains**

Cereals crops such as wheat, rice, corn, barley, rye, oats, and millet are the base of the world's food supply. Not only are they important sources of human and animal feed but they are also useful for fuel production. Over the past 50 years, cereals have emerged as rapidly evolving crops because of new technologies and advances in agronomy, breeding, biotechnology, genetics, and so on. Population growth and climate change have led to new challenges, among which are feeding the growing global population and mitigating adverse effects on the environment. One way to deal with these issues is through sustainable cereal production. This book, the second installation in a two-volume work, discusses ways to achieve sustainable production of cereals via agronomy, breeding, transcriptomics, proteomics, and metabolomics. Chapters review research, examine challenges, and present future prospects in the field.

I would like to thank all the contributing authors for their outstanding efforts and timely work in producing such fine chapters. I greatly appreciate all the reviewers for their helpful comments. I would also like to thank the staff at IntechOpen, particularly Kristina Kardum Cvitan and Lucija Tomicic-Dromgool, for their assistance, advice, and encouragement during the development of this book. Lastly, I express my heartfelt thanks to my family for their love, encouragement, and vision that unveiled in me the desire to reach the highest mountain in everything I do.

**Aakash Kumar Goyal (Ph.D.)**
CEO,
RAYN Cltivation Inc.,
Edmonton, Canada

Section 1

# Germplasm

Chapter 1

# Characterization of Wild Rice - *Oryza* Species Complexes in Sri Lanka

*Shyama R. Weerakoon*

## Abstract

Rice is the staple food crop in Sri Lanka, which occupies 34% (0.77/million ha) of the total cultivated area. Sri Lanka currently produces 2.7 million tonnes of rough rice annually and satisfies around 95% of the domestic requirement. In Sri Lanka, genus *Oryza* consists of two species complexes, *O. sativa* (AA) and *O. officinalis* (CC). These two complexes are both pan tropical and have very similar overall distribution. Five wild rice species are reported in Sri Lanka, (*O. nivara* [AA], *O. rufipogan* (AA) *O. eichengeri* [CC], *O. rhizomatis* (CC) and *O. granulate* (GG). *O. rhizomatis* has been reported only in Sri Lanka and considered endemic to Sri Lanka. Recent studies demonstrated, the reliance on single source of information could mislead results in the phylogenetic inferences due to analytical inconsistency and biological processes. Therefore, exact number of wild rice species in Sri Lanka becomes uncertain and the necessity arises to assess *Oryza* species complexes in Sri Lanka using morphological, anatomical, and molecular information to enumerate number of species within each *Oryza* complex and characterization of species and species complexes. The study revealed, characterization of wild rice species, to a certain extent, can be made through morphological and anatomical characters, specially lamina anatomical characters. Molecular information is more reliable in delimitation of wild rice species complexes in Sri Lanka. *O. rhizomatis* and *O. eichingeri* (CC) are well separated from the rest of wild rice species (AA). Molecular data revealed, *O. nivara* and *O. rufipogon* have undergone independent evolution within Sri Lanka. Well separated five wild rice species are existing in Sri Lanka. Studies on ecological resilience of morphological, anatomical, and molecular studies are very useful for species enumeration of wild rice complexes in Sri Lanka. The findings led to conclude that wild rice species in Sri Lanka are "ecological swarms" and represents allopatric or sympatric populations. A comprehensive knowledge on genetic diversity and population structure of wild rice germplasm in Sri Lanka provides useful information to include these locally adapted and evolved wild rice species in rice crop improvement/breeding.

**Keywords:** Wild rice, *Oryza* species complexes, Sri Lanka

## 1. Introduction

Rice serves as the main staple food crop of nearly half of the world's population and it is obvious that genetic improvement of rice cultivars play an important role in the rice production for fulfilling ever increasing food demand. Rice is the staple food which occupies 34% (0.77/million ha) of the total cultivated area in Sri Lanka and currently produces 2.7 million tonnes of rough rice annually and satisfies around 95% of the domestic requirement [1].

The rice genus *Oryza* L. consists of *ca.* 21 wild and two cultivated species distributed in Asia, Africa, Australia, and the America [2, 3] and these species have been categorized into ten different genome types, such as six diploids (AA, BB, CC, EE, FF, and GG) and four allotetraploid species (BBCC, CCDD, HHJJ, and HHKK) [4, 5]. Wild rice spices are important in rice breeding programs because, these species comprise traits of agronomic interest, for example, the resistance and tolerance to biotic and abiotic stresses [2, 6–8]. However, due to the sterility barriers, most of the *Oryza* germplasm is of limited use in rice breeding programs [8, 9]. Genetic resources of the AA- genome group also referred to as the *Oryza* complex, have long been a focal point of the rice breeders.

The *Oryza sativa* complex includes eight diploid species [2] and the Asian cultivated rice consists of main subspecies, *O. sativa* ssp. *Indica* and *O. sativa* ssp. *Japonica* [10–12] are of Asian origin and globally cultivated today. The two presumed wild progenitors; the perennial *O. rufipogon* (**Figure 1**) is distributed throughout tropical Asia and Oceania, whereas the annual *O. nivara* is distributed in tropical continental Asia (**Figure 2**). Another cultivated species in the genus, *O. glaberrima*, was parallelly domesticated in West Africa where it is endemic [2]. There are two additional wild species also endemic to Africa, *O. barthii* and *O. longistaminata*. The former is the annual wild progenitor of *O. glaberrima*, while the latter is a perennial, rhizomatous and partially self-incompatible grass species [13].

In Sri Lanka, the genus *Oryza* consists of two species complexes, the *O. sativa* complex that includes the AA genome species, the *O. officinalis* complex which includes the CC genome species [3, 14] and a single species *O. granulate* (GG) [15, 16]. The two complexes, *O. sativa* complex and *O. officinalis* complex are both pan tropical and have very similar overall distribution. However, only AA genome species have been cultivated and domesticated. It appears that *O. officinalis* complex species do not have the attributes that make them attractive or likely to cultivate. Of the five wild rice species reported in Sri Lanka, (*O. nivara*, *O. rufipogan* (AA); *O. eichengeri*, *O. rhizomatis* (CC); and *O. granulate* (GG)), *O. rhizomatis* grows in partially shaded areas/grass lands and has been reported only in Sri Lanka and hence considered endemic to Sri Lanka [17, 18].

(a) (b)

**Figure 1.**
O. rufipogon *(a) panicle (b) growing in a periodically drying temporary ponds.*

**Figure 2.**
O. nivara *(a) panicle (b) growing along the border of a canal in Sri Lanka.*

*O. rhizomatis* is one of the species of the *O. officinalis* complex (**Figure 3**). The taxonomy of *O. officinalis* complex in Sri Lankan has been puzzling due to insufficiency of satisfactory herbarium specimens and the living plant materials. As an attempt of resolving the problem of the morphological variation in the complex, Biswal and Sharma [19] retracted the name *O. collina* and considered this taxon to be synonymous with *O. eichingeri*. Thus, Biswal and Sharma [19] agreed with both Bor [20] and Tateoka [14] that *O. eichingeri* is the sole representative of *O. officinalis* complex in Sri Lanka (**Figure 4**). *O. offocinalis* in Sri Lanka grows in both shaded and open habitats, whereas *O. eichingeri* grows in the shade of forests in Uganda [21]. However, taxonomists were not able to give much weight to the habitat of this taxon since field notes are generally infrequent.

The new collections make known clear morphological and habitat differences in *O. eichingeri* and it is a larger taxon which occurs in the drier habitats in Sri Lanka [2]. This larger rhizomatous taxon has previously been called *O. latifolia* and *O. officinalis*. *O. latifolia* is a large non-rhizomatous tetraploid from South and Central America with broader leaves and whorled panicle branches. *O. officinalis* which usually has rhizomes, has smaller spikelets, shorter palea tip, more branches of approximately equal length from the lowest panicle node, and spikelets inserted away from the base of primary branches. *O. officinalis* is also genetically different from this Sri Lankan taxon with which it can form sterile hybrids. However, Sri Lankan taxon belongs to the same genome group as both *O. officinalis* and *O. eichingeri*, which is CC [22].

There are two diploid CC genome species in Sri Lanka, *O. eichingeri* and *O. rhizomatis* [17, 19]. Previously *O. collina* was the name used for Sri Lankan germplasm of the *O. officinalis* complex [23]. *O. collina* has been used for both *O. eichingeri* and *O. rhizomatis*. However, *O. rhizomatis* is readily distinguished from *O. eichingeri* by its larger plant stature and rhizome formation. *O. rhizomatis* appears to be intermediate between *O. officinalis* and *O. eichingeri*. Analysis of the nuclear and chloroplast genome of *O. rhizomatis* by RFLP and SSR reveals that *O. rhizomatis* differed from *O. eichingeri* and *officinalis* [24–26].

The nomenclature and the taxonomy of the elements of these complexes have been studied and nomenclatural changes have been suggested and certain de novo species was described to disentangle the problem within the complexes. Due to this reason, the exact number of wild rice species in Sri Lanka becomes uncertain and

**Figure 3.**
*(a) Panicles (b) Rhizomes (c) well-spread rhizome submerged in water of* O. rhizomatis. *(d)* O. rhizomatis *in open spaces in the dry zone (Anuradhapura District), Sri Lanka.*

detailed studies specially, on morphological, anatomical, and molecular aspect of the Sri Lankan wild rice are needed for the delimitation of *Oryza* complexes in Sri Lanka.

Several recent studies demonstrated that the reliance on single source of information possibly misleading the results in the phylogenetic inferences due to analytical inconsistency and biological processes [27, 28]. The inconsistencies among the

**Figure 4.**
*Panicles of O. eichingeri in open spaces in the forest in the dry zone, Sri Lanka.*

phylogenies have become one of the most common problems during the reconstructing molecular phylogenetics using different datasets, such as individual genes. Studies carried on the genome-wide markers have witnessed new phylogenetic reconstructions that use large quantities of genome-wide markers to illustrate former controversies on evolutionary relationships at all taxonomic levels [27–31]. In general, a gene tree does not necessarily reflect a species tree, even if the orthology of marker genes are clearly identified and employed. Therefore, many genetic markers, including unlinked loci with extensive functional representation as well as intergenic genomic regions, are needed to comprehensively track organismal history. Such a robust phylogeny will build a foundation for future insights into rice genome evolution.

Therefore, there is a need to delimit the *Oryza* species complexes in Sri Lanka using morphological, anatomical, and molecular information. The objectives of the present study are to enumerate the number of species within each *Oryza* complex (*O. sativa* complex and the *O. officinalis* complex) in Sri Lanka and characterization of species and species complexes with evidence generated from morphological, anatomical, and molecular studies.

## 2. Materials and methods

### 2.1 Seed material

A total of four wild rice species; *O. rufipogon*, *O. nivara*, *O. eichingeri* and *O. rhizomatis* were collected from different localities of the Districts, Puttlam, Anuradhapura, Vavuniya, Trincomalee, Hambantota, Matara and Ampara of Sri Lanka. The botanical names and the acronyms used were given in **Table 1**. The collected samples were used for morphological, anatomical and molecular studies.

| Wild Rice species | Acronym |
|---|---|
| *Oryza eichingeri* | Eich |
| *O. nivara* | Niva |
| *O. rhizomatis* | Rhi |
| *O. rufipogon* | Rufi |

**Table 1.**
*Botanical names of the wild rice species and acronyms used in the study.*

## 2.2 Morphological studies

The morphological characterization of each species collected was based on the Plant Genetics Resource Centre (PGRC), Sri Lanka Characterization Catalogue of Rice [32] (**Table 2**). The leaf, culm, and rhizomes if available were collected and

| Character | Abbreviations |
|---|---|
| **Morphological characters** | |
| Plant Height (cm) | PLH |
| Leaf blade length (cm) | LBL |
| Leaf blade wigth (cm) | LBW |
| Leaf blade pubescence at late vegetative stage | LBP |
| Leaf blade color at late vegetative stage | LBC |
| Basal leaf sheath color at late vegetative stage | BLSC |
| Ligule length at late vegetative stage (cm) | LiguleL |
| Ligule color at late vegetative stage | LiguleC |
| Ligule shape at late vegetative stage | LiguleS |
| collar color at late vegetative stage | CollorC |
| Auricle color at late vegetative stage | AuricleC |
| Culm length (cm) | CulmL |
| Culm angle after flowering | CulmA |
| Internode color after flowering | IINCAF |
| Culm strength | CulmS |
| Panicle length | PanicleL |
| Panicle type | PanicleT |
| Panicle excretion | Panicleex |
| Awning after full heading | AWNAFH |
| Awn color at maturity | AAWNC |
| Apiculuscolor | ApiculeC |
| Seed coat (bran) color at maturity | SeedCC |
| Leaf senescence | LeafS |
| **Lamina Anatomical characters** | |
| Vein diameter (μm) | VD |
| Inter Venial distance (μm) | IVD |

| Character | Abbreviations |
|---|---|
| Vein width (μm) | VW |
| Vein height (μm) | VH |
| Leaf thickness (μm) | LTH |
| Height of mesophyll layer (μm) | MESOH |
| Width of mesophyll layer(μm) | MESOW |
| Bundle cell length(μm) | BCLEN |
| Bundle cell width (μm) | BCWIDT |

**Table 2.**
*Characters Observed for characterization of wild rice.*

processed for micro sectioning. Temporary and permanent slides were prepared for cross sections of leaves, culm and rhizomes.

### 2.3 Anatomical studies

The free hand sections of the collected specimens were taken and observed under the light microscope. Measurements of anatomical characteristic features were made using standard methods.

### 2.4 Molecular studies

Total genomic DNA was extracted from 7-day old seedlings of wild rice species; *O.rufipogon*, *O. nivara*, *O. eichingeri* and *O. rhizomatis* respectively using Promega Plant DNA extraction kit. A total of twelve SSR primer pairs were used (**Table 3**) for molecular study. SSR markers were obtained from Gramene (http://www.gramene.org/). All SSR PCR amplification reactions were carried out in a total volume 30 μl of which consist 1 x PCR buffer, 1 mM dNTPs, 2 μM SSR primers, 2 mM

| SSR | Chr | Fw5′–3′ | Rev5′–3′ |
|---|---|---|---|
| RM11 | 7 | tctcctcttcccccgatc | atagcgggcgaggcttag |
| RM14 | 1 | ccgaggagaggagttcgac | gtgccaatttcctcgaaaaa |
| RM19 | 12 | caaaaacagagcagatgac | ctcaagatggacgccaaga |
| RM21 | 11 | acagtattccgtaggcacgg | gctccatgagggtggtagag |
| RM44 | 8 | acgggcaatccgaacaacc | tcgggaaaacctaccctacc |
| RM55 | 3 | ccgtcgccgtagtagagaag | tcccggttattttaaggcg |
| RM84 | 1 | taagggtccatccacaagatg | tgcaaatgcagctagagtac |
| RM211 | 2 | ccgatctcatcaaccaactg | cttcacgaggatctcaaagg |
| RM219 | 9 | cgtcggatgatgtaaagcct | catatcggcattcgcctg |
| RM253 | 6 | tccttcaagagtgcaaaacc | gcattgtcatgtcgaagcc |
| RM280 | 4 | acacgatccactttgcgc | tgtgtcttgagcagccagg |
| RM289 | 5 | ttccatggcacacaagcc | ctgtgcacgaacttccaaag |

**Table 3.**
*SSR markers used for the molecular studies.*

$MgCl_2$, 50 ng of genomic DNA and 0.5 Units of *Taq* DNA polymerase. SSR alleles were resolved on Poly Acrylamide Gel. The SSR banding patterns were identified using Poly Acrylamide Gel Electrophoresis (PAGE).

### 2.5 Analysis of Data

Gathered data were analyzed with univariate, bivariate and multivariate statistical procedures. Suitable statistical software was employed in the analysis of data. In addition, data mining analysis were also attempted for the data gathered from the study to reduce the noise in the data set.

Molecular data were analyzed using Genemapper 4.1 software and SSR profiles were analyzed using PowerMarker 3.25.

## 3. Results

### 3.1 Morphological studies

The mean values of the parametric morphological measurements of wild rice species are given in **Table 4**. According to the table, the species *O. rufipogon* indicated highest mean for the plant height (153.23 cm) and the culm length (94.11 cm) and minimum plant height was observed in *O. eichingeri* (99.25 cm). Similarly, the highest leaf length and breadth were found in the samples of *O. nivara* and narrow leaves were occurred in samples of *O. rufipogon*. The variation of ligule length indicated that *O. nivara* possessed a higher ligule length with respect to other species included in the study. The summary of the ANOVA carried out on the parametric lamina morphological characters are shown in **Table 5**, except ligule length, panicle length, the rest of the characters are significantly varying across the wild rice species.

The association of the non-parametric characters with wild rice species included in the study is shown in **Table 6**. The characters such as leaf blade pubescent, awn after full heading and intermodal color after full heading are not significantly differ across the species ($p > 0.05$). However, the rest of the characters are significantly associated with the wild rice species and are of potential characters in separating wild rice species.

| Species | PLHEI (cm) | LBL (cm) | LBW (mm) | LIGULEL (mm) | CULML (cm) | PANICLEL (cm) |
|---|---|---|---|---|---|---|
| Eich | 99.25 | 41.33 | 10.73 | 9.25 | 90.32 | 22.75 |
| | 10.84 | 6.50 | 1.20 | 6.15 | 11.11 | 5.97 |
| Niva | 140.30 | 52.00 | 10.50 | 11.75 | 120.78 | 26.33 |
| | 10.50 | 6.68 | 0.58 | 1.50 | 21.91 | 2.87 |
| Rhi | 116.08 | 48.75 | 5.58 | 7.00 | 119.05 | 21.40 |
| | 3.28 | 2.99 | 1.06 | 3.37 | 1.07 | 0.66 |
| Rufi | 153.23 | 38.56 | 4.01 | 6.31 | 94.11 | 23.66 |
| | 7.17 | 6.80 | 4.52 | 6.46 | 14.37 | 2.00 |

**Table 4.**
*Summary of the parametric morphological characters of the four wild rice species (Mean value and standard deviation below mean value).*

| Character | Sum of Squares | df | Mean Square | F | Sig. |
|---|---|---|---|---|---|
| PLHEI | 11727.01 | 3 | 3909.004 | 54.527 | S |
| LBL | 647.173 | 3 | 215.724 | 5.522 | S |
| LBW | 220.522 | 3 | 73.507 | 7.972 | S |
| LIGULEL | 94.823 | 3 | 31.608 | 1.065 | NS |
| CULML | 3951.715 | 3 | 1317.238 | 6.74 | S |
| PANICLEL | 53.154 | 3 | 17.718 | 1.423 | NS |

***Note: S = Significant at $p < 0.05$, NS = Not significant, $p > 0.05$.*

**Table 5.**
*Summary of the ANOVA performed on the parametric morphological characters of the four wild rice species.*

| Character | $\chi^2$ Value | df | Sig. |
|---|---|---|---|
| LBP | Constant | | |
| LBC | 23.00 | 3 | S |
| BLC | 4.97 | 3 | NS |
| LiguleC | 23.00 | 3 | S |
| LiguleS | 23.00 | 3 | S |
| CollorC | 13.55 | 3 | S |
| AuricleC | 23.00 | 3 | S |
| AWNAFH | 13.01 | 6 | NS |
| AWNC | 4.55 | 6 | S |
| ApiculeC | 7.53 | 6 | S |
| SeedCC | 18.59 | 3 | S |
| LasfS | 39.25 | 12 | S |
| CULMA | 22.85 | 6 | S |
| INCAF | 7.16 | 6 | NS |
| CulmS | 24.28 | 9 | S |
| Panilceexer | Constant | | |

***Note: S = denote statistically significant difference; NS = Not significant.*

**Table 6.**
*Result of $\chi^2$ test performed on the non-parametric morphological characters of the wild rice species included in the study.*

A total of three clusters were resulted from the cluster analysis of morphological characters (**Figure 5**) and species were grouped under each cluster with respect to their similarities. The samples of *O. nivara* and *O. rufipogon* were intermingled and separated into two groups. Meanwhile the samples of *O. eichingeri* and *O. rhizomatis* well-separated from 80% similarity level and from rest of the clusters representing two populations. However, one sample of *O. eichingeri* was grouped with O. *rhizomatis.* The phylogenetic tree (**Figure 6**) constructed by morphological characters clearly showed a well separated cluster of *O. rhizamatis*. The samples of *O. nivara* and *O. rufipogon* were intermingled and separated into four groups. Findings of the study led to conclude that wild rice species in Sri Lanka are "ecological swarms" and represents allopatric or sympatric populations.

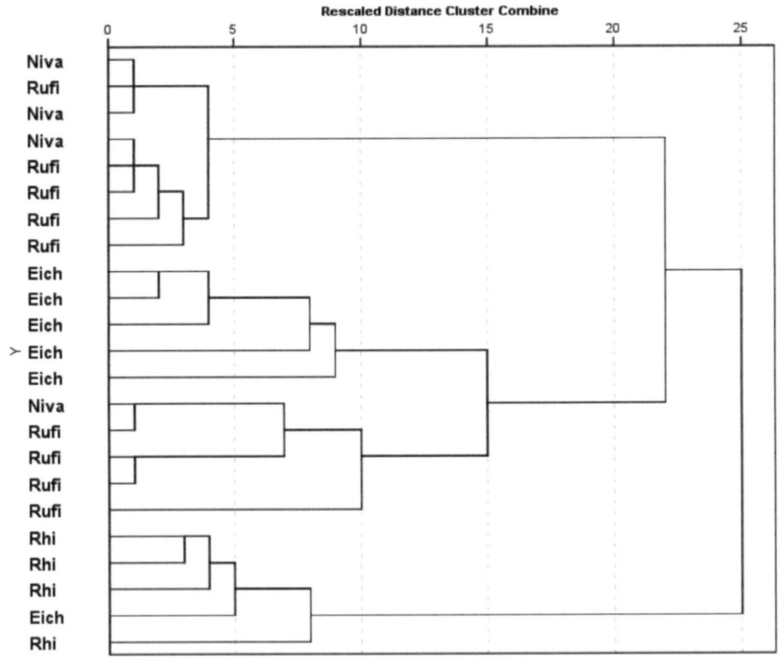


**Figure 5.**
*Dendrogram produced by cluster analysis of 22 morphological characters of wild rice species*, O. nivara, O. rufipogan, O. rhizomatis *and* O. eichingeri.

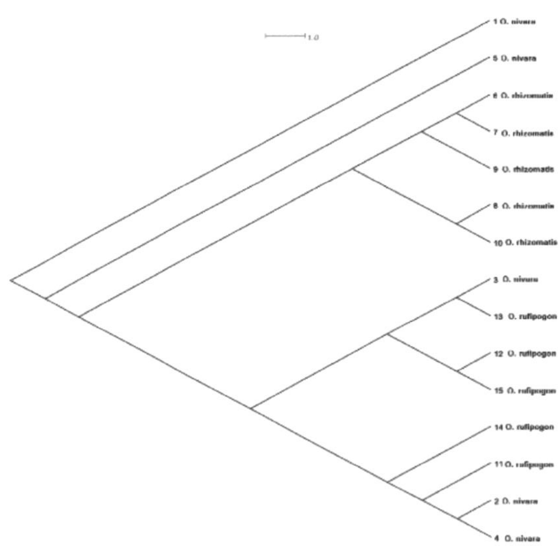


**Figure 6.**
*Phylogenetic Tree constructed by 22 morphological characters of wild rice species,* O. nivara, O. rufipogan, O. rhizomatis.

## 3.2 Anatomical studies

The variation of anatomical characters, especially the laminar anatomical features across the wild rice species are given in **Table 7**. Comparatively, the magnitude of mean values of bundle sheath cell width indicated a considerable variation between the wild rice species *O. eichingeri* (11.77 μm) and *O. rufipogon* (10.74 μm). The summary of the ANOVA (**Table 8**), indicated that the all the anatomical

| Species | VD | IVD | VW | VH | LTH | MESOH | MESOW | BCLEN | BCWIDT |
|---|---|---|---|---|---|---|---|---|---|
| Eich | 6.15 | 181.48 | 20.78 | 24.32 | 75.72 | 12.62 | 6.75 | 11.78 | 11.77 |
| | 0.80 | 11.79 | 1.53 | 2.76 | 4.43 | 0.70 | 1.88 | 0.95 | 1.70 |
| Niva | 4.33 | 207.08 | 29.65 | 37.03 | 87.68 | 12.33 | 7.65 | 10.18 | 9.85 |
| | 0.25 | 6.37 | 2.99 | 6.82 | 5.28 | 1.31 | 1.04 | 0.79 | 0.58 |
| Rhi | 5.65 | 155.53 | 25.00 | 28.48 | 68.60 | 12.10 | 3.63 | 10.55 | 8.55 |
| | 0.82 | 1.32 | 1.07 | 1.14 | 5.80 | 0.62 | 0.93 | 0.91 | 1.05 |
| Rufi | 4.91 | 210.90 | 28.91 | 34.41 | 85.68 | 12.49 | 7.97 | 11.32 | 10.74 |
| | 0.45 | 6.82 | 2.15 | 5.59 | 5.93 | 0.83 | 0.92 | 1.46 | 1.15 |

**Table 7.**
*Summary of the parametric lamina anatomical characters of the wild rice species (Mean value and standard deviation below mean value).*

| Character | Sum of Squares | df | Mean Square | F | Sig. |
|---|---|---|---|---|---|
| VD | 9.953 | 3 | 3.318 | 8.978 | S |
| IVD | 10089.31 | 3 | 3363.102 | 53.518 | S |
| VW | 295.528 | 3 | 98.509 | 23.708 | S |
| VH | 542.508 | 3 | 180.836 | 7.955 | S |
| LLTH | 1150.885 | 3 | 383.628 | 12.917 | S |
| MESOH | 0.72 | 3 | 0.24 | 0.319 | NS |
| MESOW | 55.442 | 3 | 18.481 | 11.584 | S |
| BCLEN | 7.877 | 3 | 2.626 | 1.914 | NS |
| BCWIDT | 27.049 | 3 | 9.016 | 5.856 | S |

**Note: S = denote statistically significant difference; NS = Not significant.*

**Table 8.**
*Summary of the ANOVA performed on the laminar anatomical characters of the four wild rice species.*

characters except mesophyll height and bundle sheath height. The anatomy of the culm and leaf sheath of wild rice species indicated that the characteristic features of the structures reflect the habitat conditions (**Figures** 7 and **8**).

The result of the cluster analysis of anatomical characters of wild rice species is shown in **Figure 9**. Comparatively, the Dendrogram resulted from the anatomical

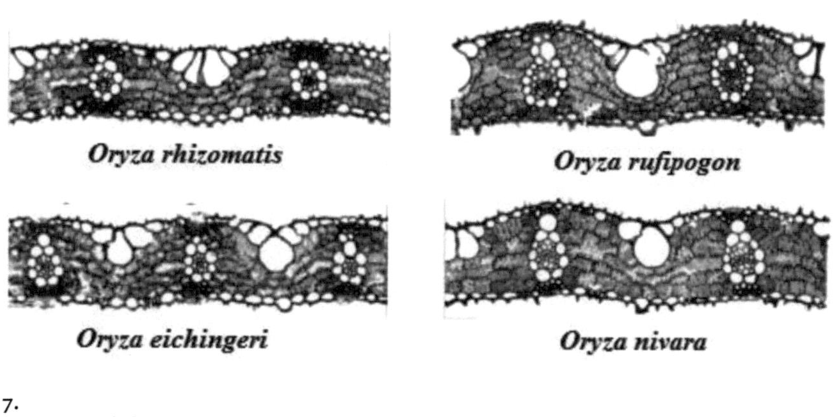


**Figure 7.**
*Laminar anatomical characters of 4 wild rice species collected during the study.*

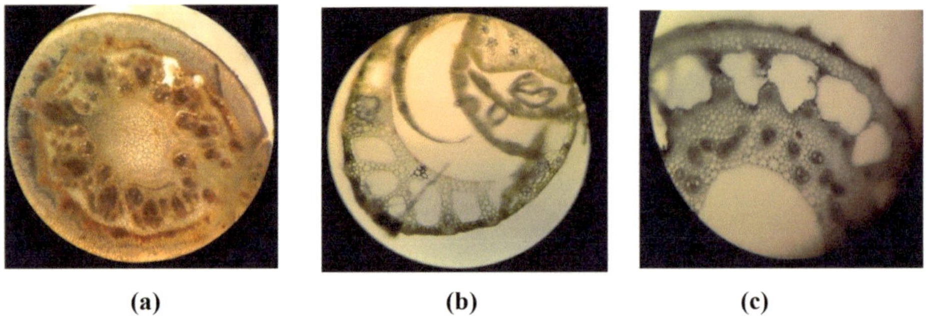


**Figure 8.**
*(a) The transverse section of culm of* O. rhizomatis *(b) the section through a portion of* O. rufipogon *culm encircled by leaf sheath (c) the culm section of* O. nivara.

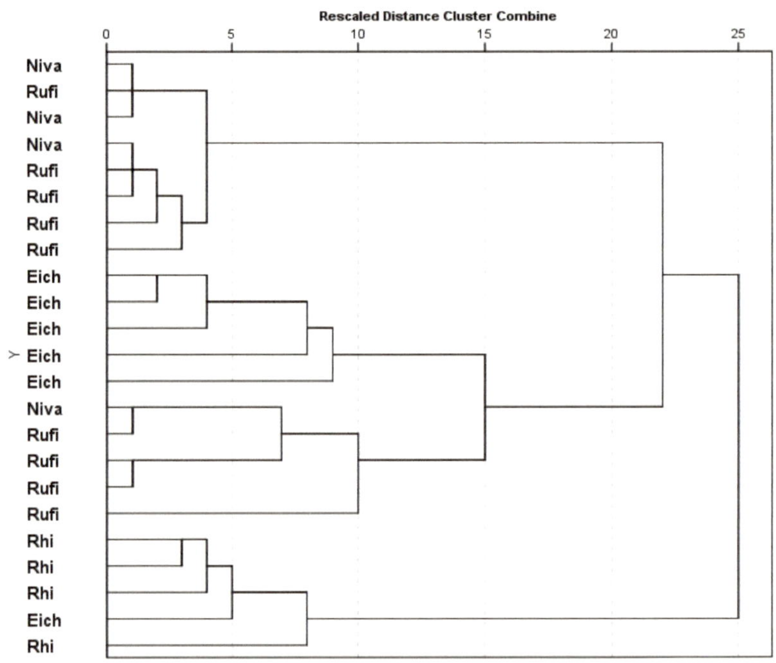


**Figure 9.**
*Dendrogram produced by cluster analysis of anatomical characters of wild rice species,* O. nivara, O. rufipogan, O. rhizomatis *and* O. eichingeri.

features indicated that anatomical characters well-separate the samples of each wild rice species. The samples of *O. rhizomatis* formed a unique group at similarity level of 80%. The pattern of the sample grouping was similar to the results obtained from the cluster analysis of morphological characters. However, samples were homogenized representing each wild rice species by pure tree branch.

The dendrogram resulted from the morphological and anatomical characters are shown in **Figure 10**. The grouping pattern of wild rice samples obtained from the analysis of morphological characters and anatomical characters reflect the same pattern observed in previously (**Figures** 5 and **9**).

### 3.3 Molecular studies

A total of three clusters were resulted from the cluster analysis of molecular data (**Figure 11**) and species were grouped under each cluster with respect to

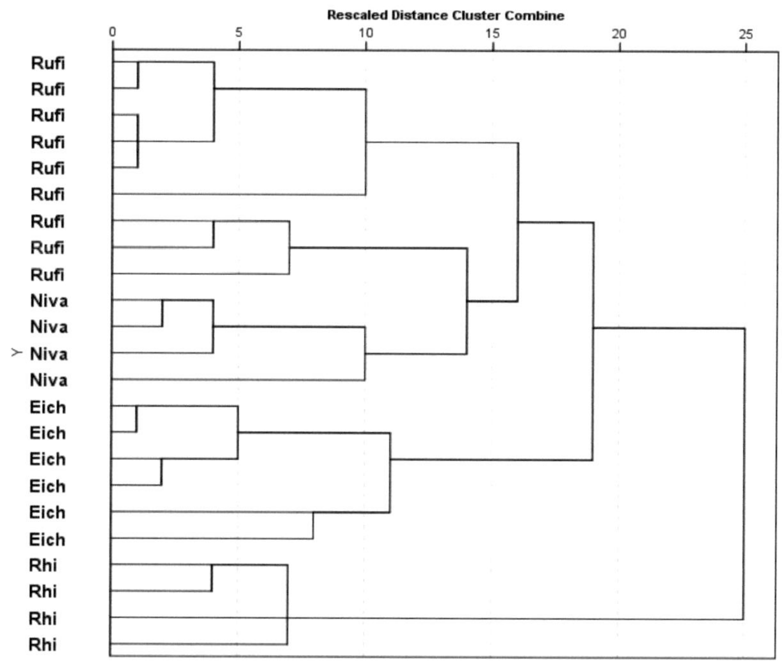


**Figure 10.**
*Dendrogram produced by cluster analysis of morphological and anatomical characters of Wild rice cultivars,* O. nivara, O. rufipogan, O. rhizomatis *and* O. eichingeri.

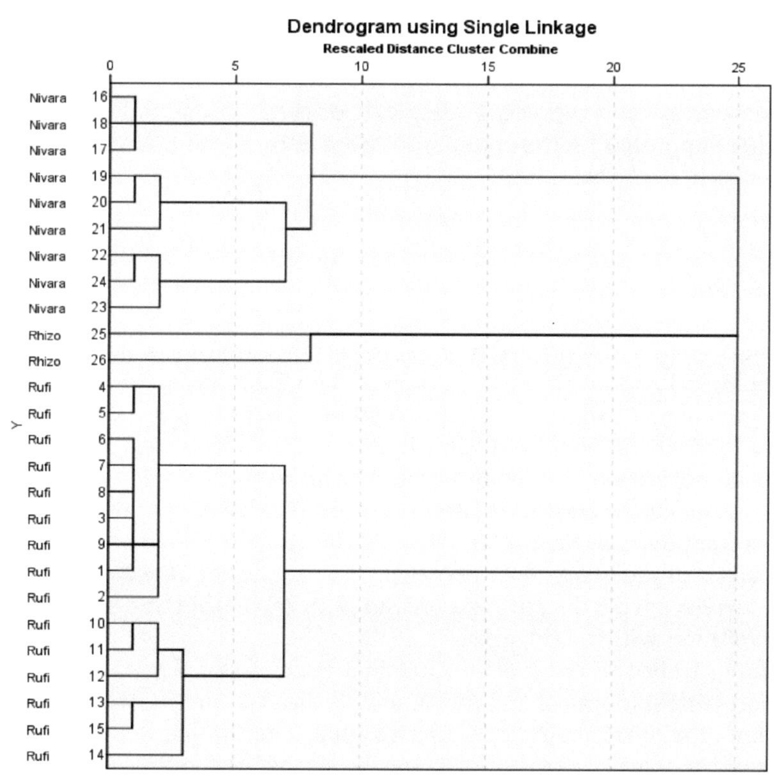


**Figure 11.**
*Dendrogram produced by cluster analysis of molecular data of wild rice species,* O. nivara, O. rufipogan *and* O. rhizomatis.

their genetic similarities. The samples of *O. nivara*, *O. rufipogon* and *O. rhizomatis* were very well separated from 40% similarity level confirming their distant relationship with each other and of independent evolution within Sri Lanka.

## 4. Discussion

The morphological and anatomical characters were investigated in relation to the species identification and delimitation of wild rice species complex in the country. The results of the morphological characters have indicated that they were useful in identification of wild rice species. However, the ecological resilience of the morphological characters is to be investigated before reaching a firm conclusion on the diagnostic value of the morphological characters. Compared to the morphological characters, the anatomical characters especially, lamina and culm anatomical characters are also indicted higher potential identification of species and delimitation of the wild rice species in each complex. Both morphological and anatomical characters can be used to separate the *O. rhizomatis* and *O. eichingeri* (CC) from the rest of wild rice species (AA). Further, based on both morphology and anatomy, *O. rhizomatis* can be distinguished from *O. eichingeri.* This finding suggests that species status of these two species deserved to maintain for further confirmation by molecular characterization. As far as the samples of two wild rice species of AA, *O. nivara* and *O. rufipogon* is concerned, there were considerable overleaps with respect to morphology and anatomy. However, the analysis of molecular data revealed that samples of *O. nivara*, *O. rufipogon* and *O. rhizomatis* have a distant relationship with each other and undergone independent evolution within Sri Lanka.

Finding of the study led to conclude that wild rice species in the island are "ecological swarms" and represents allopatric or sympatric populations. This finding is further supported by the connotations made by Nelson on the genus *Oryza* and its species in Sri Lanka [33].

## 5. Conclusions

The identification of wild rice species, to certain extent, can be made through the morphological and anatomical characters. The delimitation of the species complexes also achieved through the morphology and anatomy specially lamina anatomical characters. The nodal and culm anatomical characters are of limited value in the species identification and delimitation of wild rice species complexes.

However, molecular characterization is more reliable in characterization of wild rice species complexes in Sri Lanka.

The analysis of molecular data revealed that samples of *O. nivara*, *O. rufipogon* and *O. rhizomatis* have a distant relationship with each other and undergone independent evolution within Sri Lanka.

Therefore, studies on the ecological resilience of morphological characters in combination with anatomical and molecular studies are very useful for species enumeration of wild rice complexes in Sri Lanka. The finding led to conclude that wild rice species in Sri Lanka are "ecological swarms" and represents allopatric or sympatric populations.

A comprehensive knowledge on genetic diversity and population structure of wild rice germplasm in Sri Lanka provides useful information to include these locally adapled and evolved wild rice species in rice crop improvement and breeding programmes.

## Acknowledgements

The author wishes to thank Faculty of Natural Sciences Research Grant, The Open University of Sri Lanka for financial support. Statistical analyses of data by Dr. S. Somaratne and technical assistance by Mr. K. C. M. Liyanage is gratefully acknowledged.

## Author details

Shyama R. Weerakoon
The Open University of Sri Lanka, Nawala, Sri Lanka

*Address all correspondence to: shyamaweerakoon@gmail.com

## References

[1] Fertilizer Levels. https://www.doa.gov.lk/rrdi/index.php?option=com_sppagebuilder&view=page&id=42&lang=en. (Retrieved on 2019.09.06)

[2] Vaughan DA. 1989. The genus Oryza L. current status of taxonomy. IRRI Res. Pap. Ser. 1989; 138:21.europepmc.org › article › agr › ind90027658.

[3] Vaughan DA, Morishima M. Biosystematics of the genus Oryza. In: Smith CW, Dilday RH, editors. Rice: Origin, History, Technology, and Production. Wiley Series in Crop Science, John Wiley & Sons, Inc., New Jersey; 2002. p. 27–65.

[4] Aggarwal R, Brar D, Khush G. Two new genomes in the Oryza complex identified on the basis of molecular divergence analysis using total genomic NA hybridization. Mol. Gen. Genet. 1997;254:1–12. https://link.springer.com/article/10.1007/s004380050384.

[5] Ge S, Sang T, Lu BR, Hong DY. Phylogeny of rice genomes with emphasis on origins of allotetraploid species. Proc Natl Acad Sci USA. 1999; 96:14400–14405. DOI:10.1073/pnas.96.25.14400.

[6] Chang TT. The origin, evolution, cultivation, dissemination, and diversification of Asian and African rices. Euphytica 1976;25:425–441. https://link.springer.com/article/10.1007/BF00041576.

[7] Sitch LA, Dalmacio RD, Khush GS. Crossability of wild Oryza species and their potential use for improvement of cultivated rice. Rice Genet. Newsl. 1989; 6:58–59. https://shigen.nig.ac.jp/rice/oryzabase/asset/rgn/vol6/v6p58.html.

[8] Khush GS. Origin, dispersal, cultivation and variation of rice. Plant Mol. Biol. 1997;35: 25–34. https://pubmed.ncbi.nlm.nih.gov/9291957/.

[9] Piegu B, Guyot R, Picault N, Roulin A, Saniyal A, Kim H, Collura K, Brar DS, Jakson S, Wing RA, Panaud O. Doubling genome size without polyploidization: dynamics of retrotransposition-drivengenomic expansions in Oryza australiensis, a wild relative of rice. Genome Res. 2006;16: 1262–1269. http://www.genome.org/cgi/doi/10.1101/gr.5290206.

[10] Nayar NM. Origin and cytogenetics of rice. Advances in Genetics. 1973;17: 153–292. DOI:https://doi.org/10.1016/S0065-2660(08)60172-8.

[11] Oka HI. Origin of cultivated rice. Japan Sci. Soc. Press/Elsevier, Tokyo/Amsterdam; 1988. ISBN 0444989196.

[12] Morishima H, Sano Y, Oka HI. Evolutionary studies in cultivated rice and its wild relatives. Oxford Surveys in Evolutionary Biology. 1992; 8:135–184. https://www.google.com/search?q=Evolutionary+studies+in+cultivated+rice+and+its+wild+relatives.+Oxford+Surveys+in+Evolutionary+Biology

[13] Ghesquiere A. Evolution of Oryza longistaminata. Rice Genetics Collection Rice Genetics I. 2008. p. 15–25. DOI: 10.1142/9789812814265_0002.

[14] Tateoka T. Taxonomic studies of Oryza II. Several species Complex. Bot. Mag. Tokyo. 1962;75:455–461. www.jstage.jst.go.jp › article › jplantres1887 › _pdf.

[15] Liyanage, ASU, Hemachandra PV, Edimsinghe DK, Senevirathna SK, Takahashi J. Surveying and Mapping of Wild Species of Oryza in Sri Lanka. Jpn. J. Trop. Agr. 2002; 46(1):14–22. DOI: 10.11248/jsta1957.46.14

[16] Abhayagunasekara AVC, Bandaranayake PCG, Samarasinghe WLG, Pushpakumara DKNG. Diversity of wild

rice species in Sri Lanka: some reproductive traits. In: Proceedings of the 7th YSF Symposium, Colombo, Sri Lanka; 2018. P. 7–10. https://www.researchgate.net/publication/326998769_Diversity_of_wild_rice_species_of_Sri_Lankasome_reproductive_traits#fullTextFileContent.

[17] Vaughan DA. A new rhizomatous Oryza species (Poaceae) from Sri Lanka. Botanical Journal of the Linnean Society. 1990;103:159–163. https://doi.org/10.1111/j.1095-8339.1990.tb00182.x

[18] Somaratne S, Weerakoon SR, Siriwardana KGDI. Oryza rhizomatis Vaughan. In. Mondal TK, Henry RI, editors. The Wild Oryza Genomes, Compendium of Plant Genomes. Springer International Publishing AG; 2018. P. 263–269. DOI:10.1007/978-3-319-71997-9_23

[19] Biswal J, Sharma, SD. Taxonomy and phylogeny of Oryza collina. Oryza. 1987;24:24–29. https://agris.fao.org/agris-search/search.do?recordID=IN8800340.

[20] Bor NL. Grasses of Burma, Ceylon, India and Pakistan. Pergamon Press, London; 1960. https://www.cabi.org/isc/abstract/19611602837

[21] Tateoka T. Notes of some grasses. XVI. Embryo structure of the genus Oryza in relation to their systematics. Am J Bot. 1964; 51:539–543. DOI: 10.1002/j.1537-2197.1964.tb06667.x

[22] Jena KK, Khush GS. Introgression of genes from Oryza officinalis Well ext Watt. To cultivated rice, O. sativa L. Theoretical and Applied Genetics. 1990; 80:737–745. DOI:10.1007/bf00224186.

[23] Sharma SD, Shastry SVS. Taxonomic studies in genus Oryza L. III. O. rufipogon Griff. sensustricto and O. nivara Sharma et Shastry nom. Nov. Indian Journal of Genetics and Plant Breeding. 1965;25:157–167. grassworld.myspecies.info › content › taxonomic-studies

[24] Provan J, Corbett G, McNicol JW, Powell W. Chloroplast DNA variability in wild and cultivated rice (Oryza sativa) revealed by polymorphic chloroplast simple sequence repeats. Genome. 1997; 40:104–110. DOI: https://doi.org/10.1139/g97-014.

[25] Wang XK, Cai HW, Cheng KS. The discovery of an Est locus related to the origin, evolution and classification of Asian rice. Chinese Rice Res. Newslt. 1992;7:1–2. www.ricescience.org › abstract › abstract9333.

[26] Dally AM, Second G. Chloroplast DNA diversity in wild and cultivated species of rice (genus Oryza, section Oryza): ladistics-mutation and genetic-distance analysis. Theor. Appl. Genet. 1990;80:209–222. https://link.springer.com/article/10.1007/BF00224389.

[27] Wendel JF, Doyle JJ. Phylogenetic incongurence: Window into genome history and molecular evolution. In. Soltis DE, Soltis PS, Doyle JJ. editors. Molecular Systematics of Plants II: DNA Sequencing. Kluwer, Boston; 1998. p. 265–296. ISBN: 978–0–412-11121-1.

[28] Zou XH, Zhang FM, Zhang JG, Zang LL, Tang L, Wang J, Sang T, Ge S. Analysis of 142 genes resolves the rapid diversification of the rice genus. Genome biol. 2008;9:R49. http://genomebiology.com/2008/9/3/R49 G.

[29] Bapteste E, Brinkmann H, Lee JA, Moore DV, Sensen CW, Gordon P, Duruflé L, Gaasterland T, Lopez P, Müller M. The analysis of 100 genes supports the grouping of three highly divergent amoebae: Dictyostelium, Entamoeba, and Mastigamoeba. Proc. Natl. Acad. Sci. USA. 2002;99:1414–1419. DOI: 10.1073/pnas.032662799.

[30] Rokas A, Williams BL, King N, Carroll SB. Genome-scale approaches to

resolving incongruence in molecular phylogenies. Nature. 2003; 425:798–804. DOI:10.1038/nature02053.

[31] Cranston KA, Hurwitz B, Ware D, Stein L, Wing RA. Species trees from highly incongruent gene trees in rice. SystBiol. 2009;58:489–500. DOI: 10.1093/sysbio/syp054.

[32] PGRC. Characterization Catalogue of Rice (Oryza sativa). Department of Agriculture., Ministry of Agriculture and Lands, Sri Lanka; 1999.

[33] Harriman NA. (1994). Oryza. In: Flora of Ceylon. Dasanayake MD, Clayton WR, editors. Amerind Publishing Co. Pvt., Ltd., New Delhi, India; 1992. VIII: p. 326–330.

Chapter 2

# Economic Perspectives on Analysis of Ensuring Cereal Production and Consumption Security

*Ryusuke Oishi*

## Abstract

Cereals are essential for human nutrition. However, the ever-increasing world population makes it difficult to maintain the cereal production and consumption security. It is essential to overcome this situation by increasing cereal yield. An additional issue is the fact that while some countries suffer from hunger, a significant amount of food is discarded in others. This study analyses both production and consumption of cereals with the goal of ensuring food security. On the production side, many developing countries lack production capacity in contrast to developed countries. This is mainly due to a lack of capital, technology and human resources skills. In this study, we first theoretically demonstrated a cereal production gap between developing and developed countries. Second, we performed an empirical analysis to confirm the theoretical demonstration. On the consumption side, we focused on the cause of food loss and waste. We apply economic theories to demonstrate the situation where food loss and waste are occurring in the market. Then we introduced the related data to interpret the current world situation. Finally, we discussed the potential measures to improve cereal production and consumption.

**Keywords:** cereals, rice, wheat, maise, production, yield

## 1. Introduction

Cereals are essential for human nutrition. However, the ever-increasing world population makes it difficult to maintain the security of this food source. It is essential to overcome this situation by increasing cereal yield. An additional issue is the fact that while some countries suffer from hunger, a significant amount of food is discarded in others. This study analyses both production and consumption of cereals with the goal of ensuring food security.

On the production side, many developing countries lack production efficiency mainly due to a lack of capital and human resources skills. First, we theoretically demonstrated a cereal production gap between developing and developed countries. Second, we performed empirical analysis to confirm the theoretical demonstration. We also clarified how the production and consumption status of three typical cereals (wheat, rice and maize—or corn in United States English) are related to the regional food cultures.

On the consumption side, we investigated the cause of food loss and waste. We applied economic theories to demonstrate the market situation where food loss and waste are occurring. We then introduced the data relating to food loss and waste in G20 and the least developed countries in the world to interpret the current world situation. In addition, we also considered some cases of macroeconomic impacts on cereal productions (i.e. weather extremes and changes in commodity prices).

The remainder of this chapter is organised as follows. Section 2 provides the formal definition of food security. Section 3 provides both the theoretical and empirical analysis of cereal production. Section 4 analyses the causes of food loss and waste. Section 5 interprets the additional issues of cereal production (i.e. the macroeconomic impacts) and Section 6 concludes the chapter.

## 2. Food security

According to the Food and Agriculture Organisation of the United Nations (FAO), food security exists when all people—at all times—have physical, social and economic access to sufficient, safe and nutritious food that meets their dietary needs and food preferences for an active and healthy life [1]. The issue of food security has been raised as a major agenda at the United Nations (for example, Goal 2 of the Sustainable Development Goals: End Hunger, Achieve Food Security and Improved Nutrition and Promote Sustainable Agriculture) [2].

The above agenda is mainly due to the expected increase in the world population potentially make it difficult to ensure food security with the current level of food production. Overcoming this situation will require increasing the production capacity of nutritious foods for human consumption (especially wheat, rice and maize). However, the cereal production capacity varies across countries. In particular, there is more room for improvement in the crop yield in developing countries than in developed countries. Moreover, a significant amount of edible food is being discarded without being consumed in many developed countries whereas some developing countries concern food shortages. Eliminating such consumption waste can contribute to food security as much as increasing production capacity.

## 3. Cereal production

### 3.1 Theoretical analysis

In this section, we used a simple economic model to illustrate the problems of cereal production in developing countries and the methods for its improvement[1]. In order to highlight the problem of cereal production in developing countries, we comparatively analysed developing and developed countries. A farmer produces a product (i.e. cereals) by using labour and capital as demonstrated in Eq. (1).

$$Y_i = L_i^{\alpha_i} K_i^{1-\alpha_i} \tag{1}$$

Eq. (1) takes the form of a Cobb–Douglas production function, where $Y_i$ is an output of cereals. $i$ is a subscript that informs on whether the farmer lives in a developed or developing country (i.e. $i = R, i = P$ for developed and developing

[1] Because many readers of this paper are not expected to be specialised in economics, we will deal with a simple model that can easily be understood.

countries, respectively); $L_i$ denotes the labour in country $i$ and $K_i$ denotes the capital used to produce cereals (e.g. machinery). The degree of contribution of $L_i$ in producing cereals is measured by $\alpha_i$. We assume that $\alpha_R > \alpha_P$. Because capital is often exported from industrialised countries to the rest of the world, we assume that there is no significant difference in performance of capital between developing and developed countries. In this case, the degree of $\alpha_i$ is determined by the production capacity of $L_i$ relative to the performance of $K_i$. By considering its relationship to $K_i$, labourers in developing countries are often under trained and less productive than in developed countries.

Next, we investigated the relationship between labour and capital in production for further details. Specifically, by fixing the production of cereals at a certain level (i.e.$\overline{Y}_i$), the marginal rate of technical substitution of $i$ ($MRTS_i$) is derived as Eq. (2).[2]

$$MRTS_i = -\frac{\partial L_i}{\partial K_i} = \left(\frac{1-\alpha_i}{\alpha_i}\right)\left(\frac{\overline{Y}_i}{K_i}\right)^{\frac{1}{\alpha_i}} \tag{2}$$

According to Eq. (2), $MRTS_P$ is larger than $MRTS_R$. This indicates that, as compared to developed countries, introducing one unit of capital can replace a larger number of labourers to produce a given amount of cereal in developing countries. This is because, as compared to when there is enough farming machinery, productivity is more likely be improved by introducing one machine when there is a shortage of farming machinery.

Next, we set the farmer's budget constraints. To use labour and capital, wages and purchase/maintenance costs for capital resources were incurred, and the farmers are supposed to cover them out of their budgets. Eq. (3) shows the farmer's budget constraints.

$$B_i = L_i I_i + K_i C \tag{3}$$

$B_i$ is a farmer's budget in country $i$, $I_i$ is an income of $L_i$ and $C$ is the cost of purchasing and maintaining $K_i$. We assumed that $B_R > B_P$ and $I_R > I_P$ because, in most cases, the income level in developed countries is higher than in developing countries. In contrast, we did not distinguish $C$ in terms of subscript $i$. This is because capital is often exported from industrialised countries around the world, and hence its prices do not differ largely between countries.

By using Eqs. (1)-(3), we demonstrate the farmer's cereal production regarding to its first-order conditions. The first-order condition is derived by solving the problem with a Lagrange multiplier (Eq. (4)).

$$LF_i = L_i^{\alpha_i} K_i^{1-\alpha_i} + \lambda_i (B_i - L_i I_i - K_i C) \tag{4}$$

where $LF_i$ is the dependant variable of the Lagrange function in country $i$ and $\lambda_i$ is a Lagrange multiplier of the problem in country $i$, respectively.

To solve the problem, we first differentiate $LF_i$ with respect to $L_i$, $K_i$ and $\lambda_i$ shown in Eqs. (5)-(7), respectively.

$$\frac{\partial LF_i}{\partial L_i} = \alpha_i L_i^{\alpha_i - 1} K_i^{1-\alpha_i} - \lambda_i I_i \tag{5}$$

[2] The $MRTS_i$ measures the number of $L_i$ to be replaced by increasing one unit of $K_i$ to maintain a certain level of production (i.e. $\overline{Y}_i$).

$$\frac{\partial LF_i}{\partial K_i} = (1-\alpha_i)L_i^{\alpha_i}K_i^{-\alpha_i} - \lambda_i C \tag{6}$$

$$\frac{\partial LF_i}{\partial \lambda_i} = B_i - L_i I_i - K_i C \tag{7}$$

By using the first-order conditions of Eqs. (5)-(7), we can derive the optimal level of $L_i$ ($L_i^*$) and $K_i$ ($K_i^*$) as follows:

$$L_i^* = \frac{\alpha_i B_i}{I_i} \tag{8}$$

$$K_i^* = \frac{(1-\alpha_i)B_i}{C} \tag{9}$$

Eq. (8) specifies the optimal number of labourers to produce cereals in country $i$. By assuming that the difference between $B_i$ and $I_i$ in developing and developed countries is the same (i.e.$\frac{B_R}{I_R} = \frac{B_P}{I_P}$), the optimal number of labourers in developing countries is smaller than in developed countries.

Eq. (9) specifies the optimal amount of capital resources to produce cereals in country $i$. Due to the comparison of $B_i$ and $\alpha_i$ between developing and developed countries, we are unable to comparatively conclude the size of $K_R^*$ and $K_P^*$.

The optimal set of inputs $(L_R^*, K_R^*)$ and $(L_P^*, K_P^*)$ are illustrated in **Figures 1** and **2**, respectively.

The figures are two-dimensional, with the number of labourers in country $i$ ($L_i$) on the vertical axis and the quantity of capital in country $i$ ($K_i$) on the horizontal axis. The negative slope straight line in each figure illustrates the farmers' budget constraints in country $i$ ($B_i$).[3] The curves illustrate the farmers' production in country $i$ ($\overline{Y}_i$).[4] In this situation, the farmers' best option is to produce a certain amount of cereals by minimising the cost, which is reflected by the points where the production curve touches the budget constraint.

As illustrated in **Figures 1** and **2**, due to the small budget, the optimal amount of both inputs in developing countries is less than in their developed counterparts.[5] However, closer inspection reveals that the optimal amount of capital is relatively larger than the number of labourers in developing countries. This is due to the lack of skilled labourers and the fact that there is more room for productivity improvements through capital in the case of developing countries.[6] Even though this is the optimum input situation based on the model, the situation in reality differs. In many developing countries, development support has not sufficiently progressed, and agriculture is often carried out manually.

---

[3] The farmers are only able to purchase the set of inputs allocated inside (lower left) of the budget constraints.

[4] The curve of $\overline{Y}_i$ illustrates the sets of $L_i$ and $K_i$ to maintain the fixed amount of the farmer's production in country $i$. Moreover, slope of the curve is reflected by size of $MRTS_i$ (without the negative sign). Because $MRTS_P$ is larger than $MRTS_R$, slope of the curve in **Figure 2** is steeper than **Figure 1**.

[5] In this case, it is assumed that the difference in $B_i$ is larger than the difference in $\alpha_i$ between the countries.

[6] However, as the country develops, the optimal input of labour and capital changes. As the country develops, labourers acquire higher skills. Moreover, capital accumulation leaves less room for marginal productivity gains. Hence, the developing country's optimal condition converges toward that of developed countries.

DOI: http://dx.doi.org/10.5772/intechopen.96377

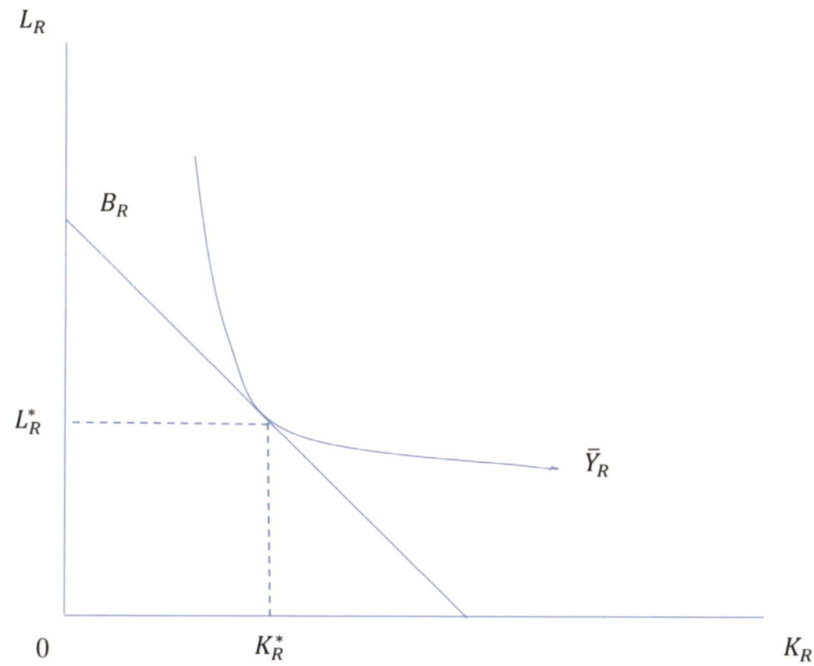


**Figure 1.**
*Optimal Set of Imputs in Developed Countries.*

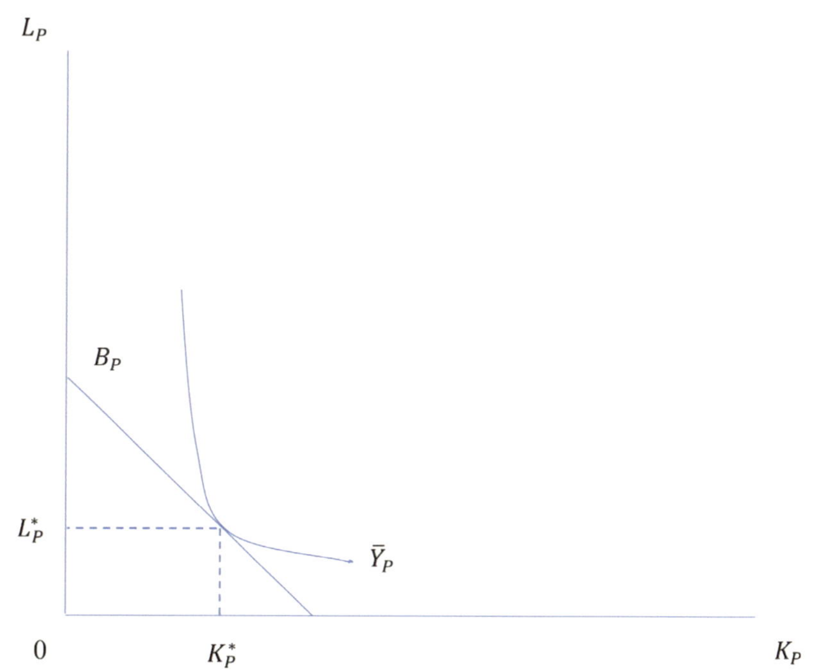


**Figure 2.**
*Optimal Set of Imputs in Developing Countries.*

### 3.2 Empirical analysis

Although we theoretically demonstrate the optimal input of cereal production in developing and developed countries, it does not reflect the actual situation. In this section, we investigate the empirical data to identify the discrepancies between the optimality condition ($L_i^*$ and $K_i^*$) demonstrated in 3.1 and reality.

*3.2.1 Data source*

The data employed for the empirical analysis was extracted from FAOSTAT.[7] Specifically, the production quantity (Element code: 5511) and domestic supply quantity (Element code: 5301) of wheat (Item code: 2511), rice (Item Code: 2805) and maize (Item code: 2514) in the group of 20 (G20) countries and the least developed countries in 2018 (Year code: 2018) are extracted from a domain of new food balance (Domain code: FBS).[8] The number of tractors (Element code: 5116 and Item code: 2455009) in the G20 and least developed countries in 2006 (Year code: 2006) is extracted from a domain of machinery (Domain code: RM).[9] The population (Element code: 511 and Item code: 3010) in the G20 and least developed countries in year 2018 were extracted from a domain of annual population (Domain code: OA). The food losses (Element Code: 5123) of wheat, rice and maize in the G20 countries and the least developed countries in 2018 are extracted from the domain of new food balance.[10]

*3.2.2 Data description*

**Tables 1** and **2** in the Appendix show the domestic production quantity and the production quantity of wheat, rice and maize, the number of tractors used for agriculture and the population in the world's least developed countries and the G20 countries. Close inspection of the tables reveals the problems related to cereal production and food security in developing countries.

First, in many developing countries, both the quantity of production and the domestic production of cereals are far below those of the G20, which indicates that developing countries lack cereal production capacity. Moreover, many G20 countries' production exceeds its domestic supply, indicating that those countries import cereals to cover shortages in their national consumptions. However, the opposite is true in the case of the least developed countries. A low level of production is indicative of threatened food security.

Second, although based on scarce information, the number of tractors in the G20 countries is significantly higher than in developing countries. As demonstrated in the theoretical analysis, abundant machinery contributes to the production of cereals. The lack of machinery likely causes the difference in the productivity between developed and developing countries.

Third, the production of wheat, rice and maize varies according to the country's region. For example, wheat production in many countries in **Table 1** is lower than that in countries in **Table 2**. This is because many developed countries are Western countries whose staple food is wheat (i.e. bread). On the contrary, many countries

---

7 FAOSTAT (http://www.fao.org/faostat/en/#data).

8 The production quantity is a value that considers import / export and change in stocks from domestic production quantity.

9 Due to a small data availability, the author decided to show the number of tractors in 2006, which reports more data than other years.

10 The author identifies G20 countries in the 'countries' and the least developed countries in the 'special group: Least Developed Countries > List) in the domains of the items (i.e. new food balance, machinery and annual population). We excluded the European Union (EU) from G20 because the latter is not a single nation.

in **Table 1** produce more rice. This is because rice is often consumed in Asia and Africa, and those regions include many countries in **Table 1**. Unlike the other two cereals, maize is actively produced in the countries listed in both **Tables 1** and **2**. This is because maize is often consumed in Africa as well as in Western countries.

### *3.2.3 Empirical analysis*

To make the discussion in 3.2.2 more reliable, we performed regression analyses. We confirmed the first (as compared to developed countries, developing countries lack cereal production capacity) and third argument (i.e. the production of wheat, rice and maize varies according to the country's region).[11]

$$\begin{aligned} Cereal_{cit} = {} & \beta_{c0} + \beta_{c1} G_{20i} + \beta_{c2} Africa_i + \beta_{c3} Asia_i + \beta_{c4} America_i + \beta_{c5} Europe_i \\ & + \sum_{t=2015}^{2018} Year_t + \varepsilon_{it} \end{aligned} \quad (10)$$

In Eq. (10), the dependent variable $Cereal_{cit}$ represents the domestic supply quantity of cereal $c$ in country $i$ at year $t$.[12] Subscript $c$ distinguishes three types of cereals (i.e. wheat, rice and maize). Subscript $t$ specifies the year in the sample period.[13] The explanatory variable $G_{20i}$ is a dummy variable that takes the value of 1 if country $i$ is a member of G20, and 0 otherwise. $Africa_i, Asia_i, America_i$ and $Europe_i$ are dummies that take the value of 1 if country $i$ is in that region and 0 otherwise.[14] $\beta_{c0}, \beta_{c1}, \beta_{c2}, \beta_{c3}, \beta_{c4}$ *and* $\beta_{c5}$ are a constant term, the coefficients for $G_{20i}, Africa_i,$ $Asia_i, America_i$ and $Europe_i$ in the case of regressing for cereal $c$, respectively. $\sum_{t=2015}^{2018} Year_t$ is the sum of the dummies between 2015 and 2018 for capturing the year fixed effect.[15] $\varepsilon_{it}$ is an error term for country $i$ at year $t$.[16]

**Table 3** presents the descriptive statistics for the data employed for the regression analysis of Eq. (10). As mentioned earlier, the G20 production of all cereals far exceeds that of the developing countries. The difference is particularly noticeable in maize production and smallest of the three (wheat, rice and maize) in rice production. This is because there are many countries that have rice as their staple food in poor regions such as Southeast Asia.

**Table 4** shows the estimation results of Eq. (10). Close inspection of the estimation results reveals some features that are consistent with the argument in 3.2.2. First, the estimated coefficients of $G_{20i}$ $(\hat{\beta}_{c1})$ in the case of all three cereal types are

---

[11] Due to the small amount of available data, we did not perform the regression analysis on the second argument (i.e. the number of tractors in the G20 countries is significantly higher than that in developing countries).

[12] We separately regressed Eq. (10) for the three cereals types (i.e. $Cereal_{Wheat\ i\ t},\ Cereal_{Rice\ i\ t}$ *and* $Cereal_{Maize\ i\ t}$). Although $i$ is used to distinguish developed or developing country (i.e. $i = R, i = P$) in subsection 3.1, in this subsection, $i$ is used to distinguish the countries in the dataset (i.e. $i =$ *the US*, $i =$ *the UK*, *etc* … ).

[13] Although **Tables 1** and **2** only show the 2018 data for each country, we decided to use the 2014–2018 data for the estimation. This is because a larger data size helps to perform a more significant quantitative analysis.

[14] To deal with collinearity, we dropped $Oceania_i$ from the regression equation.

[15] To deal with collinearity, we dropped $Year_{2014}$ from the regression equation.

[16] The data employed in subsection 3.2.3 are extracted from the source introduced in 3.2.1. The dummies were created by the author. The regional dummies $Africa_i, Asia_i, America_i, Europe_i$ and $Oceania_i$ were created by using FAOSTAT information: Regions (Africa > (List), Asia > (List), Americas > (List), Europe > (List), Oceania > (List)).

positively estimated with strong significance, indicating that G20 countries produce larger amounts of all three cereals than the least developed countries. Second, by looking at the estimation result of the regional dummies, differences in cereal production by region can be found. For example, in case of $Cereal_{Rice\ i\ t}$, the size of the estimated coefficient for $Asia_i$ $(\hat{\beta}_{Rice\ 3})$ is significantly larger than those for other regions. This finding makes sense given that rice is the staple food for people in Asian countries. Similarly, in the case of $Cereal_{Maize\ i\ t}$, the size of the estimated coefficient for $America_i$ $(\hat{\beta}_{Maize\ 4})$ is much larger than those for other regions. This is also understandable because people in the United States consume many foods made from maize. In case of $Cereal_{Wheat\ i\ t}$, estimation of the regional dummies is different from our expectation. Given that wheat-based foods are mainly consumed in Europe, the estimated coefficient for $Europe_i\,(\hat{\beta}_{Wheat\ 5})$ was expected to be larger than for other regions. However, $\hat{\beta}_{Wheat\ 3}$ is larger than that of $\hat{\beta}_{Wheat\ 5}$. This might be due to China's wheat production, that surpassed that in European countries. Finally, the low $R^2$ value in all three cases indicates that the explanatory ability of these estimation results is not high. This is because all the explanatory variables (except the constant) are dummy variables.

In order to comparatively analyse cereal production in the developed and developing countries in greater detail, in addition to Eq. (10), we conducted another estimation procedure. The production of cereals in a country depends on its population. By using per capita productivity, we can exclude the impact of the population. In order to reveal the latter issue, we set Eq. (11) as follows:

$$\begin{aligned} Cereal\ Per\ Capita_{cit} &= \beta_{c6} + \beta_{c7}G_{20i} + \beta_{c8}Africa_i + \beta_{c9}Asia_i + \beta_{c10}America_i \\ &\quad + \beta_{c11}Europe_i + \sum_{t=2015}^{2018} Year_t + \varepsilon_{it} \end{aligned} \tag{11}$$

Eq. (11) is different from Eq. (10) in that it uses $Cereal\ Per\ Capita_{cit}$ as a dependant variable. This variable was obtained by dividing $Cereal_{cit}$ by population of country $i$.[17]

**Table 5** shows the descriptive statistics for $Cereal\ Per\ Capita_{cit}$. Information in **Table 5** is significantly different from **Table 3**. Specifically, as compared from the developed country, the mean value of $Cereal\ Per\ Capita_{Rice\ i\ t}$ is higher in the least developing countries. This is probably because the large populations in developed countries used as a denominator value of $Cereal\ Per\ Capita_{Rice\ i\ t}$ lowered their value of $Cereal\ Per\ Capita_{Rice\ i\ t}$.

**Table 6** shows the estimation results of Eq. (11); the interpretation of these results requires careful consideration. In the case of wheat and maize, the coefficients of $G_{20i}$ $(\hat{\beta}_{Wheat\ 7}$ *and* $\hat{\beta}_{Maize\ 7})$ are positively significant, indicating that even if the effects of the population are eliminated, G20 countries are more productive than the least developed countries, which is consistent with our view. In contrast, in the case of rice, surprisingly, the estimation of $G_{20i}$ $(\hat{\beta}_{Rice\ 7})$ is negatively significant. This is probably because G20 countries with rice food cultures include populous countries (i.e. China and Indonesia).

In the case of rice, looking at the estimated result of the regional dummies shows that only $Asia_i$ $(\hat{\beta}_{Rice\ 9})$ is positively significant. This confirms that Asia stands out in

[17] The regression of Eq. (11) additionally uses population data of the countries in 3.2.1 (in the period between 2014 and 2018) to the data employed for the regression of Eq. (10). To distinguish from the coefficients in Eq. (10), the constant, coefficient of $G_{20i}, Africa_i, Asia_i, America_i$ *and* $Europe_i$ in Eq. (11) are expressed as $\beta_{c6}, \beta_{c7}, \beta_{c8}, \beta_{c9}, \beta_{c10}$ *and* $\beta_{c11}$, respectively.

terms of rice productivity. Similarly, in the case of maize, the estimated coefficient of $America_i$ ($\hat{\beta}_{Rice\ 10}$) is significantly larger than those of other regions, indicating that maize productivity in the American region is higher than in other regions. One unexpected observation is that, in the case of wheat, all regional dummies are estimated to be negative. Because the estimated coefficient for $Europe_i$ ($\hat{\beta}_{Wheat11}$) has the smallest negative value, we can argue that the per capita production of wheat in Europe is the largest; however, the reason for this is not clear.

Compared to **Table 4**, all the $R^2$ values are slightly higher in **Table 6**. Therefore, the explanatory power of the estimation results improved slightly in **Table 6**.

## 4. Cereal consumption

In Section 3, we describe both the theoretical and empirical analyses of the cereal supply. In this section, we identify the problems with and the methods for improving the consumption of cereals as food security issues. One might think that the only issue regarding food security is the improvement of production capacity; however, management of consumption is also important. Among many problems relating to cereal consumption, we focused on food loss and waste.

The food situation differs from country to country. While some countries suffer from poverty and lack of food, others have excessive food supply in their markets and dispose of consumable foods.

**Table 7** shows the amount of food lost in the least developed and G20 countries.[18] Not surprisingly, as compared to that in the least developed countries, the larger amount of food loss is reported in the G20 countries. Considering the larger production volume and the population, the loss in the G20 can be understood as a natural consequence of its food consumption. However, the developing countries are not completely devoid of food loss; although small comparatively, food loss is also reported in the least developed countries. However, reasons for food loss are different between the G20 and the least developed countries. In the case of the G20 countries, the cause of food loss is overstocking. For example, in Japan, consumers strongly demand food safety and quality [3]. As a result, foods that are not sold by the expiration date are discarded [3]. Moreover, consumable foods with slight scratches or incompatible size are discarded [3]. Another issue is that the main concern of business food suppliers is to make profit. When such economic agents compete with each other to form a market is a basis of capitalism. However, in order to make profits, they sometimes make decisions that are detrimental to society. For example, convenience stores in Japan often overstock foods, because they do not want to lose customers due to out of food stock [3].

In the case of developing countries, due to a lack of sufficient capital for processing and preserving food, some foods are rarely delivered to consumers while fresh. For example, in developing countries, most of the postharvest grains are stored in traditional storage structures, which cannot prevent insect infestation and mould during storage [4].

**Figure 3** illustrates an excess supply of food (caused by overstocking) in developed countries.[19] According to economic theory, if goods are traded at their market equilibrium (i.e. the intersection of the demand and supply curves), production and

[18] Because the food loss data is reported from a limited number of countries, the countries with no data were deleted from **Table 7**.

[19] $S$, $D$, $P$, $Q$, $Q_S$ and $Q_D$ in **Figures 3** and **4** represent supply, demand, price, quantity, quantity supplied and quantity demanded for a product in the market.

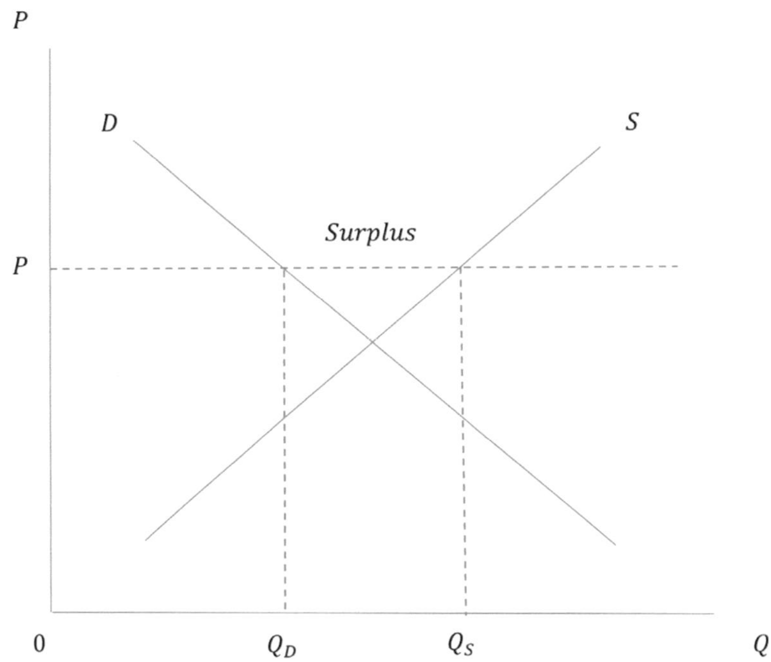


**Figure 3.**
*The Market with Oversupply of Food.*

consumption match and no goods will be left unsold. However, as shown in **Figure 3**, if the price of goods deviates from its equilibrium, supply and demand of goods do not match, and some goods are not sold.

**Figure 4** illustrates a food shortage situation in poor countries.[20] Unlike in **Figure 3**, the supply curve in **Figure 4** becomes vertical at a certain quantity ($Q_S$). This shows supplier's limit of cereal production capacity. Due to the limited production capacity in poor countries, once the supply reaches its limit, the quantity does not increase regardless of the price. Moreover, in this situation, the price of the goods deviates downward from its equilibrium. The discrepancy between supply and demand becomes the shortage.

In theory, eliminating oversupply of food and the assurance of food security in developing countries is mutually achievable. For example, transferring food surplus from developed countries to developing countries to combat the shortages would provide a solution to both problems.

## 5. Measures to ensure world food security

As discussed in Sections 3 and 4, there are the challenges to ensuring world food security both in food supply and consumption. On the supply side, the main reason for food insecurity is the lack of capital and training of human resources in developing countries. However, due to lack of funds, increasing capital and training human resources are hardly be achieved. Therefore, loans and assistance from international organisations and developed countries are required. Additionally, investment projects (i.e. foreign direct investment) are also effective in raising funds for developing countries. Furthermore, to gain an understanding of the

[20] Here, in order to focus on the issue of food security in developing countries, we ignored food loss in the developing countries.

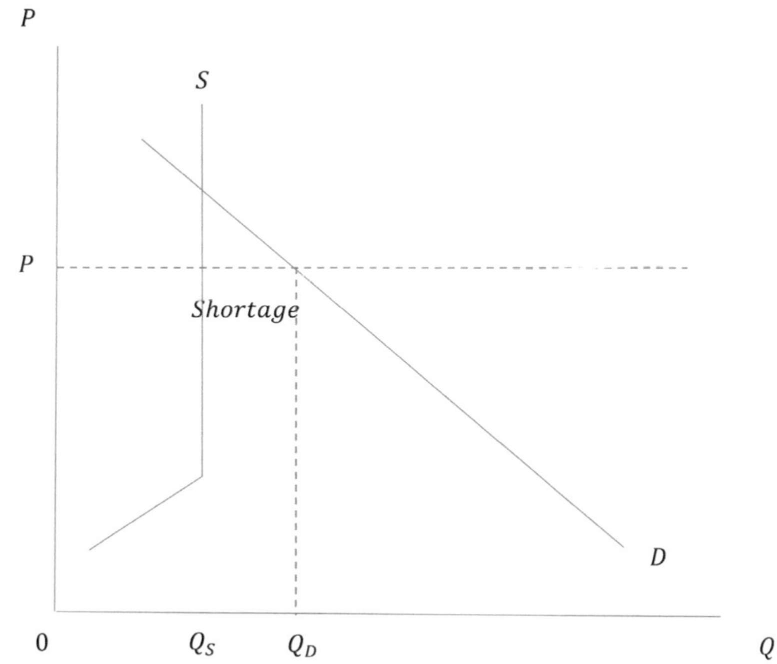


**Figure 4.**
*The Market with Shortage of Food.*

outcome of such investments is larger in developing countries than in developed countries will be important[21].

There are other points that need improvement. First, improvement in seed breeding is effective for sustainable agricultural production [5]. Second, in many cases, as compared to farmers with large operations, smallholder farmers are disadvantaged in accessing market; therefore, farming should be re-invented as an attractive local business opportunity for smallholder families [6]. Third, in the agri-food system, a few large international firms dominate the market share, and smallholders mainly trade on local short-supply chains [7]. Fourth, in recent years, production and consumption of rice in many parts of Asia has been steadily declining, whereas the opposite is true in Africa [8]. Fifth, due to economic development, the demand for maize to feed livestock has increased [9]. Six, the problem of food self-sufficiency does not apply only to developing countries. In Japan, due to the ageing population and the lack of young people willing to engage in agriculture, the self-sufficiency rate is low [10]. As a result, Japan imports a substantial amount of food from abroad (i.e. China) to cover their shortages [10].

Stabilisation of markets to protect suppliers from macroeconomic shocks is also important. For example, in 2020, the global economy was greatly affected by the spread of COVID-19. Because the main transmission route of COVID-19 is droplets from infected people, people stopped eating out. In Japan, food manufacturing and service businesses were severely damaged by people's restraints and many food manufacturers and restaurants went bankrupt [11]. This was mainly because in

21 This is because countries with small capital grow faster by capital investment than countries with large capital. However, on the other hand, we cannot ignore that investment in developing countries often involves higher risk than that in developing countries.

Japan many restaurants are small and medium-sized enterprises that lack corporate strength to withstand severe economic adversities [12]. As a result, a significant amount of ingredients that were to be delivered to restaurants went left unsold. In addition, a sharp drop in the demand for restaurant food has spurred the overstock of food, leading to the collapse of the prices of some foods. If the situation does not improve, farmers will not be able to make sufficient profits and will be forced out of business. This will lead to a decline in food production capacity and will worsen food insecurity.

In addition, inflation and instability of agricultural prices will put pressure mainly on the world's low-income group of people and an increase in labour productivity will be necessary to overcome the problem [13]. The effect of extreme weather is also a major issue for cereal production. Weather extremes in the United States in 2012 caused a sharp increase in the world maize price, and the poor countries with high maize import dependency were the most seriously affected [14].

On the consumption side, the main challenge is to reduce food loss in the developed countries. One measure to correct such market failure is government intervention in the market to internalise negative externalities. Additionally, transferring food surpluses from developed countries to developing countries to combat food shortages may improve the situation. Food banks are one of the potential measures for recycling food within a country [15]. If such food banks could be internationalised, food security in developed countries may be ensured in a way that is close to our idea. However, there are many difficulties with this potential solution. For example, some foods do not last long and are not suitable for long-distance transportation. Moreover, as discussed before, due to regional variances in food culture, some foods may be hard to replenish.[22] Many countries also impose tariffs on imported products because large amounts of imported food may adversely affect local farmers' businesses.

Additionally, food loss also occurs in developing countries, where food security is not ensured. The main cause of food loss here is lack of capital for storage; therefore, as in the supply side discussion, increasing capital through investment and assistance is the effective measure[23].

In this section, we examined measures to improve food security. Although the authors' proposed methods have the potential to improve food security, there are many challenges. In order to achieve these improvements, in addition to international efforts (i.e. international organisations and governments), changing the mindset of each economic entity (i.e. supplier and consumer) will be necessary.

## 6. Conclusion

Cereals are essential for human nutrition. However, the ever-increasing world population makes it difficult to maintain food security. It is necessary to consider various factors related to production and consumption of cereals to ensure food security. This study considers the food security issues for cereals in terms of both supply and consumption.

---

[22] South Asia, where rice is the staple food, is one of the most densely populated regions and the second poorest region in the world [16]. In order to guarantee its food security, self-sufficiency is important [16].

[23] Installation of evaporative coolers can be considered as one of the realistic measures to reduce food loss and waste [15].

On the supply side, we proposed that cereal food security can be improved by increasing production capacity in developing countries. The optimal condition of our theoretical model specifies that, as compared to the case of developed countries, farmers in developing countries should more actively utilise capital resources. However, reality largely differs from theory. As shown by the data, the use of farming machinery in developing countries is lower than in developing countries. Additionally, our regression result confirms that the production of cereals (i.e. wheat, rice and maize) in G20 countries surpasses that in the least developed countries. Furthermore, our empirical analysis also confirms that the type of cereals produced vary significantly regarding the region.

On the consumption side, we emphasised the problem of food loss. Although some countries in the world suffer from food insecurity, others discard large amounts of food. Reducing such waste can contribute to global food security. However, the causes of food loss differ between developing and developed countries. In the case of the former, the lack of capital means that food rots before it is consumed. In the case of the latter, foods are oversupplied to the market due to excessive competition between firms.

In order to mitigate these problems, we proposed the following measures. In order to increase production in developing countries, their capital resources need to increased; this requires significant investment and assistance. On the other hand, one way to prevent food loss in developed countries is to mitigate excessive competition among firms through government market intervention. It may also be effective to transport surplus food from developed countries to developing countries to fill shortages. However, there are many challenges to implementing these measures. To truly improve the global food situation, in addition to international efforts (i.e. international organisations and governments), each of us needs to change our consciousness.

## Acknowledgements

We are deeply grateful for the editor, Dr. Aakash Goyal. We are also grateful to IntechOpen and Meikai University, which supported our research in various ways.

## Appendix

| Country | Wheat | | Rice | | Maize | | Tractors | Population |
|---|---|---|---|---|---|---|---|---|
| | Product | Domestic Supply | Product | Domestic Supply | Product | Domestic Supply | | |
| Afghanistan | 3613 | 6995 | 352 | 626 | 107 | 165 | 711 | 37171.92 |
| Angola | 3 | 1111 | 10 | 725 | 2765 | 2996 | | 30809.79 |
| Bangladesh | 1099 | 6885 | 54416 | 56415 | 3288 | 4841 | | 161376.7 |
| Benin | 0 | 186 | 459 | 2131 | 1510 | 1415 | | 11485.04 |
| Burkina Faso | 0 | 277 | 161 | 523 | 1700 | 1616 | | 19751.47 |
| Cambodia | 0 | 46 | 10647 | 10006 | 604 | 692 | | 16249.79 |
| Central Africa | 0 | 5 | 11 | 11 | 90 | 86 | | 4666.368 |
| Chad | 2 | 75 | 260 | 257 | 438 | 422 | | 15477.73 |
| Djibouti | 0 | 339 | 144 | 241 | 0 | 4 | | 958.923 |

| Country | Wheat | | Rice | | Maize | | Tractors | Population |
|---|---|---|---|---|---|---|---|---|
| | Product | Domestic Supply | Product | Domestic Supply | Product | Domestic Supply | | |
| Ethiopia | 4500 | 6394 | 54 | 441 | 8350 | 8255 | | 109224.4 |
| Gambia | 0 | 84 | 2340 | 214 | 39 | 42 | | 2280.094 |
| Guinea | 0 | 420 | 171 | 3527 | 819 | 806 | | 12414.29 |
| Haiti | 0 | 278 | 183 | 735 | 260 | 288 | | 11123.18 |
| Laos | 0 | 24 | 3585 | 3436 | 982 | 737 | | 7061.507 |
| Lesotho | 9 | 59 | 258 | 19 | 100 | 292 | | 2108.328 |
| Liberia | 0 | 73 | 4030 | 598 | 0 | 1 | | 4818.973 |
| Madagascar | 6 | 365 | 112 | 4774 | 215 | 222 | | 26262.31 |
| Malawi | 1 | 145 | 3168 | 118 | 2698 | 2736 | | 18143.22 |
| Mali | 29 | 341 | 232 | 3027 | 3625 | 3226 | | 19077.75 |
| Mauritania | 7 | 551 | 134 | 295 | 16 | 23 | | 4403.313 |
| Mozambique | 21 | 691 | 27574 | 1101 | 1250 | 1315 | | 29496 |
| Myanmar | 116 | 560 | 5152 | 26176 | 1984 | 1559 | 102750 | 53708.32 |
| Nepal | 1958 | 1961 | 102 | 5700 | 2473 | 2616 | | 28095.71 |
| Niger | 5 | 66 | 120 | 309 | 30 | 60 | 375 | 22442.82 |
| Rwanda | 11 | 158 | 763 | 148 | 410 | 207 | | 12301.97 |
| Senegal | 0 | 682 | 920 | 2051 | 264 | 645 | | 15854.32 |
| Sierra Leone | 0 | 91 | 3 | 1595 | 23 | 25 | | 7650.15 |
| Sudan | 595 | 2754 | 87 | 125 | 45 | 63 | | 41801.53 |
| Togo | 0 | 144 | 145 | 222 | 887 | 905 | | 7889.093 |
| Uganda | 23 | 647 | 246 | 374 | 2773 | 2500 | | 42729.04 |
| Tanzania | 57 | 984 | 2220 | 2134 | 6273 | 5206 | | 56313.44 |
| Yemen | 106 | 3650 | 0 | 781 | 43 | 687 | | 28498.68 |
| Zambia | 114 | 160 | 43 | 56 | 2395 | 2859 | | 17351.71 |

*Note: **Table 1** is provided by the author based on the data extracted from FAOSTAT. Product and domestic supply of wheat, rice and maize are in 1,000 tonnes. Population is in 1,000 people.*

**Table 1.**
*Production of cereals, number of tractors and population in the world's least developed countries.*

| Country | Wheat | | Rice | | Maize | | Tractors | Population |
|---|---|---|---|---|---|---|---|---|
| | Product | Domestic product | Product | Domestic product | Product | Domestic product | | |
| Argentina | 18539 | 5596 | 1368 | 753 | 43462 | 19237 | | 44361.15 |
| Australia | 20941 | 8716 | 635 | 453 | 387 | 382 | | 24898.15 |
| Brazil | 5422 | 12530 | 11749 | 12273 | 82288 | 64173 | | 209469.3 |
| Canada | 32216 | 7940 | 0 | 533 | 13885 | 14215 | 733182 | 37074.56 |
| China | 131690 | 127248 | 214079 | 206919 | 257349 | 277032 | | 1459378 |
| France | 35798 | 19521 | 73 | 659 | 12667 | 8428 | | 64990.51 |
| Germany | 20264 | 16289 | 0 | 405 | 3344 | 6714 | 798700 | 83124.42 |
| India | 99700 | 95422 | 172580 | 142688 | 27820 | 23402 | | 1352642 |
| Indonesia | 0 | 9868 | 83037 | 73805 | 30254 | 31380 | | 267670.5 |
| Italy | 6933 | 11136 | 1512 | 699 | 6179 | 12034 | | 60627.29 |
| Japan | 766 | 6700 | 9728 | 11232 | 0 | 15819 | | 127202.2 |
| Mexico | 2943 | 6217 | 284 | 1352 | 27170 | 40514 | | 126190.8 |
| Korea | 26 | 4663 | 5195 | 6688 | 78 | 10037 | | 51171.71 |
| Russia | 72136 | 32416 | 1038 | 1175 | 11419 | 6447 | 439600 | 145734 |
| Saudi Arabia | 518 | 3667 | 0 | 1882 | 45 | 3160 | | 33702.76 |
| South Africa | 1900 | 3593 | 3 | 1406 | 12510 | 10510 | | 57792.52 |
| Turkey | 20000 | 19134 | 940 | 1359 | 5700 | 7071 | 1037383 | 82340.09 |
| The UK | 13555 | 16364 | 0 | 606 | 0 | 2067 | | 67141.68 |
| The US | 51398 | 33837 | 10153 | 5504 | 364262 | 292878 | | 327096.3 |

*Note:* ***Table 2*** *is provided by the author based on the data extracted from FAOSTAT. Product and domestic supply of wheat, rice and maize are in 1,000 tonnes. Population is in 1,000 people.*

**Table 2.**
*Production of cereals, number of tractors and population in G20 countries.*

| | Wheat | | Rice | | Maize | |
|---|---|---|---|---|---|---|
| | **The Least Developed Countries** | **G20 Countries** | **The Least Developed Countries** | **G20 Countries** | **The Least Developed Countries** | **G20 Countries** |
| Observation | 165 | 95 | 165 | 95 | 165 | 95 |
| Mean | 1056.667 | 22955.14 | 3775.121 | 24536.86 | 1343.521 | 42863.25 |
| Standard Deviation | 1862.218 | 30904.34 | 10106.54 | 54650.72 | 1704.479 | 83273.31 |
| Minimum Value | 5 | 2735 | −42 | 375 | 1 | 346 |
| Maximum Value | 7617 | 127248 | 56415 | 206919 | 8255 | 343651 |

*Note: **Table 3** is provided by the author based on the data extracted from FAOSTAT. $Cereal_{cit}$ is in 1,000 tonnes.*

**Table 3.**
*Descriptive statistics for* $\boldsymbol{Cereal_{cit}}$.

| **Independent Variable** | **Dependent Variable** $Cereal_{Wheat\ i\ t}$ | **Dependent Variable** $Cereal_{Rice\ i\ t}$ | **Dependent Variable** $Cereal_{Maize\ i\ t}$ |
|---|---|---|---|
| | **Estimated Coefficient** | **Estimated Coefficient** | **Estimated Coefficient** |
| $G_{20i}$ | 24978.12*** (0.000) | 32732.82*** (0.001) | 45346.42*** (0.000) |
| $Africa_i$ | 17217.26*** (0.002) | 31947.7*** (0.001) | 44958.31*** (0.000) |
| $America_i$ | 7068.021*** (0.001) | 8418.469*** (0.001) | 80329.6*** (0.000) |
| $Asia_i$ | 24711.49*** (0.000) | 54529.34*** (0.000) | 46233.9*** (0.000) |
| $Europe_i$ | 13487.44*** (0.000) | 254.84*** (0.002) | 7352.68*** (0.000) |
| **Constant** | −17861.62*** (0.003) | −32457.05*** (0.002) | −46880.16*** (0.000) |
| Number of Observations | 260 | 260 | 260 |
| Prob > F | 0.000 | 0.000 | 0.000 |
| $R^2$ | 0.312 | 0.309 | 0.247 |

*Note: This table shows estimation results for Eq. (10). P-values are presented in parentheses. The estimation is conducted with robust standard errors. Superscripts ***, ** and * indicate statistical significance at 1%, 5% and 10%, respectively.*

**Table 4.**
*Estimation results of Eq. (10).*

| | Wheat | | Rice | | Maize | |
|---|---|---|---|---|---|---|
| | The Least Developed Countries | G20 Countries | The Least Developed Countries | G20 Countries | The Least Developed Countries | G20 Countries |
| Observation | 165 | 95 | 165 | 95 | 165 | 95 |
| Mean | 0.0502536 | 0.1482661 | 0.1270291 | 0.0540263 | 0.0583697 | 0.20179 |
| Standard Deviation | 0.1062425 | 0.0939006 | 0.162305 | 0.0677363 | 0.0550903 | 0.2146537 |
| Minimum Value | 0.0010715 | 0.0291697 | −0.0019442 | 0.0045624 | 0.0002075 | 0.0146632 |
| Maximum Value | 0.8501159 | 0.3557684 | 0.61645 | 0.2757308 | 0.2365934 | 1.063882 |

*Note: **Table 5** is provided by the author based on the data extracted from FAOSTAT.*

**Table 5.**
*Descriptive statistics for* ***Cereal Per Capita***$_{cit}$

| Independent Variable | Dependent Variable *Cereal Per Capita*$_{Wheat\ i\ t}$ | Dependent Variable *Cereal Per Capita*$_{Rice\ i\ t}$ | Dependent Variable *Cereal Per Capita*$_{Maize\ i\ t}$ |
|---|---|---|---|
| | Estimated Coefficient | Estimated Coefficient | Estimated Coefficient |
| $G_{20i}$ | 0.0388*** (0.001) | −0.150*** (0.000) | 0.132*** (0.000) |
| $Africa_i$ | −0.225*** (0.000) | −0.090*** (0.002) | 0.177*** (0.000) |
| $America_i$ | −0.211*** (0.000) | −0.014 (0.103) | 0.399*** (0.000) |
| $Asia_i$ | −0.209*** (0.000) | 0.123*** (0.000) | 0.132*** (0.000) |
| $Europe_i$ | −0.066*** (0.000) | −0.011*** (0.000) | 0.087*** (0.000) |
| **Constant** | 0.270*** (0.000) | 0.169*** (0.000) | −0.119*** (0.000) |
| Number of Observations | 260 | 260 | 260 |
| Prob > F | 0.000 | 0.000 | 0.000 |
| $R^2$ | 0.340 | 0.403 | 0.530 |

*Note: This table shows estimation results for Eq. (11). P-values are presented in the parentheses. The estimation is conducted with robust standard errors. Superscripts ***, ** and * indicate statistical significance at 1%, 5% and 10%, respectively.*

**Table 6.**
*Estimation results of Eq. (11).*

| The Least Developed Countries | | | | G20 | | | |
|---|---|---|---|---|---|---|---|
| | Wheat | Rice | Maize | | Wheat | Rice | Maize |
| Afghanistan | 542 | 25 | 16 | Argentina | 316 | 53 | 633 |
| Angola | 0 | 0 | 307 | Australia | 209 | 6 | 2 |
| Bangladesh | 238 | 3104 | 250 | Brazil | 247 | 1181 | 8321 |
| Benin | 0 | 115 | 378 | China | 2901 | 8624 | 11806 |
| Chad | 0 | 10 | 36 | France | 326 | 3 | 112 |

| The Least Developed Countries | | | | G20 | | | |
|---|---|---|---|---|---|---|---|
| Ethiopia | 168 | 3 | 252 | India | 5987 | 4654 | 2785 |
| Madagascar | 0 | 403 | 10 | Italy | 49 | 34 | 11 |
| Malawi | 0 | 5 | 529 | Japan | 163 | 190 | 4 |
| Mali | 21 | 127 | 218 | Mexico | 197 | 53 | 4571 |
| Mauritania | 22 | 7 | 1 | Korea | 23 | 481 | 205 |
| Mozambique | 1 | 5 | 75 | Russia | 433 | 21 | 115 |
| Myanmar | 24 | 861 | 90 | Saudi Arabia | 35 | 0 | 92 |
| Nepal | 195 | 486 | 248 | South Africa | 85 | 0 | 569 |
| Niger | 0 | 4 | 1 | Turkey | 2133 | 30 | 202 |
| Rwanda | 0 | 3 | 49 | The US | 2334 | 398 | 17864 |
| Senegal | 6 | 30 | 33 | | | | |
| Sierra Leone | 0 | 120 | 1 | | | | |
| Sudan | 48 | 2 | 4 | | | | |
| Uganda | 20 | 5 | 175 | | | | |
| Tanzania | 3 | 30 | 738 | | | | |
| Zambia | 3 | 2 | 72 | | | | |

*Note: **Table 7** is provided by the author based on the data extracted from FAOSTAT. Food loss of wheat, rice and maize in the least developed and G20 countries are in 1,000 tonnes.*

**Table 7.**
*The amount of food loss in the least developed and G20 countries.*

## Author details

Ryusuke Oishi
Faculty of Economics, Meikai University, Urayasu City, Chiba Prefecture, Japan

*Address all correspondence to: r-o@meikai.ac.jp

## References

[1] Food and Agriculture Organization of the United Nations, Food Security Statistics [Internet]. Available from: h ttp://www.fao.org/economic/ess/ess-fs/en/ [Accessed: 2020-12-25]

[2] United Nations, Department of Economic and Social Affairs, Suitable Development, Food Security and Nutrition and Sustainable Agriculture [Internet]. Available from: https://sdgs.un.org/topics/food-security-and-nutrition-and-sustainable-agriculture [Accessed: 2021-01-04].

[3] Oishi R. Food loss and waste in Japan. New Food Industry. 2019;61:908-914.

[4] Kumar D, Kalita P. Reducing postharvest losses during storage of grain crops to strengthen food security in developing countries. Foods. 2017;6:8. DOI:10.3390/foods6010008

[5] Union of Concerned Scientists. Seeds of the Future: How Investment in Classical Breeding Can Support Sustainable Agriculture [Internet]. Union of Concerned Scientists Fact Sheet. 2015. Available from: www.jstor.org/stable/resrep17294 [Accessed: 2021-01-04].

[6] Co-chairs of the Sustainable Development Solutions Network, Dobermann, A., Nelson, R. Opportunities and Solutions for Sustainable Food Production. Background Paper for the High-Level Panel of Eminent Persons on the Post-2015 Development Agenda. 2013.

[7] Oehen B, Home R. Sustainable Food Systems. In: van der Meulen N., Wiesel, J. Culinary Turn, Aesthetic Practice of Cookery. Transcript Verlag. 2017. p. 87-90.

[8] Timmer CP. The changing role of rice in Asia's food security. ADB Sustainable Development Working Paper Series. 2010;15.

[9] Shiferaw B, Prasanna BM, Hellin J, Bänziger M. Crops that feed the world 6. past successes and future challenges to the role played by maize in global food security. Food Security. 2011;3:307-327. DOI:10.1007/s12571-011-0140-5

[10] Oishi R. Current trend and future prospects in Japanese food imports from China. New Food Industry. 2019; 61:19-24.

[11] Oishi R. Effect of new coronavirus (COVID-19) on food and restaurant industry. New Food Industry. 2020;62: 801-806.

[12] Oishi R. Impact of novel coronavirus (COVID-19) countermeasures (self-restraint) on small- and medium-sized enterprises (SMEs) in Japan. New Food Industry. 2020;62:432-435.

[13] Dorward A. Agricultural labour productivity, food price and sustainable development impacts and indictors. Food Policy. 2013;39:40-50. DOI: 10.1016/j.foodpol.2012.12.003

[14] Chung U, Gbegbelegbe S, Shiferaw B, Robertson R, Yun JI, Tesfaye K, Hoogenboom G, Sonder K. Modeling the Effect of a heat wave on maize production in the USA and its implication on food security in the developing world. Weather and Climate Extremes. 2014;5-6:67-77. DOI:10.1016/j.wace.2014.07.002

[15] Lipinski B, Hanson C, Lomax J, Kitinoja L, Waite R, Searchinger T. Reducing food loss and waste. Working Paper, Instalment 2 of Creating a Sustainable Food Future. Washington, DC: World Resource Institute. 2013.

[16] Bishwajit G, Sarker S, Kpoghomou MA, Gao H, Jun L, Yin D, Ghosh S. Self-sufficiency in rice and food security: A South Asian perspective. Agriculture & Food Security. 2013;2:10. DOI:10.1186/2048-7010-2-10

Chapter 3

# Present Status and Future Prospects of Drought Tolerance in Rice

*Veerendra Jaldhani, Ponnuvel Senguttuvel, Bathula Srikanth, Puskur Raghuveer Rao, Desiraju Subrahmanyam and Durbha Sanjeeva Rao*

## Abstract

Rice is an important staple food crop across the world. It is mainly cultivated under irrigated lowland and also rain-fed upland conditions where drought stress is often noticed. Global climate change predicts an intensification of drought stress in future due to uneven rainfall which was witnessed for the last few years. Confronting drought stress can deliver fruitful crop returns in rice and scope for research extents. Drought stress affects the overall plant growth and yield. A prominent improvement has been made during last two decades in our understanding of the mechanisms involved in adaptation and tolerance to drought stress in rice. In order to achieve the marked crop returns from rainfed areas, there is a requisite of drought tolerant rice varieties, and genetic improvement for drought tolerance should be a prime area of concern in the future. A huge rice germplasm is available and good number of the germplasm possess drought tolerance and these genomic regions have been exploited in developing some drought tolerant rice varieties. The application of available genotyping methodologies, the identification of traits of interest, and key genetic regions associated with the drought tolerance have opened new prospects to successfully develop new drought tolerant varieties. This chapter deals with the importance of drought tolerance in rice crop followed by the evolution of molecular markers and breeding techniques in identifying drought tolerant QTL's/genes and their utilization in the improvement of drought tolerant rice varieties.

**Keywords:** Rice, Drought tolerance, Markers, QTLs, Varieties

## 1. Introduction

Rice is a staple cereal consumed by more than half of the world's population. It is cultivated in wide agro-ecological conditions including rain-fed conditions where drought stress is often evident due to erratic rainfall. Rice crop consumes around 3000 L water to produce 1 Kg rice. Drought stress is one of the major factors that leads to decreased rice production [1] and it can expand to above 50% of the global arable land by 2050 [2] due to loss of ground water, global climate change leading to decreased water sheds. Drought stress was noted in approximately 42 Mha of rice-producing area [3]. It is of two types, terminal and intermittent [4].

Drought stress is usually a dry condition where the water availability is less than a threshold level which causes damage to plants [5].

Lack of water for a long time leading to the death of plants is terminal drought while lack of water for a short time that leads to improper growth is intermittent drought [6].

Drought stress tolerance varies among the plant species and is defined as the ability of a plant to grow, develop and produce significant yield as well as economic benefit [7]. It is also defined as the ability of plant to survive at minimum water level in the fresh tissue (23%) [6].

Drought stress leads to morphological, physiological and biochemical changes in plants which ultimately lead to decreased yields. In response, plants synthesize reactive oxygen species (ROS), proteins and osmolytes to maintain turgor pressure. This osmotic adaptation provides dehydration tolerance to tissues [8–10]. But, this did not show yield benefit in rice [11].

The first landmark achievement in drought rice tolerance study was the identification of a small region on chromosome 8 homologous to chromosome 7 of wheat [12]. A marker based selection was proposed and this region was worked out to identify QTL's for vegetative leaf rolling and root traits- thickness and root to shoot ratio [13].

Historically, upland-adapted germplasm was of Japonica type while lowland-adapted germplasm was of Indica type [14]. In general, Japonica genotypes were dehydration avoidant while Indica types were dehydration tolerant. Geographically separated evolution and sterility problems have limited hybridization between the two types [15]. If breeding for both osmotic adjustment and rooting capacity is considered desirable, then the linkage between high osmotic adjustment and poor root traits needs to be broken.

RG1 QTL was identified while working with 52 recombinant inbred (RI) lines (F 7), a randomly sampled subset of a population originally developed to study the genetics of resistance to rice blast (*Pyricularia oryzae*) [16].

## 2. First generation markers

### 2.1 Restriction fragment length polymorphism (RFLP)

Rice RFLP maps developed at Cornell University [17, 18] and Japan [19, 20] is the basis of gene mapping research. In the 1988 wet season at IRRI, Co39 (maternal), a lowland, Indica cultivar developed in India, and Moroberekan, a traditional upland, Japonica cultivar originally developed in Guinea. Moroberekan is considered to be resistant to drought while Co39 is drought susceptible. About 50 $F_1$ seeds were obtained from the cross and only 15 $F_1$ seeds were randomly chosen and grown in a greenhouse to obtain an $F_2$ population. About 300 $F_2$ seeds were randomly selected and planted in the Rapid Generation Advance (RGA) [21] greenhouse from $F_2$ to $F_6$ using single seed descent (SSD). All panicles were bagged at each generation and $F_7$ seeds were used for genotype analysis [16].

DNA was extracted from the leaves of the two parents and digested with restriction enzymes DraI, EcoRI, EcoRV, HindIII and ScaI. The digested DNAs were electrophoresed on 0.9% agarose gels and transferred to Hybond $N^+$ membranes (Amersham Corp., Chicago) according to the manufacturer's instructions. 280 DNA clones distributed throughout the 12 chromosomes of rice (184 rice genomic clones, coded RG; 62 rice cDNAs, coded Rz; and 28 oat cDNAs, coded CDO) were linearized and labeled with $^{32}$PdCTP by the random hexamer method [22]. Hybridized filters were washed once in 1.5 X SSPE and once in 0.5 X SSPE at 65° for

15–20 min. Filters were exposed to X-ray film −80° at with one intensifying screen for 1–4 days. 127 informative probes were used for segregation analysis of the RI lines using the procedures outlined above. Mapmaker [23] and Map Manager [24] were used to establish the RFLP map.

Since the *indica* × *japonica* crosses suffer from sterility, distorted segregation, etc., [16], breeders chose to work on the accessions of *indica or japonica* sub-species, RFLP was considered laborious over RAPD which gained popularity among researchers [25].

### 2.2 Random amplified polymorphic DNA (RAPD)

Prior to the availability of complete rice genome sequence, RAPD markers were useful in developing drought tolerant varieties. It is the simplest, cheaper, sensitive and useful technique for the genotype identification, population and pedigree analysis, phylogenetic studies, genetic mapping [26] and analysis of genetic fidelity of commercially micropropagated plants [27]. It is simple since it requires less DNA and simultaneously doesn't require southern blotting and radioactive labeling [28]. Considering the usage of RAPD in wheat, maize, pea-nut, broccoli and cauliflower, RAPD was applied in rice to detect diversity in the low land and upland rice varieties with an aim to identify drought-resistant loci [29] as given below.

Thirteen rice cultivars were grown in a growth chamber at 28°-day temperature, 25°-night temperature and with an irradiance of 800 μmoles $m^{-2}$ $s^{-1}$ for 12 h a day. At six leaf stage, leaves were collected and stored in liquid nitrogen and genomic DNA was extracted and purified [22]. Forty-two GC-rich 10 bp random primers were used as RAPD markers. DNA amplification was done in a PCR (Perkin Elmer Cetus) programmed for 45 cycles of1min at 94°C,1 min at 37°C and 2 min at 72°C. The reaction conditions include 25 μl total reaction volume having 10 mM Tris-HC1 (pH 8.3), 50 mM KC1, 2mMMgC$1_2$, 50 mM each of dNTP's, 10 ng of a single random primer, 25 ng of genomic DNA and 2 units of Ampli Taq DNA polymerase (Perkin Elmer Cetus) and 50 ml of sterilized mineral oil.

At the end of amplification, 10 μl of each amplification mixture was loaded in either 1.4% agarose (0.5 to 4 kb) or 5% polyacrylamide (<500 bp) gels for electrophoresis in 1 x TBE (89 mM Tris, 89 mM boric acid and 2 mM EDTA). Gels were stained with ethidium bromide and photographed under UV light. Pair-wise comparisons of genotypes, based on the presence (score 1) or absence (score 2) of each marker similarity coefficients were calculated which were used to construct UPGMA tree. One-to-twelve DNA amplicons were observed from each genomic DNA sample. A total of 260 DNA fragments were amplified and 208 (80%) of these showed polymorphisms. Upland cultivars and lowland cultivars were classified into two main clusters-*japonica* and *indica* with seven (Azucena, Rikuto Norin 21, Moroberekan, IAC25, IRAT13, OS4, and 63–83) six (BPI-76 NS, IR20, IR36, CO39, MGL-2 and Salumpikit) cultivars respectively. Later, they screened 2074 rice varieties for drought tolerance where upland varieties were identified with higher score for drought tolerance and were recommended for donors for breeding drought tolerant varieties as well as for developing molecular markers.

However, like RFLP, RAPD is also disadvantageous due to low polymorphism among *japonica* rice, need for hundreds of markers to locate markers in the QTL and absence of RAPD markers for some regions of the chromosomes [30].

### 2.3 AFLP (amplified fragment length polymorphism)

Restriction enzymes were reported from bacteria. This enzyme identifies foreign DNA in bacterial cells based on the target site and generally, cuts the DNA within this site if the site is un-methylated. Eventually, the broken DNA cannot show its effect on the host bacterial cell and hence, the natural function of restriction enzymes is to restrict the growth of foreign DNA. Restriction enzymes differ in the length of their target sites (4–8). It is universally accepted that the probability of finding the smaller length of sequence is manifold higher than longer sequences i.e., in other words if the length of the target site is less, then, it can be repeated more frequently in the DNA. In the above example, a 6-base target is rarer to find than the 4-base target site. Adapter /adaptor/linker is a short, chemically synthesized (known sequence), single-stranded or double-stranded oligonucleotide that can be ligated to the ends of other DNA/RNA molecules to convert them to sticky ends of desired sequence.

DNA was isolated from 80 plants of $F_2$ population of the cross 'Labelle' × 'Black Gora' [30] and AFLP was conducted [31]. AFLP involves cutting of genomic DNA with two restriction enzymes that need different lengths of target sites, a 6-base (or "rare") cutter (EcoRI) and a 4-base (or "frequent") cutter (MseI), ligating adapters to the fragment ends, amplifying MseI-EcoRI fragments with primers that match the adapter and contain additional selective nucleotides at the 3′ end and separating the fragments on denaturing polyacrylamide gels. EcoRI and MseI adapters, T4 DNA ligase, DTT and water were added to the restriction reaction-mixture which was incubated for 3 h at 37°C. Then, pre-amplification of DNA was done with primers which were labeled subsequently with spermidine. The amplicons were separated on 4.5% denaturing (urea) polyacrylamide gels for 90 to 120 min at 120 W. The gels were dried and exposed to X-ray film for 4–7 days.

Bands showing clear polymorphism were scored as present ("1") or absent ("0"). Genetic similarity was computed as the number of common bands divided by the total number of bands of both accessions, and genetic distance was computed as 1 minus this value [32]. These distances were used to construct cluster diagrams by UPGMA method (SAS, PROC CLUSTER) [33]. AFLP is cheaper than RAPD. AFLP is useful to screen small number of samples with large number of markers.

## 3. Second generation markers

### 3.1 Inter simple sequence repeats (ISSR)

ISSR markers are well distributed in the eukaryotic genome [34] more feasible and reproducible than RAPD [35] highly polymorphic, less expensive and independent of sequence information [36]. Polymorphism of 73.02% with RAPD markers and 90.91% with ISSR markers was observed between six rice lines [37]. Seventeen ISSR primers- (8 based on $(AG)_8$, 8 on $(GA)_8$, and 1 on $(GATA)_4$ were used to screen 12 cultivars as presented below [38].

Genomic DNA was isolated from freshly harvested young leaves of each cultivar by Mini prep method. ISSR-PCR was conducted and the amplified products were resolved in 1.8% agarose gel in the presence of ethidium bromide and were documented under ultraviolet light. Polymorphism information content (PIC) was calculated based on the presence (score 1) or absence (score 0) of band. Similarity scores were calculated and un-weighted pair-group method with arithmetic average (UPGMA) dendrogram was generated sub-program of NTSYS-PC software version 2.0at 1000 bootstrap. The three drought tolerant varieties formed one sub-cluster by (GA)8YG primer.

### 3.2 Simple sequence repeat (SSR) and SNPs (third generation markers)

Rice crop was recognized with more than 20,000 SSR markers and over one million SNPs and Indels, which include both functional and non-functional markers [39–41]. This has opened up huge opportunities for the use of molecular markers in diversity analysis, mapping genes/QTLs for various agronomic traits under drought, and their use in marker-assisted breeding (MAB) and also in positional cloning of QTLs to identify candidate genes for complex traits [42].

The accessibility of complete rice genomic sequence information, rice linkage maps, and molecular marker technology has made it possible to dissect complex traits into individual quantitative trait loci (QTLs) [43–46]. Linkage mapping, association mapping, nested association mapping, marker-aided recurrent selection (MARS), and genome-wide selection (GWS) are different approaches currently followed for the mapping and introgression of QTLs for drought tolerance. There are several sources of genomic variation such as QTL main effects, QTL × QTL interactions, and QTL × environment interactions. A thorough understanding of this variation is very important before embarking on marker-assisted selection (MAS) of drought QTLs. Several major-effect QTLs for grain yield under drought have been identified and are being used in marker-assisted breeding/pyramiding. Courtois et al. [47] identified the meta-QTLs for root traits under drought stress. Meta-analysis of 53 grain yield associated QTLs identified from 15 previous studies bring about in 14 meta-QTLs. In general, rice varieties with deep and high-volume root system exhibits better adaptability under drought conditions [48]. Among the root traits, root length, volume, thickness and root growth angle (RGA) are playing key role in mitigating drought tolerance [49]. In turn, RGA determines the root depth. The deeper and profuse root architecture support plants to extract water from deeper soil layers. The three major QTLs governing root growth angle (RGA) in rice were reported by Uga et al. [49–51]. Among them, DEEPER ROOTING 1 (DRO1) is the significant one and this QTL has been fine mapped and the underlying gene, an early auxin responsive factor, has been cloned using IR64, a shallow rooted variety and Kinandang Patong (KP), a deeply rooted variety. The variation in the DRO1 gene in the major Indian rice genotypes used in drought tolerant breeding plans and their association with RGA is reported [52]. So far, 675 root QTLs and more than 85 genes related to 29 different root parameters have been reported in rice [47]; https://snp-seek.irri.org/). The introduction of deep rooting traits in high yielding varieties is a resourceful way of enlightening drought tolerance in rice.

## 4. Tissue culture

Polyethylene glycol (PEG) can decrease moisture in tissues [53] by osmosis and eventually reduce callus growth. Manually dehusked brown rice of four rice varieties- Pusa Basmati 1, Taraori Basmati, Pant Sugandh Dhan 17 and Narendra 359were washed with detergent (teepol), washed with sterile distilled water, surface sterilized by 70% ethanol followed by 1% sodium hypochlorite and 0.1% mercuric chloride and were inoculated on to MS medium [54] having 30% sucrose and 8% agar. One month old calli were transferred onto the MS medium having a series of concentrations (0 (control), 10, 20, 30, 40, 50, 60 and 70 g/L) of PEG. After 30-day incubation, healthy calli at 70 g/L PEG were identified as drought tolerant and were subjected to shoot and root inductions separately followed by the development of plant lets. Loss of moisture content was least in Narendra 359 (2.99%) and highest in Pusa Basmati 1 (20.64%) and it indicates the drought response variation among the rice genotypes. Proline content in calli of all varieties increased with the increase in PEG concentration [55].

## 5. Gene expression

Transcriptomic analysis identified stress responsive genes like transcription factors, genes encoding for osmolyte production, reactive oxygen species (ROS) scavenging and other metabolic pathways etc. which help in developing drought tolerant varieties [56, 57]. They are divided into signaling and functional groups [58]. Despite many techniques are available for transcriptome studies, micro-chips developed from quality rice genome sequence were used to identify the regulating reproductive development, hormone signaling and abiotic stress response [59, 60].

Seven-day-old seedlings of Dhagaddeshi (DD) and IR20 cultivars were subjected to drought stress and microarray hybridization of the RNA isolated from samples collected after 3 h and 6 h along with that of control seedlings, was carried out as per manufacturer's instructions (GeneChip® 3' IVT Express Kit User Manual, 2008, Affymetrix). The number of probe sets expressing differentially after 3 h stress is almost double for DD (10,901) than IR20 (5,502) in comparison with the control. However, this difference was less after 6 h with 8,601 for IR20 and 11,041 for DD. Despite the initial delay in sensing drought stress by IR20, differences in transcript levels were more or less mitigated at the 6 h time point. Fructose-bisphosphate aldolase (LOC_Os01g67860), OsVP1 (LOC_Os01g68370), auxin response factor 2 (LOC_Os01g70270) showed high expression levels along with other conserved genes and those of unknown function in DD. Drought stress is known to induce accumulation of osmolytes like proline, glycinebetaine that help in the prevention of dehydration in plants. A significant increase in the accumulation of free proline was observed in both cultivars as the stress duration progressed [61]. The gene expression analysis of DRO1 gene elucidates structural variation and this information is very crucial for breeding rice for drought tolerance in future [52].

## 6. Drought tolerance varieties

Development of tolerant varieties is the strategy chosen across the field crops including rice. The following are the list of ways by which drought tolerance varieties were developed (**Table 1**). Grain yield was used as a trait to develop drought tolerant varieties and presently, physiological traits are on focus [62]. Most of the characters are influenced by numerous loci termed as Quantitative trait loci (QTL) and they have only minor influence on the trait [63–65]. Marker assisted selection (MAS) is the integration of molecular genetics with artificial selection.

Conventional breeding involves the art of hybrid cross to develop new and improved cultivars. It includes the identification of drought tolerant genetic variants followed by introduction of these traits into popular varieties [66]. It was accepted among the breeders that the existence of drought tolerance variation in the germplasm indicates the presence of stress tolerance genes [67].

However, in conventional breeding programs, only a few parents are involved and their use efficiency in rice accessions remains low because of the inefficient cross-pedigree breeding method, resulting in a narrow genetic base of the developed cultivars [68]. In addition, 15 drought tolerant rice landraces were identified with stable yield under the drought stress while screening both under net house and laboratory evaluation [69].

| S. No. | Varieties | Key trait | Country |
|---|---|---|---|
| 1 | Shabhagidhan | Possess qDTY12.1 | India |
| 2 | Birsa Vikas Dhan 111 | Root QTL from Azucena | India |
| 3 | DRR Dhan 42 | IR 64 NIL with qDTY4.1 & qDTY2.2 | India |
| 4 | DRR Dhan 50 | Samba Mahsuri sub 1 with qDTY 2.1 & 3.1 | India |
| 5 | CR Dhan 801 | Swarna sub 1 with qDTY 2.1 & 3.1 | India |

**Table 1.**
*List of drought tolerant varieties released with drought QTLs through Marker assisted breeding.*

## 7. Recurrent selection (RS)

It involves multiple parents which is an ideal breeding approach to steadily improve the level of quantitative traits in a breeding population. RS was first applied in cross-pollinated crops, maize [70]. In rice, Virmani et al. [71] showed random mating composite population facilitated by IR36ms having recessive genic male sterility for the improvement of restorers and maintainers. However, these two methods were cumbersome and inefficient. In 2001, a mutant of "Sanming Dominant Genic Male Sterile Rice" was found from an F2 population of a cross between SE21S and Basmati370 [72]. The male sterility of this mutant was controlled by a dominant gene and it was fine mapped on chromosome 8 [73]. Further, by multiple backcrosses, they introduced this dominant male sterile (DMS) allele into the genetic background of rice cultivar Jiafuzhan (known as Jiabuyu), which was used to develop 12 drought-tolerant lines through RS [74].

## 8. Marker assisted back crossing

In backcrossing a donor and recurrent parents are used. Donor parent contains the gene or QTL of interest and the recurrent parent is mega variety or line that is improved by adding the gene or QTL of interest. The donor parent is crossed to the recurrent parent. The progeny of this cross is then crossed back to the recurrent parent (back cross). The progeny of this cross is selected for the added trait and

| S.No. | QTL | Chr | Parentage | Reference |
|---|---|---|---|---|
| 1 | 1 QTL | | CO39 × Moroberekan | Lilley et al. [75] |
| 2 | 39 QTLs | | CO39 × Moroberekan | Ray et al. [76] |
| 3 | 39 QTLs | - | IR64 × Azucena | Yadav et al. [77] |
| 4 | 18 QTLs | | Azucena × Bala | Price et al. [78] |
| 5 | 17 QTLs | | Bala × Azucena | Price et al. [78] |
| 6 | 28 QTLs | | IR58821 × IR52561 | Ali et al. [79] |
| 7 | QCMS1.1 | 1 | IR62266 × CT9993 | Tripathy et al. [80] |
| 8 | QCMS3.1 | 3 | IR62266 × CT9993 | Tripathy et al. [80] |
| 9 | QCMS7.1 | 7 | IR62266 × CT9993 | Tripathy et al. [80] |
| 10 | QCMS8.1 | 8 | IR62266 × CT9993 | Tripathy et al. [80] |
| 11 | QCMS8.2 | 8 | IR62266 × CT9993 | Tripathy et al. [80] |
| 12 | QCMS9.1 | 9 | IR62266 × CT9993 | Tripathy et al. [80] |
| 13 | QCMS9.2 | 9 | IR62266 × CT9993 | Tripathy et al. [80] |

| S.No. | QTL | Chr | Parentage | Reference |
|---|---|---|---|---|
| 14 | QCMS11.1 | 11 | IR62266 × CT9993 | Tripathy et al. [80] |
| 15 | QCMS12.1 | 12 | IR62266 × CT9993 | Tripathy et al. [80] |
| 16 | 15 QTLs | | IR64 × Azucena | Hemamalini et al. [81] |
| 17 | 5 QTLs | | CT9993 × IR62266 | Zhang et al. [82] |
| 18 | 23 QTLs | | IR1552 × Azucena | Zheng et al. [83] |
| 19 | qgy3.1 | 3 | CT9993-5-10-1-M × IR62266-42-6-2 | Lanceras et al. [84] |
| 20 | qgy4.3 | 4 | CT9993-5-10-1-M × IR62266-42-6-2 | Lanceras et al. [84] |
| 21 | qGY-2b | 2 | Zhenshan 97B × IRAT109 | Zou et al. [85] |
| 22 | qDTY1.1 | 1 | CT9993-5-10-1-M × IR62266-42-6-2 | Kumar et al. [86] |
| 23 | qDTY12.1 | 12 | Way Rarem × Vandana | Bernier et al. [87] |
| 24 | qGy10 | 10 | Tequing × Lemont | Zhao et al. [88] |
| 25 | 13 QTLs | - | Azucena × Bala | Khowaja, F. S., & Price, A. H. [89] |
| 26 | 7 QTLs | - | Indica × Azucena | Zheng et al. [90] |
| 27 | 1 QTL | | Apo/2 × Swarna | Venuprasad et al. [91] |
| 28 | QDS_9.1 | 9 | IR64 × IR77298-5-6-B-18(Aday Sel) | BP et al. [92] |
| 29 | 1 QTL | | R77298 × Sabitri | Yadav et al. [93] |
| 30 | qDTY 3.4 | 3 | Danteshwari × Dagaddeshi | Verma et al. [94] |
| 31 | qPN-6-2 | 6 | Xiaobaijingzi × Kongyu131 | Xing et al. [95] |
| 32 | qDTY3.1 | 3 | TDK1 × IR55419-04 | Dixit et al. [96] |
| 33 | qDTY6.1 | 6 | TDK1 × IR55419-04 | Dixit et al. [96] |
| 34 | qPSS8.1 | 8 | IR64 × IRAT177 | Trijatmiko et al. [97] |
| 35 | qGPP8.2 | 8 | IR64 × IRAT177 | Trijatmiko et al. [97] |
| 36 | 1 QTL | | IR64 × Cabacu | Trijatmiko et al. [97] |
| 37 | QSnp1b | | Teqing × Lemont | Wang et al. [98] |
| 38 | QSnp3a | | Teqing × Lemont | Wang et al. [98] |
| 39 | QSnp11 | | Teqing × Lemont | Wang et al. [98] |
| 40 | QGyp2a | | Teqing × Lemont | Wang et al. [98] |
| 41 | QSf8 | | Teqing × Lemont | Wang et al. [98] |
| 42 | qDTY3.2 | | Swarna × WAB | Saikumar et al. [99] |
| 43 | qPDL1.2 | 1 | Appo × Moroberekan | Sellamuthu et al. [100] |
| 44 | qHI3 | 3 | Appo × Moroberekan | Sellamuthu et al. [100] |
| 45 | qDTY2-2 | 2 | MRQ74 cultivar | Shamsudin et al. [101] |
| 46 | qDTY3-1 | 3 | MRQ74 cultivar | Shamsudin et al. [101] |
| 47 | qDTY1-3 | 1 | Dular × IR62-21 | Catolos et al. [102] |
| 48 | qDTY8-1 | 8 | Dular × IR62-21 | Catolos et al. [102] |

**Table 2.**
*List of QTLs reported for drought tolerance associated traits in rice.*

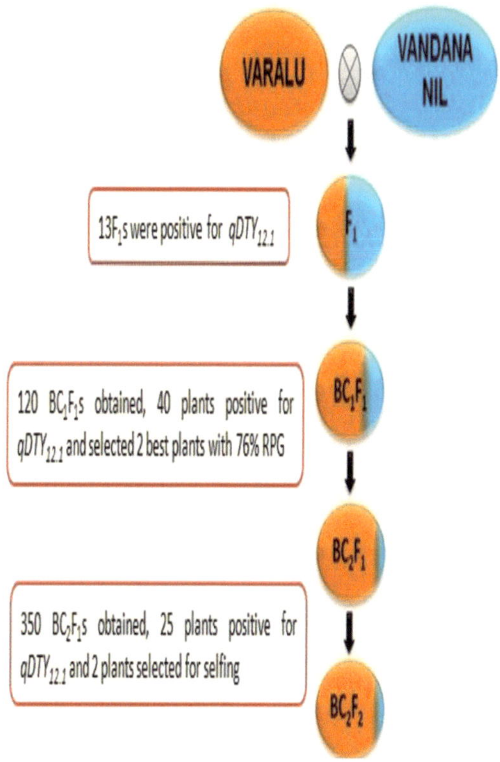


**Figure 1.**
*Back crossing flow chart to introduce drought QTL) (adopted from Balija et al. [103].*

again subjected to back cross with the recurrent parent. This process is repeated to obtain a line as identical as possible to the recurrent parent with the addition of the gene of interest that has been added through breeding.

Among the QTLs for drought stress tolerance (**Table 2**), qDTY12.1 offers significant yield under reproductive-stage drought stress by contributing 51% genetic variance [87] and is available in Vandana NIL (near isogenic line). Hence, this QTL was introduced into Varalu (WGL 14377× CR-544-1-2) which is a popular variety cultivated in upland areas of India by back crossing method (**Figure 1**) [103].

Later, responsible genomic regions have been identified and are popularly known as molecular markers. They were used to develop drought tolerant varieties by a process known as marker assisted selection or breeding [66].

## Author details

Veerendra Jaldhani[1], Ponnuvel Senguttuvel[1], Bathula Srikanth[1,2], Puskur Raghuveer Rao[1], Desiraju Subrahmanyam[1] and Durbha Sanjeeva Rao[1*]

1 Indian Institute of Rice Research, Hyderabad, India

2 Professor Jayashankar Telangana State Agricultural University, Hyderabad, India

*Address all correspondence to: sraodurbha@gmail.com

## References

[1] Swamy BM, Kumar A. Genomics-based precision breeding approaches to improve drought tolerance in rice. Biotechnology advances. 2013;31(8):1308-18.

[2] Singhal P, Jan AT, Azam M, Haq QM. Plant abiotic stress: a prospective strategy of exploiting promoters as alternative to overcome the escalating burden. Frontiers in Life Science. 2016;9(1):52-63.

[3] Yang X, Wang B, Chen L, Li P, Cao C. The different influences of drought stress at the flowering stage on rice physiological traits, grain yield, and quality. Scientific Reports 2019;9:1-12.

[4] Polania J, Rao IM, Cajiao C, Grajales M, Rivera M, Velasquez F, Raatz B, Beebe SE. Shoot and root traits contribute to drought resistance in recombinant inbred lines of MD 23-24× SEA 5 of common bean. Frontiers in Plant Science. 2017;8:296.

[5] Ishaku GA, Tizhe DT, Bamanga RA, Afolabi ET. Biotechnology and drought stress tolerance in plants. Asian Plant Research Journal. 2020;22:34-46.

[6] Oladosu Y, Rafii MY, Samuel C, Fatai A, Magaji U, Kareem I, Kamarudin ZS, Muhammad II, Kolapo K. Drought resistance in rice from conventional to molecular breeding: a review. International journal of molecular sciences. 2019 Jan;20(14):3519.

[7] Farooq M, Aziz T, Wahid A, Lee DJ, Siddique KH. Chilling tolerance in maize: agronomic and physiological approaches. Crop and Pasture Science. 2009;60(6):501-16.

[8] Hsiao TC, Acevedo E, Fereres E, Henderson DW. Water stress, growth and osmotic adjustment. Philosophical Transactions of the Royal Society of London. B, Biological Sciences. 1976;273(927):479-500.

[9] Jones MM, Turner NC. Osmotic adjustment in expanding and fully expanded leaves of sunflower in response to water deficits. Functional Plant Biology. 1980;7(2):181-92.

[10] Morgan JM. Osmoregulation and water stress in higher plants. Annual review of plant physiology. 1984;35:299-319.

[11] Lilley JM, Ludlow MM, McCouch SR, O'toole JC. Locating QTL for osmotic adjustment and dehydration tolerance in rice. Journal of Experimental Botany. 1996;47(9):1427-36.

[12] Ahn S, Anderson JA, Sorrells ME, Tanksley S1. Homoeologous relationships of rice, wheat and maize chromosomes. Molecular and General Genetics MGG. 1993;241(5):483-90.

[13] Champoux MC, Wang G, Sarkarung S, Mackill DJ, O'Toole JC, Huang N, McCouch SR. Locating genes associated with root morphology and drought avoidance in rice via linkage to molecular markers. Theoretical and Applied Genetics. 1995 Jun 1;90(7-8):969-81.

[14] Glaszmann JC. A varietal classification of Asian cultivated rice (*Oryza sativa* L.) based on isozime polymorphism. In rice genetics. InProceeding of the International Rice Genetics Symposium 1985 (pp. 27-31).

[15] Chang TT. The origin, evolution, cultivation, dissemination, and diversification of Asian and African rices. Euphytica. 1976;25(1):425-41.

[16] Wang GL, Mackill DJ, Bonman JM, McCouch SR, Champoux MC, Nelson RJ. RFLP mapping of genes

conferring complete and partial resistance to blast in a durably resistant rice cultivar. Genetics. 1994;136(4):1421-34.

[17] McCouch SR, Kochert G, Yu ZH, Wang ZY, Khush GS, Coffman WR, Tanksley SD. Molecular mapping of rice chromosomes. Theoretical and Applied Genetics. 1988;76(6):815-29.

[18] Causse MA, Fulton TM, Cho YG, Ahn SN, Chunwongse J, Wu K, Xiao J, Yu Z, Ronald PC, Harrington SE. Saturated molecular map of the rice genome based on an interspecific backcross population. Genetics. 1994 Dec 1;138(4):1251-74.

[19] SAITO A, YANO M, KISHIMOTO N, NAKAGAWA M, YOSHIMURA A, SAITO K, UKAI Y, KAWASE M, NAGAMINE T, YOSHIMURA S, IDETA O. Linkage map of restriction fragment length polymorphism loci in rice. Japanese Journal of Breeding. 1991 Dec 1;41(4):665-70.

[20] Kurata N, Nagamura Y, Yamamoto K, Harushima Y, Sue N, Wu J, Antonio BA, Shomura A, Shimizu T, Lin SY, Inoue T. A 300 kilobase interval genetic map of rice including 883 expressed sequences. Nature genetics. 1994 Dec;8(4):365-72.

[21] Vergara BS, Patena G, Lopez FSS. Rapid Generation of rice at the International Rice Research Institute. 1982. IRRI Research Paper Series No. 84. Los Baños: International Rice Research Institute.

[22] Feinberg AP. A technique for radiolabelling DNA restriction endonuclease fragments to high specific activity. Ann Biochem. 1984;137:266-7.

[23] Lander ES, Green P, Abrahamson J, Barlow A, Daly MJ, Lincoln SE, Newburg L. MAPMAKER: an interactive computer package for constructing primary genetic linkage maps of experimental and natural populations. Genomics. 1987;1(2):174-81.

[24] Manly KF. A Macintosh program for storage and analysis of experimental genetic mapping data. Mammalian Genome. 1993;4(6):303-13.

[25] Mackill DJ. Rainfed lowland rice improvement. Int. Rice Res. Inst.; 1996.

[26] Williams JG, Kubelik AR, Livak KJ, Rafalski JA, Tingey SV. DNA polymorphisms amplified by arbitrary primers are useful as genetic markers. Nucleic acids research. 1990;18(22):6531-5.

[27] Verma P, Srivastava P, Singh SC, Mathur A. Genetic fidelity of long-term micropropagated plantlets of Valeriana wallichii–an endangered medicinal plant. Current Science. 2010;99(4):436-8.

[28] Mackill DJ, Zhang Z, Redona ED, Colowit PM. Level of polymorphism and genetic mapping of AFLP markers in rice. Genome. 1996; 39 (5): 969-77.

[29] Yu LX, Nguyen HT. Genetic variation detected with RAPD markers among upland and lowland rice cultivars (*Oryza sativa* L.). Theoretical and Applied Genetics. 1994 Jan 1;87(6):668-72.

[30] Redona ED, Mackill DJ. Mapping quantitative trait loci for seedling vigor in rice using RFLPs. Theoretical and Applied Genetics. 1996 Mar;92(3):395-402.

[31] Vos P, Hogers R, Bleeker M, Reijans M, Lee TV, Hornes M, Friters A, Pot J, Paleman J, Kuiper M, Zabeau M. AFLP: a new technique for DNA fingerprinting. Nucleic acids research. 1995;23(21):4407-14.

[32] Nei M, Li WH. Mathematical model for studying genetic variation in terms

of restriction endonucleases. Proceedings of the National Academy of Sciences. 1979;76(10):5269-73.

[33] SAS Institute Inc. SAS/STAT user's guide, version 6. 4th ed, vol 2. Cary, North Carolina: SAS Institute Inc., 1989.

[34] Tautz D, Renz M. Simple sequences are ubiquitous repetitive components of eukaryotic genomes. Nucleic acids research. 1984;12(10):4127-38.

[35] Godwin ID, Aitken EA, Smith LW. Application of inter simple sequence repeat (ISSR) markers to plant genetics. Electrophoresis. 1997;18(9):1524-8.

[36] Bornet B, Muller C, Paulus F, Branchard M. Highly informative nature of inter simple sequence repeat (ISSR) sequences amplified using tri-and tetra-nucleotide primers from DNA of cauliflower (*Brassica oleracea var. botrytis* L.). Genome. 2002 Oct 1;45(5):890-6.

[37] Alzohairy AM. Molecular markers for new promising drought tolerant lines of rice under drought stress via RAPD-PCR and ISSR markers. Journal of American science. 2010;6(12).

[38] Reddy CS, Babu AP, Swamy BM, Kaladhar K, Sarla N. ISSR markers based on GA and AG repeats reveal genetic relationship among rice varieties tolerant to drought, flood, or salinity. Journal of Zhejiang University Science B. 2009 Feb;10(2):133-41.

[39] Raghuvanshi S, Kapoor M, Tyagi S, Kapoor S, Khurana P, Khurana J, Tyagi A. Rice genomics moves ahead. Molecular breeding. 2010 Aug;26(2):257-73.

[40] McNally KL, Childs KL, Bohnert R, Davidson RM, Zhao K, Ulat VJ, Zeller G, Clark RM, Hoen DR, Bureau TE, Stokowski R. Genomewide SNP variation reveals relationships among landraces and modern varieties of rice. Proceedings of the National Academy of Sciences. 2009 Jul 28;106(30):12273-8.

[41] McCouch SR, Zhao K, Wright M, Tung CW, Ebana K, Thomson M, Reynolds A, Wang D, DeClerck G, Ali ML, McClung A. Development of genome-wide SNP assays for rice. Breeding Science. 2010;60(5):524-35.

[42] Salvi S, Tuberosa R. To clone or not to clone plant QTLs: present and future challenges. Trends in plant science. 2005 Jun 1;10(6):297-304.

[43] Temnykh S, DeClerck G, Lukashova A, Lipovich L, Cartinhour S, McCouch S. Computational and experimental analysis of microsatellites in rice (*Oryza sativa* L.): frequency, length variation, transposon associations, and genetic marker potential. Genome research. 2001;11(8):1441-52.

[44] McCouch SR, Teytelman L, Xu Y, Lobos KB, Clare K, Walton M, Fu B, Maghirang R, Li Z, Xing Y, Zhang Q. Development and mapping of 2240 new SSR markers for rice (*Oryza sativa* L.). DNA research. 2002 Jan 1;9(6):199-207.

[45] Tuberosa R, Salvi S. QTLs and genes for tolerance to abiotic stress in cereals. InCereal genomics 2004 (pp. 253-315). Springer, Dordrecht.

[46] Tuberosa R, Salvi S. Genomics-based approaches to improve drought tolerance of crops. Trends in plant science. 2006 Aug 1;11(8):405-12.

[47] Courtois B, Ahmadi N, Khowaja F, Price AH, Rami JF, Frouin J, Hamelin C, Ruiz M. Rice root genetic architecture: meta-analysis from a drought QTL database. Rice. 2009 Sep;2(2):115-28.

[48] Kim Y, Chung YS, Lee E, Tripathi P, Heo S, Kim KH. Root response to drought stress in rice (*Oryza sativa* L.). International journal of molecular sciences. 2020 Jan;21(4):1513.

[49] Uga Y, Okuno K, Yano M. Dro1, a major QTL involved in deep rooting of rice under upland field conditions. Journal of experimental botany. 2011 May 1;62(8):2485-94.

[50] Uga Y, Sugimoto K, Ogawa S, Rane S, Ishitani M, Hara N, Kitomi Y,Inukai Y, Ono K, Kanno N, Inoue H, Takehisa H, Motoyama R, Nagamura Y, Wu J, Matsumoto T, Takai T, Okuno K, Yano M(2013) Control of root system architecture by DEEPERROOTING 1 increases rice yield under drought conditions.Nat Genet 45(9):1097-1102.

[51] Uga Y, Kitomi Y, Yamamoto E, Kanno N, Kawai S, Mizubayashi T, Fukuoka S (2015) A QTL for root growth angle on ricechromosome 7 is involved in the genetic pathway of DEEPERROOTING 1. Rice 8:8.

[52] Singh BK, Ramkumar MK, Dalal M, Singh A, Solanke AU, Singh NK, Sevanthi AM. Allele mining for a drought responsive gene DRO1 determining root growth angle in donors of drought tolerance in rice (*Oryza sativa* L.). Physiology and Molecular Biology of Plants. 2021 Feb 26:1-2.

[53] Adkins SW, Kunanuvatchaidach R, Godwin ID. Somaclonal Variation in Rice-2 Drought Tolerance and Other Agronomic Characters. Australian Journal of Botany. 1995;43(2):201-9.

[54] Murashige T, Skoog F. A revised medium for rapid growth and bio assays with tobacco tissue cultures. Physiologia plantarum. 1962;15(3):473-97.

[55] Joshi R, Shukla A, Sairam RK. In vitro screening of rice genotypes for drought tolerance using polyethylene glycol. Acta Physiologiae Plantarum. 2011;33(6):2209.

[56] Wang H, Zhang H, Gao F, Li J, Li Z. Comparison of gene expression between upland and lowland rice cultivars under water stress using cDNA microarray. Theoretical and Applied Genetics. 2007;115(8):1109.

[57] Ray S, Dansana PK, Giri J, Deveshwar P, Arora R, Agarwal P, Khurana JP, Kapoor S, Tyagi AK. Modulation of transcription factor and metabolic pathway genes in response to water-deficit stress in rice. Functional & integrative genomics. 2011;11(1):157-78.

[58] Todaka D, Shinozaki K, Yamaguchi-Shinozaki K. Recent advances in the dissection of drought-stress regulatory networks and strategies for development of drought-tolerant transgenic rice plants. Frontiers in plant science. 2015;6:84.

[59] Arora R, Agarwal P, Ray S, Singh AK, Singh VP, Tyagi AK, Kapoor S. MADS-box gene family in rice: genome-wide identification, organization and expression profiling during reproductive development and stress. BMC genomics. 2007;8(1): 1-21.

[60] Jain M, Khurana JP. Transcript profiling reveals diverse roles of auxin-responsive genes during reproductive development and abiotic stress in rice. The FEBS journal. 2009;276(11): 3148-62.

[61] Borah P, Sharma E, Kaur A, Chandel G, Mohapatra T, Kapoor S, Khurana JP. Analysis of drought-responsive signalling network in two contrasting rice cultivars using transcriptome-based approach. Scientific reports. 2017;7(1):1-21.

[62] Anwaar HA, Perveen R, Mansha MZ, Abid M, Sarwar ZM, Aatif HM, ud din Umar U, Sajid M, Aslam HM, Alam MM, Rizwan M. Assessment of grain yield indices in response to drought stress in wheat (*Triticum aestivum* L.). Saudi journal of

biological sciences. 2020 Jul 1;27(7):1818-23.

[63] Breese EL, Mather K. The organisation of polygenic activity within a chromosome in Drosophila. Heredity. 1957 Dec;11(3):373-95.

[64] Thoday JM. Location of polygenes. Nature. 1961;191(4786):368-70.

[65] Lande R, Thompson R. Efficiency of marker-assisted selection in the improvement of quantitative traits. Genetics. 1990;124(3):743-56.

[66] Kiriga AW, Haukeland S, Kariuki GM, Coyne DL, Beek NV. Effect of Trichoderma spp. and Purpureocillium lilacinum on Meloidogyne javanica in commercial pineapple production in Kenya. Biological Control. 2018;119:27-32.

[67] Aguado-Santacruz GA, García-Moya E, Aguilar-Acuña JL, Moreno-Gómez B, Solís-Moya E, Preciado-Ortiz ER, Jiménez-Bremont JF, Rascón-Cruz Q. In vitro plant regeneration from quality protein maize (QPM). In Vitro Cellular & Developmental Biology-Plant. 2007 Jun 1;43(3):215-24.

[68] Li ZK, Zhang F. Rice breeding in the post-genomics era: from concept to practice. Current opinion in plant biology. 2013 May 1;16(2):261-9.

[69] Tu DX, Huong NT, Giang LT, Thanh LT, Khanh TD, Trung KH, Nhan DT, Tuan NT. Screening Drought Tolerance of Vietnamese Rice Landraces in the Laboratory and Net House Conditions. Advanced Studies in Biology. 2021;13(1):21-8.

[70] Bolaños J, Edmeades GO. Eight cycles of selection for drought tolerance in lowland tropical maize. II. Responses in reproductive behavior. Field Crops Research. 1993 Jan 15;31(3-4):253-68.

[71] Virmani, S. S., B. C. Viraktamath, C. L. Casal, R. S. Toledo, M. T. Lopez and J. O. Manalo (1997). Hybrid rice breeding manual, International Rice Research Institute, Philippines.

[72] Huang XB, Tian ZH, Deng ZQ, Zheng JT, Lin CB, Tang JX. Preliminary identification of a novel Sanming dominant male sterile gene in rice (*Oryza sativa* L.). Acta Agron Sin. 2008;34(10):1865-8.

[73] Yang ZM, Xie XF, Huang XB, Wang FQ, Tong ZJ, Duan YL, Lan T, Wu WR. Mapping of Sanming dominant genic male sterility gene in rice. Yi Chuan= Hereditas. 2012;34(5):615-20.

[74] Pang Y, Chen K, Wang X, Xu J, Ali J, Li Z. Recurrent selection breeding by dominant male sterility for multiple abiotic stresses tolerant rice cultivars. Euphytica. 2017;213(12):268.

[75] Lilley J, Ludlow M, McCouch S, O'Toole J. Locating QTL for osmotic adjustment and dehydration tolerance in rice. Journal of Experimental Botany.1996;47:1427-1436.

[76] Ray, J.D.; Yu, L.; McCouch, S.R.; Champoux, M.C.; Wang, G.; Nguyen, H.T. Mapping quantitative trait loci associated with root penetration ability in rice (*Oryza sativa* L.). Theor. Appl. Genet. 1996;92, 627-636.

[77] Yadav R, Courtois B, Huang N, McLaren G. Mapping genes controlling root morphology and root distribution in a doubled-haploid population of rice. Theoretical and Applied Genetics.1997;94:619-632.

[78] Price AH, Townend J, Jones MP, Audebert A, Courtois B. Mapping QTLs associated with drought avoidance in upland rice grown in the Philippines and West Africa. Plant Molecular Biology.2002;48:683-695.

[79] Ali M, Pathan M, Zhang J, Bai G, Sarkarung S, Nguyen H. Mapping QTLs for root traits in a recombinant inbred population from two indica ecotypes in rice. Theoretical and Applied Genetics.2000;101:756-766.

[80] Tripathy JN, Zhang J, Robin S, Nguyen TT, Nguyen HT. QTLs for cell-membrane stability mapped in rice (*Oryza sativa* L.) under drought stress. Theor. Appl. Genet. 2000; 100, 1197-1202.

[81] Hemamalini GS, Shashidhar HE, Hittalmani S. Molecular marker assisted tagging of morphological and physiological traits under two contrasting moisture regimes at peak vegetative stage in rice (*Oryza sativa* L.). Euphytica 2000;112, 69-78.

[82] Zhang J, Zheng H, Aarti A, Pantuwan G, Nguyen T, Tripathy J, ... Nguyen BD. Locating genomic regions associated with components of drought resistance in rice: comparative mapping within and across species. Theoretical and Applied Genetics 2001; 103:19-29.

[83] Zheng BS, Yang L, Zhang WP, Mao CZ, Wu YR, Yi KK, Wu P. Mapping QTLs and candidate genes for rice root traits under different water-supply conditions and comparative analysis across three populations. Theor. Appl. Genet. 2003; 107, 1505-1515.

[84] Lanceras JC, Pantuwan G, Jongdee B, Toojinda T. Quantitative trait loci associated with drought tolerance at reproductive stage in rice. Plant Physiology. 2004; 135:384-399.

[85] Zou G, Mei H, Liu H, Liu G, Hu S, Yu X, ... Luo L. Grain yield responses to moisture regimes in a rice population: association among traits and genetic markers. Theoretical and Applied Genetics.2005; 112:106-113.

[86] Kumar R, Venuprasad R, Atlin G. Genetic analysis of rainfed lowland rice drought tolerance under naturallyoccurring stress in eastern India: heritability and QTL effects. Field Crops Research 2007;103:42-52.

[87] Bernier J, Kumar A, Ramaiah V, Spaner D, Atlin G. A large-effect QTL for grain yield under reproductive-stage drought stress in upland rice. Crop Science. 2007;47(2):507-16.

[88] Zhao XQ, Xu JL, Zhao M, Lafitte R, Zhu LH, Fu BY, ... Li ZK. QTLs affecting morph-physiological traits related to drought tolerance detected in overlapping introgression lines of rice (*Oryza sativa* L.). Plant Science. 2008;174:618-625.

[89] Khowaja, F.S.; Price, A.H. QTL mapping rolling, stomata conductance and dimension traits of excised leaves in the Bala×Azucena recombinant inbred population of rice. Field crop res. 2008;106, 248-257.

[90] Zheng BS, Yang L, Mao CZ, Huang YJ, Wu P. Mapping QTLs for morphological traits under two water supply conditions at the young seedling stage in rice. Plant Science. 2008; 175:767-776.

[91] Venuprasad R, Dalid C, Del Valle M, Zhao D, Espiritu M, Cruz MS, ... Atlin G. Identification and characterization of large-effect quantitative trait loci for grain yield under lowland drought stress in rice using bulk-segregant analysis. Theoretical and Applied Genetics. 2009;120:177-190.

[92] BP MS, Ahmed HU, Henry A, Mauleon R, Dixit S, Vikram P, ... Mandal NP. Genetic, physiological, and gene expression analyses reveal that multiple QTL enhance yield of rice mega-variety IR64 under drought. PloS One. 2013;8:e62795.

[93] Yadav, R.B.; Dixit, S.; Raman, A.; Mishra, K.K.; Vikram, P.; Swamy, B.M.;

Kumar, A. A QTL for high grain yield under lowland drought in the background of popular rice variety Sabitri from Nepal. Field Crop Res.2013; 144, 281-287.

[94] Verma SK, Saxena RR, Saxena RR, Xalxo MS, Verulkar SB. QTL for grain yield under water stress and non-stress conditions over years in rice (*Oryza sativa* L.). Australian Journal of Crop Science. 2014;8:916.

[95] Xing W, Zhao H, Zou D. Detection of main-effect and epistatic QTL for yield-related traits in rice under drought stress and normal conditions. Canadian Journal of Plant Science.2014; 94:633-641.

[96] Dixit S Singh A, Kumar A. Rice breeding for high grain yield under drought: a strategic solution to a complex problem. International Journal of Agronomy 2014;1-16.

[97] Trijatmiko KR, Prasetiyono J, Thomson MJ, Cruz CMV, Moeljopawiro S, Pereira A. Meta-analysis of quantitative trait loci for grain yield and component traits under reproductive-stage drought stress in an upland rice population. Molecular Breeding. 2014;34:283-295.

[98] Wang, Y.; Zhang, Q.; Zheng, T.; Cui, Y.; Zhang, W.; Xu, J.; Li, Z. Drought-tolerance QTLs commonly detected in two sets of reciprocal introgression lines in rice. Crop Pasture Sci. 2014; 65, 171-184.

[99] Saikumar, S.; Gouda, P.K.; Saiharini, A.; Varma, C.M.K.; Vineesha, O.; Padmavathi, G.; Shenoy, V.V. Major QTL for enhancing rice grain yield under lowland reproductive drought stress identified using an O. sativa/*O. glaberrima* introgression line. Field Crop Res. 2014;163, 119-131.

[100] Sellamuthu R, Ranganathan C, Serraj R. Mapping QTLs for reproductive-stage drought resistance traits using an advanced backcross population in upland rice. Crop Science. 2015;55:1524-1536.

[101] Shamsudin NA, Swamy BPM, Ratnam W, Cruz MTS, Sandhu N, Raman AK, Kumar A. Pyramiding of drought yield QTLs into a high quality Malaysian rice cultivar MRQ74 improves yield under reproductive stage drought. Rice 2016;9:21.

[102] Catolos M, Sandhu N, Dixit S, Shamsudin NAA, Naredo MEB, McNally KL, ... Kumar A. Genetic loci governing grain yield and root development under variable rice cultivation conditions. Frontiers in Plant Sciences.2017.

[103] BALIJA V, BANGALE U, PONNUVEL S, BARBADIKAR KM, MADAMSHETTY SP, DURBHA SR, YADLA H, MAGANTI SM. Improvement of Upland Rice Variety by Pyramiding Drought Tolerance QTL with Two Major Blast Resistance Genes for Sustainable Rice Production. 水稻科学. 2021:3.

Chapter 4

# Heterosis and Heterotic Grouping among Tropical Maize Germplasm

*Richard Olutayo Akinwale*


## Abstract

Maize (*Zea mays* L.) is the most important staple cereal cultivated in sub-Saharan Africa but its productivity is considerable low due to several factors. Development and deployment of maize hybrids have been reported as one of the crucial options in achieving sustainable maize production in sub-Saharan Africa. Information on the heterotic response among available genetic materials in a breeding program is valuable before commencement of any hybrid development program. Unlike the temperate germplasm, maize tropical germplasm is characterized with wide genetic base and genetic complexities and thus, proper organization of the pools, populations, varieties and inbreds that can serve as parental materials for hybrid development through identification of a distinct heterotic groups and patterns among tropical germplasm becomes very essential. This paper reviewed past research efforts at characterizing heterotic response among tropical maize genetic materials with a view to point out merits and demerits in the methods used and future direction towards achieving sustainable hybrid cultivation and enhancing food security in the sub-region.


**Keywords:** combining ability, gene action, heterotic grouping, hybrids, tropical maize

## 1. Introduction

The term 'heterosis', as first introduced by Shull in 1909, was used to describe the phenomenon when the mean of any character or characters in a hybrid exceeds the mean of its descendants obtained by any system of close inbreeding. Four hypotheses were proposed to explain this; the dominance hypothesis which postulates that the increase in vigor after crossing results from the combination of different dominant alleles contributed by each parent [1]. The heterozygosis hypothesis attributes the increase in vigor to the existence of loci at which the heterozygous state is superior to either homozygotes [2, 3]; the pseudo-overdominance hypothesis that attributes the hybrid vigor to the effect of tightly linked genes with favorable dominant alleles in repulsion phase in the parental lines resulting in an apparent overdominance when combined in the hybrid [4] and epistasis hypothesis which explains the increased vigor in the light of the interaction of favorable alleles from two parents at different loci that show additive, dominant and/or overdominant action [5]. Among these hypotheses, heterozygosis gained prominence. Milborrow [6] asserted, from physiology view point, that even though the growth of a plant may be limited by the genes that regulate certain metabolic pathway down to a lower level than the maximum possible, heterozygous plants may partially escape the growth regulation, thereby giving them advantage over the homozygous individuals. Brieger [7] explained that heterosis is easily obtained when the parents from which the hybrids are produced are inbreds or purelines and that heterosis does not affect the individual plant as a whole, but the expression of each of the traits that are heterotic. For instance, characters in maize that are affected by heterosis include plant and ear heights, size of leaves, intensity, size and strength of root system, amount of pollen shed, number and size of kernels and response to biotic and abiotic stresses [7]. Characters such as earliness to maturity, row number of the ear, plant and kernel color are not heterotic characters.

Two types of heterosis have been described in literatures. Falconer and Mackay [8] defined mid-parent heterosis as the difference between the hybrid and the mean of the two parents. They also defined high- or best-parent heterosis as the difference between the hybrid mean and the mean of either of the parent. Mid-parent heterosis value has been of more importance because it provides the basis for the identification of heterotic patterns among a fixed set of populations/inbred lines [9]. Melani and Carena [10] asserted that the utilization of mid-parent values is an effective practical method to identify heterotic responses among parents.

A heterotic group has been defined as a collection of germplasm that, when crossed with germplasm from an external group, tends to exhibit a higher degree of heterosis (on the average) than when crossed with a member of its own group [11]. Melchinger and Gumber [12] also defined heterotic group as a collection of related or unrelated genotypes from the same or different populations, which display similar combining ability and heterotic response when crossed with genotypes from other genetically distinct germplasm groups. Heterotic pattern refers to a specific pair of two heterotic groups, which express high heterosis and consequently high hybrid performance in their cross. Melchinger and Gumbler [12] recommended the following criteria for the choice of heterotic pattern in hybrid breeding; (i) high mean performance and large genetic variance in the hybrid population; (ii) high *per se* performance and good adaption of the parent population to the target regions; and (iii) low inbreeding depression, if hybrids are produced from inbred lines. Establishing heterotic pattern is of prime importance in the development of a successful maize hybrid program [13].

## 2. Heterotic grouping methods for maize germplasm.

After establishing significant genetic variability among parental materials to use, plant breeders employ several methods for classifying the parents into heterotic groups. The methods include morphological traits, pedigree method, multivariate technique, genetic methods involving mating designs and the use of molecular markers. At advanced stage of breeding, genetic and molecular methods are preferred because of their high level of precision since their results are minimally influenced by environmental factors. Among several mating designs in plant breeding, three are prominent for classifying parents into heterotic groups. Where proven testers exist in a breeding program, a line x tester mating design is embraced in which each tester represent a heterotic group. Where there is no proven testers, diallel method and North Carolina Design II become better alternatives. In studies where such designs are employed, information on heterotic groups as well as identification of testers are usually the prime objectives. The advent of molecular markers has offered a less-stressful, faster, smarter and somewhat cheaper alternative through the use of genetic distance. Examples of markers for popularly used this purpose are Amplified Fragment Length Polymorphism (AFLPs), Simple Sequence Repeats (SSRs) and Single Nucleotide Polymorphism (SNPs) markers. The qualities of these markers that make them suitable for this purpose include the following: high throughput, they are highly reproducible and they are relatively easy to assay. In more recent times, SNPs markers has become the most popular and marker technologies include Microarray and DarT and DarTSeq have been developed on the basis of SNPs.

## 3. Classifying tropical maize germplasm into heterotic groups and identifying heterotic pattern

In temperate maize germplasm, distinct heterotic groups and establishing clear heterotic patterns such as European flint x US Lancaster, which is commonly used in Europe and Reid Yellow Dent x Lancaster extensively exploited in US, China [14, 15] and many parts of the world ([10, 16]). Paterniani [17] noted that there is lack of information on the heterotic response among tropical maize germplasm and that might be due to the existence of the large number of races and cultivars that were yet to be studied, thus making it difficult to have well-defined heterotic patterns in tropical maize. By then, only a few studies have been conducted to establish heterotic patterns of the tropical maize germplasm [18]. An earlier study reported a combination of Tuxpeno with Eto and other Carribean flints as a promising heterotic pattern [19]. The study also established Tuson x Tuxpeno, Cuban flint × Tuxpeno, and Suwan 1 x Tuxpeno as promising combinations for hybrid production in the tropics. Previous studies on the heterotic pattern using late/intermediate maturing inbred lines from IITA could not establish clear heterotic patterns [20–22]. The reason adduced for the results was that the inbred lines were derived from source populations formed by mixing different germplasm without taking into consideration the need to maintain heterotic groups intact [22]. Menkir *et al.* [20] recommended a combination of divergent testers with molecular markers as a better alternative to classifying tropical maize inbred lines. On this basis, Menkir *et al.* [23] attempted to classify 38 tropical maize inbred lines into heterotic groups using two testers (TZMI102 and TZMI 1407) and molecular markers. The testers successfully classified 23 out of 38 inbreds into two heterotic groups. The results of the classification based on AFLP and SSR markers were found to be largely consistent with each other but the molecular markers classified the same inbreds into groups different from those classified by the testers. The authors concluded that the line x testers method used was found to be more efficient in classifying the inbreds than the molecular markers and recommended that the molecular marker-based grouping might at best serve as a basis for designing and carrying out combining ability studies in the field for tropical maize germplasm. However with the advent of more efficient markers for genetic diversity assessment, the result appears more promising.

A similar study by Barata and Carena, [13] in their comparative analysis of heterotic grouping of maize inbreds using diallel analysis method and SSR markers corroborated the above findings and recommended extensive field evaluation as being more appropriate in assigning unrelated maize inbred lines into heterotic groups.

Breeders at national and international research institutes in sub-Saharan Africa have developed thousands of inbred lines over years and several efforts have been made to identify defined heterotic groups that can be utilized in the sub-region.

Badu-Apraku *et al.* (2005) used multivariate techniques to classify 47 inbreds based on morphological traits and 4 groups were identified. They however, considered the grouping to be preliminary since morphological traits can be greatly influenced by environmental factors. Badu-Apraku et al. [24] also selected promising inbred parents based on multiple morphological traits under stress and non-stress environments using genotype main effect and genotype by environment interaction (GGE) biplot.

Wu *et al.* [25] classified 27 maize inbreds into four distinct heterotic groups using North Carolina Design II. Agbaje *et al.* [26] used a line x tester method to classify 35 early maturing yellow endosperm inbred lines into heterotic groups with two testers, TZi 4001 and Ku1414 evaluated under *Striga*-infested and *Striga*-free conditions at Mokwa and Abuja and in *Striga*-free environment at Ile-Ife, Nigeria. None of the inbred lines could be classified into heterotic groups under any of the evaluation environments, evidently because the testers were not sufficiently effective to discriminate among the inbreds. Furthermore, Badu-Apraku *et al.* [27] could neither identify definite heterotic groups, nor identify ideal testers in a diallel study among nine yellow-grained early maize inbreds using the genotype main effect plus genotype-by-environment interaction (GGE) biplot analysis. The reason they adduced for this was the overdominating effect of SCA relative to GCA effects. Nevertheless, distinct tester groups were identified. In their study of heterosis and genetic distance among 17 lowland white-grained tropical maize under drought stress and non-stress conditions using diallel and RFLP markers, Betran *et al.* [28] found that the degree of inbreeding of the parental lines could affect their response to stress. They also observed that the environment significantly affected the correlations of genetic distance with lower values observed under more stressed conditions. They equally opined that optimal nonstress environments where grain yield is maximal could be more appropriate to measure SCA effects and the predictive value of genetic distance.

Conventionally in quantitative genetics, SCA effects of inbreds have always been used to classify genetic materials into heterotic groups. This is based on the assumption that SCA of two lines from different heterotic groups is greater than those from the same group. However, the reliability of this method is dependent on the number of materials investigated and adequate sampling of the genetic background. Fan *et al.* (2008) used heterotic group's Specific and General Combining Ability (HSGCA), a combination of SCA and GCA effects, to assign some tropical maize inbreds into heterotic groups instead of the traditional method involving SCA effects only. This proposition was on the basis that SCA effects were often greatly influenced by the interaction between two inbred lines and between hybrids and environments, which often times lead to assigning the same inbred line into different heterotic groups under different studies (Fan *et al.,* 2008). Results of the study showed that HSGCA method was more effective than the use of SCA and molecular marker methods in classifying tropical maize germplasm into distinct heterotic groups. Akinwale et al. [29] attempted to classify 28 tropical inbred lines into heterotic groups using SCA yield, HSGCA and SSR-based molecular markers and reported that HSGCA was most efficient. They also reported that classifying inbreds based on SCA-yield under non-stress environment was closely related to the groups established by SSR markers.

Because yield is a complex trait and possesses low heritability, improvement progress based on direct selection is usually very slow. Most of the methods used for heterotic grouping are based on single trait, yield. Therefore, Badu-Apraku et al. [30] devise another method, heterotic groups based on General Combining Ability of Multiple Traits (HGCAMT), which integrate general combining ability effects of multiple traits especially where additive gene effects are predominant over non-additive effects for such traits. Comparing the HGCAMT method with other grouping methods, it was reported that results obtained were consistent with those of HSGCA, yield-SCA and SNP marker-based genetic distance under stress environment and even more effective than other methods across multiple stress environments [31, 32]. Badu-Apraku and Akinwale [33] in another line by tester study among 63 lines by 4 testers using GGE biplot concluded that the GGE biplot method was efficient in classifying the inbreds.

It should be noted that heterotic grouping of tropical maize germplasm is greatly influenced by factors. The amount of genetic diversity among parental lines evaluated is a major factor. Badu-Apraku et al. [27] reported in a diallel study among nine yellow maize inbreds that the genetic diversity was small and therefore distinct heterotic groups could not be identified. The inbreds could only be classified into tester groups. In another similar study with white inbreds, significant genetic diversity was recorded and two clear heterotic groups were identified among a set of 9 inbreds [34]. Another important factor affecting heterotic grouping is the type of gene action preponderant in the set of parents under study. Heterotic groups are clearly identified when both additive and non-additive gene action are significant and there is preponderant of additive gene action over non-additive gene action [31, 32, 35]. In any study where these conditions are not met, distinct heterotic groups cannot be identified. The third factor that affects heterotic grouping among tropical germplasm is the type of mating design and heterotic grouping method used in the study. Heterotic grouping are conventionally based on combining ability effects, which are obtained from different mating designs employed in various genetic studies. It should be noted that among all mating designs, cross classification/factorial mating designs are the only type of design that are useful for heterotic grouping combining ability effects of the parents and hybrids can only be estimated using this type of mating designs. Examples of these designs are diallel, North Carolina Design II and line x tester design. Among these designs, diallel mating design has proved to be the most valuable design and most popularly used especially when evaluating a sizable number of parents for the purpose of heterotic grouping and identification of testers ([29]: [27, 34]). North Carolina design II added the advantage of being able to classify more parents more efficiently. Based on the type of combining ability effect used in the classification, three grouping methods are commonly used; yield's specific combining ability effects (SCA) [20], heterotic group's general and specific combining ability effects (GCA + SCA) [36] and heterotic group's general combining ability effects of multiple traits (HGCAMT) [30]. Each of these grouping methods gives different heterotic groups and the efficiency of these methods differ depending on the mode of gene action prevalent for the trait(s) under study. For instance, the use of HGCAMT (which is predominantly based on additive gene action for classification) is grossly inappropriate where non-additive gene action is prevalent. The fourth factor affecting heterotic grouping is the environmental complexes including environmental stresses such as drought, striga infestation, insect infestation, and low soil N, which characterize production environment in the sub-Saharan Africa. Different groups, in terms of number of groups and constitution of each group, are created among the same set of parents but under different research environments ([29]: [31, 32, 35]: [33]). These factors and their interaction are very important in grouping tropical maize germplasm and to establish a clear heterotic pattern with broad application to tropical germplasm, which in turn will greatly facilitate development and deployment of superior maize hybrid in the sub-region. It is therefore recommended that research efforts should be intensified to study these factors and how some of them interact in order to decipher the genetic complexities and environmental complexes that characterize maize production and productivity in sub-Saharan Africa.

## Author details

Richard Olutayo Akinwale
Department of Crop Production and Protection, Obafemi Awolowo University, Ile-Ife, Nigeria

*Address all correspondence to: akinrichie2002@yahoo.com

## References

[1] Bruce, A.B. (1910). The mendelian theory of heredity and the augmentation of vigour. Science 32: 627- 628.

[2] East, E.M. and Hayes, H.K. (1912). Heterozygosis in evolution and in plant breeding. Bullettin of US Department of Agriculture. 243: 7-58.

[3] Shull, G.F. (1911). Hybridization methods in corn breeding. American Breeders Magazine 1: 98-107.

[4] Crow, J. F., (1952). Dominance and overdominance. In Gowen, J.W. (ed.) Heterosis. Iowa State

[5] Schnell, F.W., and Cockerham, C.C. (1992). Multiplicative vs. arbitrary gene action in heterosis. Genetics 131:461-469

[6] Milborrow, B. (1998) A biochemical mechanism for hybrid vigour. Journal of Experimental Botany 49: 1063-1071

[7] Brieger, F.G. (1950). The genetic basis of heterosis in maize. Genetics 35: 420-445.

[8] Falconer, D.S. and Mackay, T.F.C. (1996) Introduction to quantitative genetics. 4th Edition. Longman Technical, Essex, UK

[9] Hallauer, A.R. Miranda, J.B. and Carena J. (2010) Quantitative genetics in maize breeding. Iowa State University, Ames, Iowa, U.S.A.

[10] Melani, M.D. and Carena, M.J. (2005) Alternative maize heterotic patterns for the northern corn belt. Crop Science 45: 2186-2194.

[11] Lee, M., (1995). DNA markers and plant breeding programmes. Advances in Agronomy 55:265-344.

[12] Melchinger, A.E. and Gumber, R. K. (1998). Overview of heterosis and heterotic groups in agronomic crops. In Lamkey, K. R. and Staub, J. E. (eds), *Concept and Breeding of Heterosis in Crop Plants*. CSSSA-SP, Publication no. 25, Madison, WI, pp. 29-44.

[13] Barata, C. and Carena, M. J. (2006). Classification of North Dakota maize inbred lines into heterotic groups based on molecular and testcross data. Euphytica 151: 339-349.

[14] Fan, X.M., Tan, J., Yang, J.Y., Liu, F., Huang, B.H., and Huang, Y.X. (2002). Study on combining ability for yield and genetic relationship between exotic tropical, subtropical maize inbreds and domestic temperate maize inbreds. Science Agriculture Sin. 35: 743-749.

[15] Huang, Y.Q., and Li, J.S. (2001). Classification of heterotic groups with RFLPs among 45 maize inbred lines. Agric. Sin. China 34: 244-250

[16] Vasal, S.K., Srinivasan, G., Han, G.C., and Gonzalez, F. (1992). Heterotic patterns of ninety-two white tropical CIMMYT lines. Maydica 37: 259-270.

[17] Paterniani, E., (1990). Maize breeding in the tropics. CRC Critical review in plant science. 9: 125-154

[18] Paliwal, R.L., Granados, G., Lafitte, H.R., and Violic, A.D. (2000). Tropical maize improvement and production. FAO, Rome, Italy.

[19] Goodman, M.M. (1985) Exotic maize germplasm: status, prospects and remedies. Iowa State Journal of Research 59: 497-527

[20] Menkir, A., Melake-Berhan, A., Thé, C., Ingelbrecht, I. and Adepoju, A. (2004). Grouping of tropical mid-altitude maize inbred lines on the basis of yield data and molecular markers. Theoretical and Applied Genetics 108: 1582-1590.

[21] Kim, S.K., and Ajala, S.O. (1996) Combining ability of tropical maize germplasm in West Africa II: Tropical vs temperate x tropical origins. Maydica 41: 135-141.

[22] Kim, S.K., Ajala, S.O., and Brewbaker, J.L. (1999). Combining ability of tropical maize germplasm in West Africa III. Tropical maize inbreds. Maydica 44: 285-291

[23] Menkir, A., Kling, J.G., Badu-Apraku, B. and Ingelbrecht, I. (2005). Molecular marker-based genetic diversity assessment of *Striga*-resistant maize inbred lines. Theoretical and Applied Genetics 110(6): 1145-1153.

[24] Badu-Apraku, B., Fakorede, M.A.B., Oyekunle, M. and Akinwale, R.O. (2011c). Selection of extra-early maize inbreds under low N and drought at flowering and grain- filling for hybrid production. Maydica 56: 1721-1735.

[25] Wu, J.C., Xu, C.X., Chen, H.M., Tan, J., Han, X.R., Huang, B.H. and Fan, X.M. (2007). Studies on combining ability and heterotic grouping of 27 quality protein maize inbreds and four temperate representative inbreds of Chinese major heterotic groups. Science in Agriculture Sin 40: 1288-1296.

[26] Agbaje, S. A., Badu-Apraku, B. and Fakorede, M.A.B. (2008). Heterotic patterns of early maturing maize inbred lines in *Striga*-free and *Striga* infested environments. Maydica 53: 87-96.

[27] Badu-Apraku, B., Lum, A.F., Akinwale, R.O. and Oyekunle, M. (2011a). Biplot analysis of diallel crosses of early maturing tropical yellow maize inbreds in stress and nonstress environments. Crop Science 51: 173-188.

[28] Betrán, F.J., Ribaut, J.M., Beck, D., Gonzalez de Leon, D. (2003). Genetic diversity, specific combining ability, and heterosis in tropical maize under stress and nonstress environments. Crop Science 43: 797-806.

[29] Akinwale R.O., Badu-Apraku, B., Fakorede, M.A.B. and Vroh-Bi I. (2014) Heterotic grouping of tropical early-maturing maize inbred lines based on combining ability in *Striga*-infested and *Striga*-free environments and the use of SSR markers for genotyping. Field Crop Research 156:48-62

[30] Badu-Apraku, B., M. Oyekunle, R.O. Akinwale and M. Aderounmu (2013). Combining ability and genetic diversity of extra-early white maize inbreds under stress and nonstress environments. Crop Science 53: 9-26 DOI: 10.2135/cropsci2012.06.038

[31] Badu-Apraku, B., Fakorede, M.A.B., Gedil, M., Annor, B., Talabi, A. O., Akaogu, I. C., Oyekunle, M., Akinwale, R. O. and Fasanmade, T. Y. (2016a). Heterotic patterns of IITA and CIMMYT early-maturing yellow maize inbreds under contrasting environments. Agronomy Journal 108: 1321-1336.

[32] Badu-Apraku, B., Fakorede, M.A.B., Talabi, A.O., Oyekunle, M., Akaogu, I.C., Akinwale, R.O., Annor, B., Melaku G., Fasanmade, Y., and Aderounmu, M. (2016b). Gene action and heterotic groups of early white quality protein maize inbreds under multiple stress environments. Crop Science 56: 183-199 doi.org/10.2135/cropsci2015.05.0276

[33] Badu-Apraku, B. and Akinwale, R.O. (2019) Biplot analysis of line × tester data of maize (*Zea mays* L.) inbred lines under stress and nonstress environments Cereal Research Communications 47(3): 518-530 DOI: 10.1556/0806.47.2019.250133-3720/$20.00

[34] Badu-Apraku, B., Oyekunle, M., Akinwale, R.O. and Fontem Lum, A. (2011b) Combining ability and heterotic groups of early-maturing tropical white maize inbred lines under stress and nonstress environments. Agronomy Journal 103: 544-557.

[35] Badu-Apraku, B., Fakorede, M.A.B., Melaku G., Talabi, A.O., Annor, B., Oyekunle, M., Akinwale, R.O., Fasanmade Y. and Akaogu I.C. (2015). Heterotic response among crosses of IITA and CIMMYT early white maize inbred lines under multiple stress environments. Euphytica 206: 245-262 DOI 10.1007/s10681-015-1506-0

[36] Fan, X.M., Zhang, Y.M., Yao, W.H., Chen, H.M., Tan, J., Xu, C.X., Han, X.L., Luo, L.M. and Kang, M.S. (2009). Classifying maize inbred lines into heterotic groups using a factorial mating design. Agronomy Journal 101 (1): 106-112.

Section 2

# Agronomy and Breeding

Chapter 5

# Physiological, Ecological and Genetic Interactions of Rice with Harmful Microbes

*Yulin Jia and Melissa H. Jia*


## Abstract

Rice is one of the most important food crops for mankind and suffers significant crop loss annually due to rice diseases. Availability of genome sequences of rice has served as a springboard to utilize its innate immunity to prevent rice diseases. Knowledge on interactions of rice and rice pathogens has rapidly accumulated. Effective resistance genes have been identified from cultivated, weedy species of rice, and wild rice relatives and their roles in plant innate immunity have been uncovered. Presently, rice diseases are being managed using host resistance genes and pesticides in diverse culture systems around the globe. This chapter presents a simple review of interactions of rice with harmful microbes causing the two major damaging diseases, rice blast and sheath blight. The review is written to target new readers in life sciences. Knowledge and critical literatures on physiological, genetic, and ecological aspects of host-pathogen interactions are presented to gain insights leading to sustainable disease management systems.


**Keywords:** rice, disease, resistance genes, innate immunity, crop protection

## 1. Introduction

Rice originated 130 million years ago to become the annual cereal crop that provides 20% of the essential calories needed to feed more than one-half of the world population [1]. Major producing countries of rice are China and India; most of the rice produced by both countries is domestically consumed. In the USA, rice has been under cultivation for over three hundred years; however, today less than 2 percent of world rice is being produced in the USA. The rice crop in the USA is known for its high rough rice yield, and excellent milling and cooking quality that occupies the top 10 in the international marketplace. The earliest rice seeds were discovered over 7,700 years ago in the Hangzhou area, eastern China [2] establishing China as the first ancient civilized country to grow rice. The domestication and agronomic improvement of rice began with the exploitation of wild rice and land race varieties during ancient times. Since then, the primary breeding objective has been to increase the yield potential to meet the rapidly increasing demands of human consumption. In the 1950s, the semi-dwarf gene *sd1* was discovered and used to develop semi-dwarf varieties with high grain yield but without lodging. The rapid utilization of this technology began the green revolution in rice. In 1960, the principles and techniques of hybrid production developed by American corn breeders were adapted for hybrid rice breeding in China [3]. During the 1970s, indica types of hybrid rice were rapidly deployed in major rice producing areas and, currently, hybrid rice is more than 50% of the rice production in China. Hybrid rice production has expanded to other countries and in 2020, occupied approximately 35% of US rice acreage. Globally rice crops have been well protected against diseases; however, throughout this intense agronomic selection for yield enhancing genes, the corresponding genetic diversity needed for effective rice disease control has decreased [4]. Increasing yield through hybrid rice is one way to increase the total rice production. However, rapid extension of hybrid rice worldwide will present a new challenge for the control of rice diseases such as rice blast and sheath blight diseases since limited germplasm can be used for hybrid seed production. An insignificant race of the southern leaf blight fungus *Bipolar maydis* under favorable conditions resulted in 6 billion crop loss when maize hybrids with cytoplasm (cms-T) were heavily deployed in the southern USA (for example, [5]). Understanding mechanisms of interactions of rice with harmful microbes are therefore critical for food security.

Most diseases of rice are caused by harmful fungi such as blast disease caused by [*Magnaporthe oryzae* (anamorph: *Pyricularа oryzae*)], sheath blight [*Rhizoctonia solani* (telomorph: *Thanatephorus cucumeris*)], brown spot [*Cochliobolus miyabeanus* (anamorph: *Bipolaris oryzae*)], false smut (*Ustilaginoidea virens*), kernel smut [*Tilletia barclayana* (*Neovossia horrida*)], Narrow brown leaf spot [*Cercospora janseana* (telomorph: *Sphaerulina oryzina*)], crown sheath rot (*Gaeumannomyces graminis* var. *graminis*), downy mildew (*Phytophthora macrospora*), aggregate sheath spot [*Rhizoctonia oryzae-sativae* (Sawada) Mordue and *R. oryzae* Ryker & Gooch)], eyespot [*Drechslera gigantea* (Heald et Wolf) S. Ito], leaf smut (*Entyloma oryzae*), leaf scald (*Microdochium oryzae*), seedling blight (*Pythium*, *Fusarium*, *Diplodia*, *Rhizoctonia*, and *Penicillium* spp), stem rot (*Phytophthora sojae*), bakanae [*Fusarium moniliforme* (syn. F. verticilloides), teleomorph: *Gibberella fujikuroi* (syn. *Gibberella moniformis*)], respectively. The next most common diseases are caused by harmful bacteria such as leaf blight caused by *Xanthomonas oryzae* pv. *oryzae* (*X. campestris* pv. *oryzae*), bacterial panicle blight [*Burkholderia glumae* and *Burkholderia gladioli* (Severin)], bacterial leaf streak (*Xanthomonas oryzae* pv. *Oryzicola*), bacterial foot rot (*Dickeya zeae*); sheath brown rot (*Pseudomonas fuscovaginae*), bakanae [*Fusarium moniliforme* (syn. *F. verticilloides*), teleomorph: *Gibberella fujikuroi* (syn. *Gibberella moniformis*)], respectively [6].

Any of the above mentioned rice diseases can result in severe yield loss when ideal circumstances favor disease infection and development. Among them, blast is the most devastating rice disease worldwide [6–8]. Disease symptoms of blast are more pronounced on rice under high nitrogen based nutrients and drought stress conditions, and often seen on rice plants grown on levees and the edges of rice paddies. Blast disease annually significantly reduces the crop in upland and often in flood irrigated rice production as well. Three notable crop damages are 1) the widespread destruction of the cultivar 'Newbonnet' in 1980s in the US; 2) 45% of rice affected by blast in 1993 in Japan; and 3) blast disease caused significant damage in 106 million hectares from 1982 to 2005 in China [9, 10]. Presently, occurrence and severity of blast are becoming more widespread in the Southern US and California. Sheath blight causes damage on all rice cultivars, especially on semidwarfs [11]. Sheath blight disease is the second most damaging disease after blast worldwide. Sheath blight was considered a minor disease for many years but has become more destructive under intensified high input production systems. The damage due to sheath blight was estimated to be from 20–42% in a simulation study [12] and 50% of crop loss under favorable conditions in the USA [13]. To date, sheath blight reduces the crop more than that of rice blast in the USA.

Understanding impacts of rice exposed to the above mentioned harmful microbes under changing climates is a never ending challenge. Rapidly evolving technologies such as genome sequencing, computational biology and genome editing methods have been used to study how nucleotide sequence changes influence the outcomes of host-pathogen interactions. The aim of this chapter is to update the complex interactions between rice plants with the harmful microbes causing diseases. Emphasis is placed on some aspects of the physiological, ecological, and genetic interactions between the rice plant and harmful pathogens such as rice blast (*M. oryzae*) and sheath blight (*R. solani*).

## 2. Physiological responses of rice plants to pathogens

Rice plants have cuticles, silica and cell walls that often serve as the first passive defense to contain pathogens [14–16]. Subsequent failed active resistance leads to visible pathogen damages. For rice blast disease, typically a diamond shaped symptom on the leaf is called leaf blast, and dark brown on the neck as panicle blast, respectively (**Figure 1**). Both leaf and panicle blast can result in significant yield reduction.

When they encounter rice plants, asexual spores of blast fungus, *M. oryzae* germinate and initiate their life cycle by penetrating the cells with infection pegs from tightly adhered swollen hyphal tips with the highest biological turgor pressures known for a biological organism [17]. During penetration *M. oryzae* absorbs nutrients by producing cutinases to degrade cutin in the cuticle and pectolytic enzymes. *M. oryae* secretes heat labile molecules such as endo ß-1,4-D-xylanase to solubilize rice cell wall fragments to kill cells [18, 19]. After penetration *M. oryzae* develops invasive hyphae that are in direct contact with the membrane of the live cells. Soon after that within approximately 48 hrs *M. oryzae* colonizes the host tissues and releases asexual spores at the end of invasion [20, 21]. *M. oryzae* is thus classified as a hemibiotrophic pathogen.

**Figure 1.**
*Typical symptoms of rice blast disease. A. Leaf blast disease after an artificial inoculation under greenhouse conditions at USDA ARS Dale Bumpers National Rice Research Center, Arkansas, USA, and B. Panicle blast disease in a rice field under natural infections in Puerto Rico (photograph credit: Miss Adriana Rivera).*

**Figure 2.**
*Typical sheath blight disease on sheath and leaves in a rice cultivar at the booting stage in Stuttgart, Arkansas, USA.*

Sheath blight disease can be found on young rice seedlings but usually does not begin vertical development until the plant is at the reproductive growth stage. Typical symptoms are found on the lower leaf sheaths of rice plants at late tillering or the booting stage (**Figure 2**). The sheath blight disease at flag leaf often results in significant yield reduction.

Sheath blight lesions appear as circular, or ellipsoid, green-gray, water-soaked spots at about 1–3 cm long. As a lesion develops it can enlarge to 2 cm in width and 3–10 cm in length and becomes bleached with an irregular purple-brown border at the center [13]. The fungus *R. solani* grows on the surface of the leaf sheath upward and produces new side branches approximately at 45- and 90-degree angles 5–6 mm from the growing tips of mycelia. The continued growing of side branches results in the formation of an infection cushion that is attached to the epidermis often with mucilage like materials. Penetration peg is then formed from the flattened cells at the base of the cushion. The penetration peg then penetrates the inner epidermal cells to obtain nutrients from the inner and later the outer epidermal cells of the leaf sheath. Within approximately, 48 hrs after inoculation new infection hyphae is produced in an epidermal cell lumen [22, 23]. During infection *R. solani* also produces toxin [24]. Such toxins are called host specific toxins and are pathogenicity factors [25]. Toxins are known to be toxic to plants by inhibiting host defense responses.

## 3. Ecological interactions of rice plants with pathogens

The three major factors necessary for disease to occur in plants: 1) A pathogen that can cause disease, 2) a host plant that is susceptible to a pathogen, and 3) an environment that favors the pathogen infection. The environment includes temperature, humidity, light intensity, surrounding areas, and/or human intervention. These three-factors referred as the disease triangle contribute to the severity of disease. After diseases occur plant pathogens are typically disseminated within the same field and are often transmitted by wind and/or insects to fields far away from the diseased plants. In nature, rice pathogens find rice and survive with and/or without rice after their infection and amplification [26]. In the tropics the climates are relatively warm year round. Hence, overwintering is not an issue for

*M. oryzae*. The overwintered conidia and mycelia on alternative hosts and/or rice often serve as a source of primary infection in the disease cycle. *M. oryzae* species is a pathogen of over 50 grass species including crops such as wheat (*Tritium aestivum* L. [27]), Barley (*Hordeum vulgare* L. [28]) and finger millet (*Eleusine coracana*; L. [29]). However, each isolate of *M. oryzae* is often limited to attack a small group of grass species [29]. The causal agent for blast disease in one group of grass species is often different from similar groups that attack other grass species. Cross infections between the species observed under controlled conditions suggest that grass weeds, *Rottboellia exaltata*, *Echinochloa colona*, *Leersia hexandra* and *Alopecurus carolinianus* could be alternative hosts for *M. oryzae* [30, 31].

Even though *M. oryzae* is not a good competitor among saprophytes, rice seeds and diseased rice residues are considered as the primary sources of *M. oryzae* [32, 33]. Infectious *M. oryzae* were purified from foundation, certified and grower seeds in Arkansas (for example, [32]). The spores of *M. oryzae* produced by the contaminated rice seeds infected seedlings from 2 to 4 leaf stages [33]. The infected seedlings are then served as an inoculum for nearby healthy plants that develop blast symptoms later. Often *M. oryzae* infection from booting to flowering/immature panicle results in *M. orzyae* contaminated rice seeds. Diseased residues are considered as another primary source of inocula [34]. Infected rice residues up to 18 months from surface mulch were the sources of *M. oryzae* that caused leaf blast under field conditions (for example, [34]).

*R. solani* infecting rice belongs to an anamosis subgroup (AG)1-IA [35]. This type of pathogen typically lives on dead tissues, and often changes hosts during alternative growing seasons. Infection of *R. solani* usually begins from sclerotia and mycelia on debris from previous crops in the soil. Sclerotia usually accumulate around rice plants at the water and plant interface. Once sclerotia germinate, they develop mycelia that grow upward on rice plants. Depending upon humidity and moisture conditions, disease development is rapid at early booting to heading and grain-filling stages [36]. Infections often occur near the waterline after the establishment of permanent flood. Lesions on the upper parts of rice plants can be developed to the entire leaves and leaf sheaths (**Figure 2**). Lesions with mycelia on rice plants change from white or gray with brown borders to brown where sclerotia are loosely attached. At maturity, sclerotia are separated from rice plants for overwintering in soil. Under favorable conditions such as high humidity (≥95%) and temperature (28–32°C), *R. solani* spreads rapidly to upper rice plants including rice leaves, grains, and to adjacent plants [13]. *R. solani* causing sheath blight diseases is a broad host pathogen which has multiple anamosis subgroups that specialize on several plant species causing other plant diseases [37].

## 4. Genetic interactions of rice plants with pathogens

Blast pathogen *M. oryzae* is known to reproduce asexually under field conditions and it has been a challenge to perform sexual crosses under laboratory conditions. The genome sequences of 50 isolates from different times and places showed that they belong to six lineages including isolates from two pandemics on japonica and indica rice [38]. The *de novo* DNA sequences also revealed that these lineages diverged about a millennium ago. Genome sequences of one lineage uncovered evidences of sexual transmission and alleles from multiple lineages. In the USA over the past 6 decades *M. oryzae* races have become more diverse and virulent [39, 40].

Genetic interaction of rice with *M. oryzae* follows the gene for gene theory where a resistance (*R*) gene is effective in preventing pathogen *M. oryzae* strains that contain the corresponding aviruelnce (*AVR*) gene [41, 42]. Presently, 40 *AVR* genes have been identified, 11 of which have been cloned. *AVR* genes in *M. oryzae* are random secreted molecules predicted to play important roles in pathogenicity and fitness [43]. Rice *R* genes are mutable that may generate more *R* genes. *R* gene polymorphism is thus a significant source of complexity of the interactions of rice with different rice pathogens [44, 45]. Most *R* genes in rice are members of a small gene

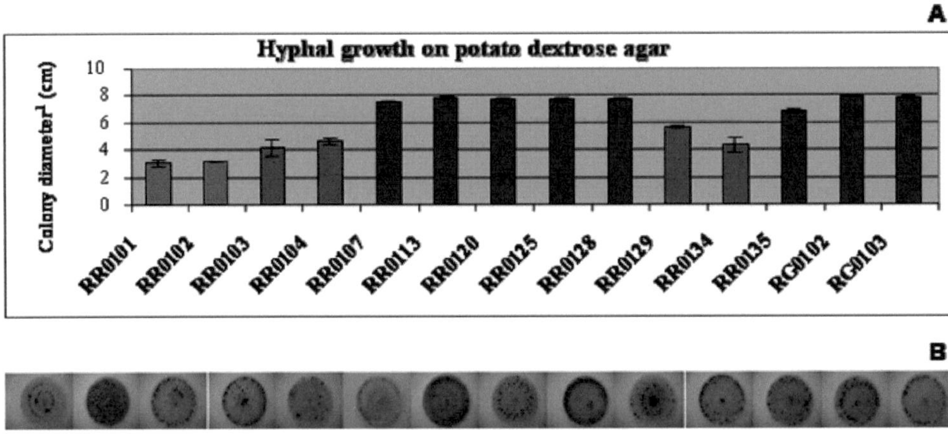


**Figure 3.**
*Physiological characterization of rice sheath blight fungus Rhizoctonia solani. A. hyphal growth on potato dextrose agar of each hyphae from indicated isolates, and B. different morphologies of sclerotia of indicated isolates in a.*

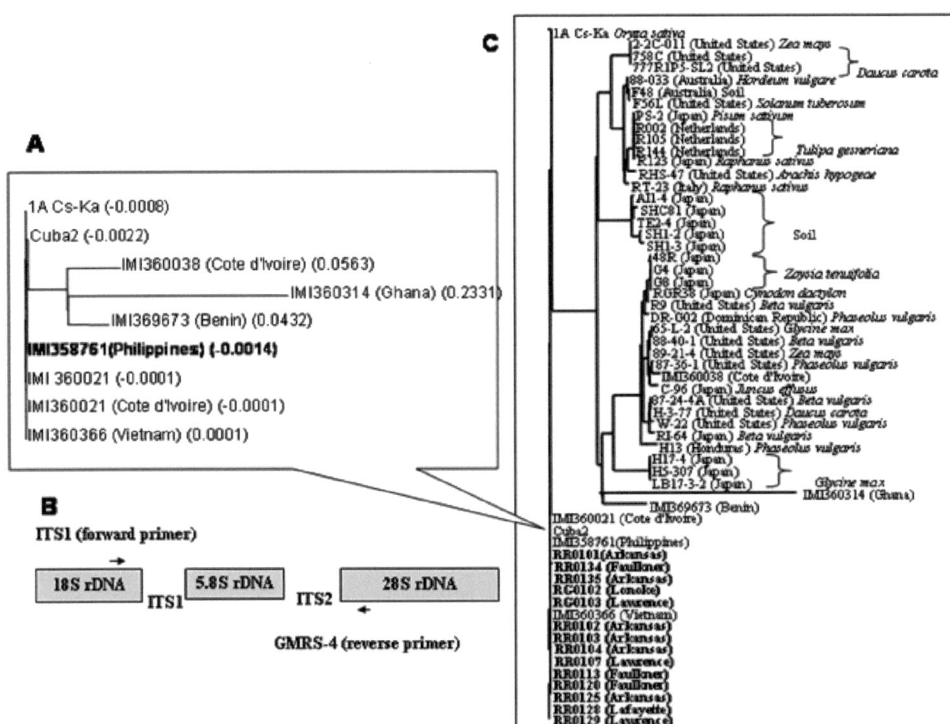


**Figure 4.**
*Molecular characterization of rice sheath blight fungus Rhizoctonia solani. A to C describing region and phylogenetic relations (http://www.ncbi.nlm.nih.gov/nuccore, genbank accession numbers, AY185104 to AY185115 of 14 isolates from indicated counties in Arkansas).*

family and are predicted to encode cytoplasmic NLR proteins with nucleotide binding site (NBS) and leucine rich repeats (LRR) [45, 46]. The rice genome (430 Mb) has 480 such NBS-LRR genes that can be the sources of *R* genes to different rice pathogens [47, 48].

Details on the genetic interaction of rice with *R. solani* has lagged far behind that of with *M. oryzae*. A major *R* gene to *R. solani* has not been discovered yet. Minor *R* genes such as *qSHB9–2* in rice cultivar Jasmine 85 have been identified [49, 50]. Further genetic and functional analyses suggest that the *ABC* transporter genes involved in rapid nutrient transportation is responsible for 25% of genetic resistance [51]. *R. solani* (AG)1-IA contains heterogenic multinuclei, and complete genome sequence of *R. solani* (AG)1-IA has been difficult due to the challenges on genome assembly [52]. Differences in morphology, speed of hyphal growth have been noticed and some isolates do display less aggressiveness (**Figure 3**).

These isolates have been recommended for controlled inoculations and genome sequencing [35]. The length of ribosomal DNA internal transcribed spacer (rDNA-ITS) can distinguish subspecies of *R. solani* from *R. oryzae* and *R. oryzae-sativae* [53], minor variation of DNA sequence, hyphal growth on potato dextrose agar and morphology of sclerotia can be seen among the isolates collected. All isolates tested so far in the USA were clustered into one clade, (**Figure 4**) [35, 54].

## 5. Remarkable features of rice plant innate immunity

Like other plants, rice cannot move to escape from pathogen attack and must evolve an efficient defense system [55]. The plant passive defense system is often initiated by cell wall, and cuticles by releasing pathogen associated molecular patterns (PAMPs, [56]). After sensing these PAMPs plants activate a variety of early defense responses including stomatal closure, transcriptional reprogramming responses and callose deposition that is called pattern-triggered immunity (PTI) [57]. More active defense response is often elicited by NLR *R* gene products. Upon the detection of pathogen *AVR* gene products NLR proteins reorganize and transduce defense signaling often resulting in programed host cell death that is called elicitor triggered immunity (ETI) [58]. Exactly how PTI and ETI lead to effective resistance response is still largely unclear (**Figure 5A**).

*R* genes are also known to be under fast evolution diversification [58] and have also evolved efficient methods to detect the unstable products of *AVR* genes in *M. oryzae*. A single amino acid of each of major blast *R* genes, *Pi-36*, *Pi-d2*, and *Pi-ta* has been found to determine its efficacy of resistance (for example, [59–61]) (**Figure 5B**). *Pi-d2* is a single copy gene encoding a predicted novel B-lectin receptor kinase with an extracellular domain of a bulb-type mannose specific binding lectin (B-lectin) and an intracellular serine–threonine kinase domain. A single amino acid difference at position 441 Isoleucine to Methionine (R to S) of Pi-d2 distinguishes resistant from susceptible allele. *Pi-d2* was localized in plasma membrane [60]. *Pi-36* is a single copy gene encoding an NBS-LRR protein. A single amino acid at the position 590 Aspartic acid to Serine (R to S) was found to associate with the resistance phenotype [61]. *Pi-ta* encodes an NLR protein with imperfect LRR [59]. Surveys in rice germplasm have identified only one *Pi-ta* allele conferring resistance and thirteen *pi-ta* alleles conferring susceptibility [62–68]. In most cases, a single nucleotide substitution results in a functional polymorphism distinguishing between resistance and susceptibility [59, 67, 68]. All resistant Pi-ta proteins have alanine at position 918 and all susceptible pi-ta proteins have serine at position 918 [62, 65].

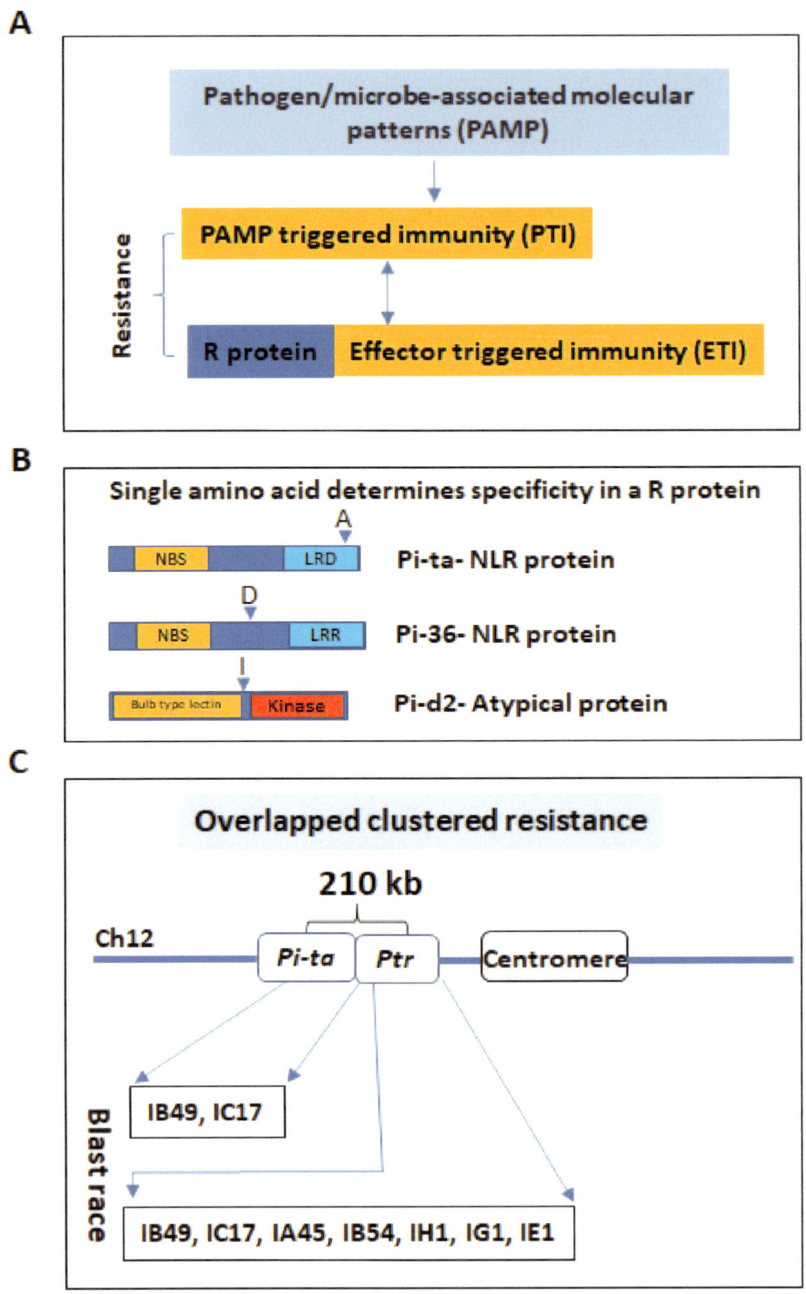


**Figure 5.**
*Diagram shows two significant mechanisms of blast R genes. A. Showing effective resistance is a result of pathogen/microbe-associated molecular pattern triggered immunity (PTI) and effector triggered immunity (ETI) mediated by R protein. B. Showing three blast R proteins with a single amino acid determining recognition specificities. Single letter code was used. C. Showing the location and resistance spectra of blast R genes Pi-ta and Ptr near the centromere of rice chromosome 12. This genomic region has been transferred as a linkage block into diverse rice germplasm due to suppressed recombination. Rice varieties with Pi-ta are resistant to the blast races IB49 and IC17 and with Ptr are resistant to the blast races IB49, IC17, IA45, IB45, IB54, IH1, IG1, and IE1. Graphics were not drawn in proportion.*

Another efficient method of plant innate immunity is a plausible failsafe mechanism. *R* genes and helper genes in plant immunity are often found in a short physical interval that can be easily passed on to the next generation (**Figure 5C**).

Rice varieties with *Pi-ta* is resistant to the blast races IB49 and IC17 [63]. *Pi-ta* was predicted to require another gene *Ptr* to be more effective [69]. The *Ptr* gene referred as *Pi-ta2* is 210 kb from *Pi-ta* on chromosome 12 [70, 71]. The *Ptr* gene was identified using a genetic screen of a mutant population created by fast neutrons and was cloned using map-based cloning approach and resistant function was validated by CRISPR-CAS 9 [70–72]. *Ptr* is a broader-spectrum blast *R* gene independent to *Pi-ta* predicted to encode a protein with 4 armadillo repeats [70]. Rice varieties with *Ptr* are resistant the blast races IB49, IC17, IA45, IB45, IB54, IH1, IG1 and IE1. Resistance spectra of both *Pi-ta* and *Ptr* were overlapped for both races IB49 and IC17. Rice genes with armadillo repeats are known to be involved in a wide range of biological functions suggesting that the *Ptr* gene in rice is a failsafe for disease resistance [73]. It remains to be determined the role of *Pi39 (t)/Pi42(t)* (LOC_Os12g1837412), 12 kb from *Pi-ta* and 198 kb from *Ptr* [74, 75] in blast resistance [76]. Resulting knowledge can aid in blast resistant breeding.

## 6. Genetic improvement of rice for enhanced resistance

Disease resistance has been one of the major breeding strategies in rice breeding programs worldwide. Continued investigation of the mechanism of plant innate immunity can accelerate disease resistant breeding efforts (**Figure 6**).

A donor or several donors for *R* genes are used for crossing or triple crossing, either resistant progeny are selected for further evaluation or these $F_2$ are advanced 2 to 3 generations and then evaluated for their disease reactions. Early breeding involved the use of field conditions to evaluate the disease reactions of landrace varieties or breeding lines. More recently, rice breeding lines and other germplasm are evaluated for disease reactions under greenhouse conditions for both blast [77, 78] and sheath blight [79]. In many countries evaluations of disease reaction are conducted under field conditions because there often exist conducive environmental conditions (for example, in Colombia, [80]). However, it is difficult to determine if observed resistance is due to any particular *R* gene due to overlapped resistance such as *Pi-ta* and *Ptr* for blast races IB49 and IC17. Marker assisted selection (MAS) uses genetic markers that are linked to the *R* genes or derived from portions of *R* genes [81]. This allows the DNA of progeny to be rapidly examined

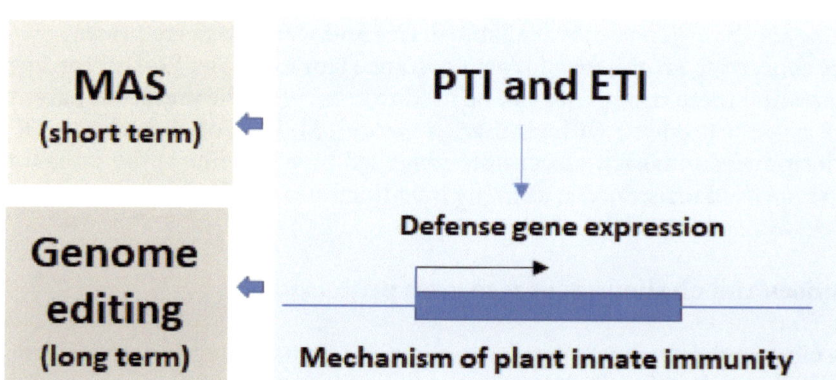


**Figure 6.**
*Understanding interplays of pattern-triggered immunity (PTI) and elicitor triggered immunity (ETI) in activating robust defense gene expression results in a short term benefit - breeding for disease resistance using marker assisted selection (MAS) approach and a long term benefit- engineering durable effective resistance using genome editing.*

for the presence of *R* genes without growing the plants to observe their reaction to pathogens. Breeders can monitor these tagged *R* genes during crossing, selection and incorporation [82–86]. MAS is a promising technology for increasing precision of selection for *R* genes and increasing speed of breeding for resistance.

One bottleneck for the use of MAS in classical breeding is the limitation of the number of markers that can be used to accurately tag resistance in any given breeding line because genetic backgrounds of breeding lines are relatively uniform in comparison with most markers developed using diverse genetic crosses of indica with japonica. Thus, the improvements of published markers are needed using local breeding lines. Gene specific DNA markers for *Pi-ta* and *Pi-b* and SSR markers for *Pikm/Piks* and *Piz* were developed based on genomic differences of local breeding lines, and are effective for MAS in the USA [85–88]. Both *Pi-ta* and *Pi-b* markers have also been adapted for blast resistant breeding programs worldwide. Another limitation for MAS is that increased numbers of markers may create a situation in which the segregating population would need to be large enough to ensure the incorporation of all *R* genes to reach the durability of resistance to the pathogens in addition to other breeding objectives. Finally, there often exist linkage blocks where disease resistance genes co-locate with inferior genes involved in quality and productivity [89, 90].

Genetic engineering is a tool that can overcome the above mentioned limitations of classical plant breeding with MAS. This approach will eliminate the inferior genes due to the linkage block in the final products. Any gene should be able to function in any other organism if genetic components for its expression are intact. Genes can be available from any organism, not just from a plant of the same species, as is the case with classical plant breeding with MAS. After genes are added to a genome, their copy numbers, location and their expression may be controlled by other genes. The common method is to insert a new gene into a plant using the natural genetic engineer, bacterium *Agrobacterium tumefaciens* [91]. The plants are regenerated from a single cell because each cell of the engineered plant must contain the new gene. Another method is biolistic transformation using the "gene gun", which bombards protoplasts with metal particles coated with the foreign genes [92]. The gene gun is much less efficient than the Ti plasmid for gene transfer because of multiple copy integrations. In rice, genetic engineering has been demonstrated to improve resistance to sheath blight and blast in various laboratories worldwide [60, 93]. One obvious advantage of genetic engineering is that genes from other organisms might give plants defenses that it never had before. The isoflavone synthase gene is not available in rice and was transferred from soybean into rice conferring an enhanced blast resistance (for example, [94]). Transgenic rice expressing these transgenes can be used to cross with the recurrent parent to produce cisgenic products without marker genes [95]. Each of cloned major *R* genes can be introduced into each susceptible advanced breeding line at the same time to develop improved resistance in as many breeding lines as possible.

## 7. Promises and challenges of rice crop protection

The most environmentally benign method for human intervention is using host *R* genes integrated with cultural management practices. General effective strategies to protect rice crop are prevention of introduction of pathogen; removal of established pathogen from infected rice plants; prevention of pathogen from infecting susceptible plants by growing rice under unfavorable climate for pathogen; broadening genetic basis; introduction of *R* genes with overlapping resistance spectra. As we better understand exactly how pathogenesis occurs, we can try to interfere with

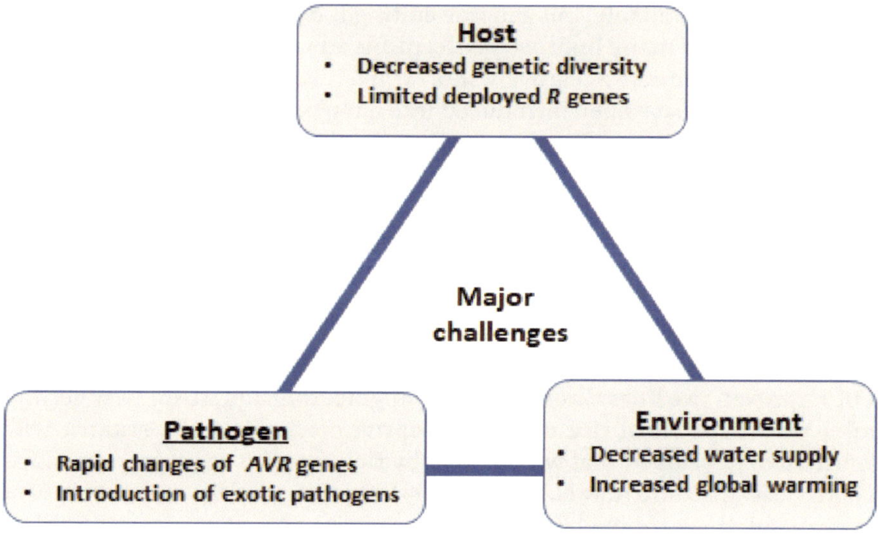


**Figure 7.**
*Major challenges of crop protection imposed by host-pathogen and environment.*

how pathogens find their hosts or inactivate important pathogen enzymes or toxins. Major *R* genes are not available to control rice sheath blight, however, sheath blight tolerant cultivars with suitable architecture and plant growth can reduce yield losses [96]. In the absence of *R* genes, fungicides are often used. The reduction in disease progress can be achieved when rice fields are treated with efficacious fungicides at the proper growth stage. A simulation study shows $43 million increase through sheath blight resistant rice production that is enough to feed 1.7 million people in the Mid-south [97].

Presently there still exist at least six major challenges for rice crop protection (**Figure** 7). 1). The intense demand for yield and quality decreases the genetic diversity of cultivated rice needed for basal defenses, such as expansion of hybrid rice that have put rice at a genetic disadvantage relative to the genetic changes of the pathogen. 2). Only limited *R* genes can be deployed locally for preventing diseases due to clonal amplification of the pathogen populations. 3). Pathogen populations can change through time and pathogen genotypes can interact with specific host genotypes leading to the "breakdown" of resistance within very short periods of time and the pathogen can adapt to new environments by rapid alteration of the *AVR* genes to create virulent races of the pathogen [98]. 4). Introduction of unwanted exotic pathogens into rice production areas through seeds. 5). Decreased water supply would increase the incidence and severity of blast disease, and 6) increased global warming not only reduces water supply but also is more dangerous if increased temperature and CO2 concentration create favorable environment for disease and weeds which can be potential sources of alternative hosts.

## 8. Future perspectives

In the future, the technique of genetic engineering should become easy and inexpensive to use, and social and economic concerns of Genetic Modified Organism (GMO) will be resolved. Genetic engineering of resistance will certainly enhance our capacity to prevent the crop loss due to diseases. Designing rice plants with novel resistance to a wide range of pathogens will be possible using genome editing mediated by CRISPR-Cas system [99]. At the present time, no commercial

GMO rice variety is available, but genome editing is expected to become the breeder's choice since many improved susceptible advanced breeding lines can be engineered with *R* genes in a relatively short time. It is important to monitor if new pathogen genotypes have been introduced into a region and at what frequencies certain pathogen genotypes change over time. On site information of the structure of pathogen populations is useful for the development and implementation of effective disease control strategies, and also provides insights into the evolution of pathogen populations in response to challenges imposed by host *R* genes. Therefore, the study of co-evolutionary mechanisms controlling the interaction of rice and pathogens should allow the application of these discoveries to the construction of more resistant plants. Novel approaches applied to study interactions of rice with *R. solani* has also begun to generate useful knowledge that will lead to the development of improved rice lines through genetic engineering and MAS. New germplasm including weedy species of rice and their adaptive mechanism of resistance will be identified or developed that will be used by rice breeders to incorporate novel sources of resistance into new cultivars [100–102]. Genetic mutants, mutant and mapping populations are available for uncovering important genes to control major rice diseases using methods of forward and reverse genetics [101–104]. More user friendly molecular markers will be identified to accelerate the development of improved rice cultivars through MAS worldwide and more robust *R* genes will be characterized for their deployment either using genetic engineering or MAS. Finally, the development of improved crop management programs to allow increased crop genetic heterogeneity can also be a solution to reduce crop damages due to rice diseases [105].

## Acknowledgements

The authors thank the past and current staff members of USDA ARS Dale Bumpers National Rice Research Center University of Arkansas Rice Research and Extension Center and Agricultural Experiment Station, Lajas Substation, Puerto Rico for technical supports. For critical reviews we thank Drs. Santosh Sharma (DB NRRC, Stuttgart, AR) and Yanbo Huang (USDA ARS Mississppi State, MS). This work was in part supported by NSF grant PGR1947609. USDA is an equal opportunity employer and provider.

## Author details

Yulin Jia* and Melissa H. Jia
USDA-ARS Dale Bumpers National Rice Research Center, Stuttgart, AR, USA

*Address all correspondence to: yulin.jia@usda.gov

## References

[1] Webpage: Available from: http://www.knowledgebank.irri.org/ericeproduction/Importance_of_Rice.htm.

[2] Zong Y, Chen Z, Innes JB, Chen C, Wang Z, Wang H. Fire and floor management of coastal swamp enabled first rice paddy cultivation in east China. Nature. 2007;449:4-59-462.

[3] Yuan L. P. 2000. Chinese can be self-supported relying on the advancement of science. Farmers daily, Nov 28, 2000. (in Chinese)

[4] Peng YL. Genome-wide identification of genes controlling hyphal growth of *Magnaporthe oryzae*. In: The 4th International Rice Blast Conference, October 9-14, 2007; Changsha, Hunan, China.

[5] Brun HA. Southern corn leaf blight: a story with retelling. Agronomy Journal. 2017;109:1-7.

[6] Ou SH. Rice Diseases, 2nd ed. Commonwealth Agricultural Bureaux Slough UK: 1985. 380p.

[7] Couch BC, Kohn LM. A multilocus gene genealogy concordant with host preference indicates segregation of a new species, *Magnaporthe oryzae*, from *M. grisea*. *Mycologia*. 2002;94:683-693.

[8] Nalley L, Tsiboe F, Durand-Morat A, Shew A, Thoma G. Economic and environmental impact of rice blast pathogen (*Magnaporthe oryzae*) alleviation in the U.S. PLoS One. 2016;11 (12). DOI:10.1371/journal.pone.0167295

[9] Kawasaki S, editor. Rice Blast: Interaction with rice and control. In: Proceedings of the 3rd International Rice Blast Conference. Springer Netherlands, 2004. 302pp. DOI:10.1007/978-0-306-48582-4

[10] Khush GS, Jena KK. Current status and future prospects of research on blast disease in rice (*Oryza sativa*). In: Proceedings of the 4th International Rice Blast Conference, October 9-14, 2007; Changsha, Hunan, China.

[11] Savary S, Castilla N, Elazegui F, McLaren C, Ynalvez M, Teng P. Direct and indirect effects of nitrogen supply and disease source structure on rice sheath blight spread. Phytopathology. 1995;85:959-965.

[12] Cu R, Mew T, Cassman K, Teng P. Effect of sheath blight on yield in tropical, intensive rice production system. Plant Disease. 1996;80:1103-1108.

[13] Lee FN, Rush MC. Rice sheath blight: A major rice disease. Plant Disease. 1983;67:829-832.

[14] Hematy K, Cherk C, Somerville S. Host–pathogen warfare at the plant cell wall. Current Opinion in Plant Biology. 2009;12:406-413. DOI: 10.1016/j.pbi.2009.06.007

[15] Seebold KW, Kucharek TA, Datnoff LE, Correa-Victoria FJ, Marchetti MA. The influence of silicon on components of resistance to blast in susceptible, partially resistant, and resistant cultivars of rice. Phytopathology. 2001;91:63-69.

[16] Yeats TH, Rose JK. The formation and function of plant cuticles. Plant Physiology. 2013;163:5-20.

[17] Howard RJ, Ferrari MA, Roach DH, Money DH. Penetration of hard substances by a fungus employing enormous turgor pressures. Proceedings of the National Academy of Sciences of the United States of America. 1991;88:11281-11284.

[18] Bucheli P, Doares SH, Albersheim P, Darvill A. Host-pathogen interactions.

XXXVI. Partial purification and characterization of heat-labile molecules secreted by the rice blast pathogen that solubilize plant cell wall fragments kill plant cells. Physiological and Molecular Plant Pathology. 1990;36:159-173.

[19] Wu SC, Kauffmann S, Darvill AG, Albersheim P. Purification, cloning and characterization of two xylanases from *Magnaporthe grisea*, the rice blast fungus. Molecular Plant-microbe Interactions. 1995;8:506-514.

[20] Wang Z, Lin H, Valent B, Rutger JN, Jia Y. Cytological and molecular analyses of disease resistance to the rice blast fungus. Chinese Journal of Rice Science. 2007;21:335-340.

[21] Jia Y, Zhou E, Lee S, Bianco T. Co-evolutionary dynamics of rice blast resistance gene *Pi-ta* and *Magnaporthe oryzae* avirulence gene *AVR-Pita1*. Phytopathology. 2016;106: 676-683.

[22] Matssura K. Scanning electron microscopy of the infection process of *Rhizoctonia solani* in leaf sheaths of rice plants. Phytopathology. 1986;76:811-814.

[23] Hirooka T, Mayagi Y, Araki F, Kunoh H. Biological mode of action of flutolanil in its systemic control of rice sheath blight. Phytopathology. 1989;79:1091-1094.

[24] Brooks S.A. Sensitivity to a Phytotoxin from *Rhizoctonia solani* correlates with sheath blight susceptibility in rice. Phytopathology. 2007;97:1207-1212.

[25] Vidhyasckaran P, Ruby-Ponmalar R, Samiyappan R, Velazhahan R, Vimala R, Ramanatham A, Paranidharan V, Muthukrishnan S. Host specific toxin production by *Rhizocotnia solani*, the rice sheath blight pathogen. Phytopathology. 1997;87:1258-1263.

[26] Mew TW, Fabellar D, Elazequi FA. Ecology of rice sheath blight pathogen: parasitic survival. International Rice Research Newsletter. 1980;5:15-16.

[27] Urashima AS, Igarashi S, Kato H. Host range, mating type, and fertility of *Pyricularia grisea* from wheat in Brazil. Plant Disease. 1993;77:1211-1216.

[28] Yaegashi H. Inheritance of blast resistance in two-rowed barley. Plant Disease. 1988;72:608-610.

[29] Kato H. Biological and genetic aspects in the perfect state of rice blast fungus, *Pyricularia oryzae* Cav. and its allies. Gamma Field Symposium. 1978;17:1-22.

[30] Mackill AO, Bonman JM. New hosts of *Pyricularia oryzae*. Plant Disease. 1986; 70:125-127.

[31] Jia Y, Gealy D, Lin MJ, Wu L, Black *H. Carolina* Foxtail (*Alopecurus carolinianus*): Susceptibility and suitability as an alternative host to rice blast disease (*Magnaporthe oryzae* [formerly *M. oryzae*]). Plant Disease. 2008;92:504-507.

[32] Guerber C, TeBeest DO. Infection of rice seed grown in Arkansas by *Pyricularia grisea* and transmission to seedlings in the field. Plant Disease. 2006;90:170-176

[33] Faivre-Rampant O, Genies L, Piffanelli P, Tharreau D. Transmission of rice blast from seeds to adult plants in a non-systemic way. Plant Pathology. 2013;62:879-887.

[34] Raveloson H, Ratsimiala Ramonta, Tharreau D, Sester M. Long term survival of blast pathogen in infected rice residues as major source of primary inoculum in high altitude upland ecology. Plant Pathology. 2018;67:610-618.

[35] Wamishe Y, Jia Y, Singh P, Cartwright RD. Identification of field isolates of *Rhizoctonia solani* to detect quantitative resistance in rice under

greenhouse conditions. Frontiers of Agri. in China. 2007;1:361-367.

[36] Lee FN. Number, viability, and buoyancy of Rhizoctonia solani sclerotia in Arkansas rice fields. Plant Disease. 1980;64:298-300.

[37] Guo CJ, Chen ZY, Wang FM. Pathogenic variability in *Thanatephorus cucumeris* (Frank) Donk and techniques for identifying varietal resistance. *Scientia Agricultura Sinica* 1985;5:50-57.

[38] Gladieux P, Condon B, Ravel S, Soanes D, Maciel JLN, Nhani A, Chen L, Terauchi R, Lebrun MH, Tharreau D, Mitchell T, Pedley KF, Valent B, Talbot NJ, Farman M, Fournier E. Gene flow between divergent cereal- and grass-specific lineages of the rice blast fungus *Magnaporthe oryzae*. mBio. 2018; 9:e01219-17. doi:10.1128/mBio.01219-17.

[39] Marchetti MA. Race specific and rate reducing resistance to rice blast in US rice cultivars. In: Zeigler RS, Leong SA, Teng PS, editors. In: Rice Blast Disease Oxon (UK):CAB International;1994. p. 231-244.

[40] Wang X, Jia Y, Wamishe Y, Jia MH, Valent B. Dynamic changes in the rice blast population in the USA over six decades. Molecular Plant-Microbe Interactions. 2017;30:803-812. doi:10.1094/MPMI-04-17-0101-R.

[41] Flor HH. Current status of the gene-for-gene concept. Annual Review of Phytopathology. 1971;9:275-296.

[42] Silue D, Notteghem JL, Tharreau D. Evidence for a gene-for-gene relationship in the *Oryze sativa-Magnaporthe grisea* pathosystem. Phytopathology. 1992;82:577-580.

[43] Leach JE, Vera Crus CM, Bai J, Leung H. Pathogen fitness penalty as a predictor of durability of disease resistance genes. Annual Review of Phytopathology. 2001;39:187-224.

[44] Liu J, Wang X, Mitchell T, Hu Y, Liu X, Dai L, Wang G-L. Recent progress and understanding of the molecular mechanisms of the rice-*Magnaporthe* interaction. Molecular Plant Pathology. 2010;11:419-427. DOI: 10.1111/J.1364-3703.2009.00607.X

[45] Wang X, Lee S, Wang J, Ma J, Bianco TA, Jia, Y. Current advances on genetic resistance to rice blast disease. In: Rice-Germplasm, Genetics and Improvement. Yan W, Bao, J. editors. IntechOpen; 2014. p.195-217. DOI: 10.5772/56824

[46] Webpage. Available from: http://prgdb.crg.eu/wiki/Main_Page

[47] International Rice Genome Sequencing Project, Sasaki, T. The map-based sequence of the rice genome. Nature. 2005;436:793-800. https://doi.org/10.1038/nature03895

[48] Zhou T, Wang Y, Chen J-Q, Araki H, Jing Z, Jiang K, Shen J, Tian D. Genome-wide identification of NBS genes in japonica rice reveals significant expansion of divergent non-TIR NBS-LRR genes. Molecular Genetics and Genomics. 2004;271:402-415.

[49] Liu G, Jia Y, Correa-Victoria F, Prado GA, Yeater KM, McClung A, Correll JC. Mapping quantitative trait loci responsible for resistance to sheath blight in rice. Phytopathology. 2009;99:1078-1084.

[50] Liu G, Jia Y, McClung A, Oard JH, Lee FN, Correll JC. Confirming QTLs and finding additional loci responsible for resistance to rice sheath blight disease. Plant Disease. 2013;97:113-117.

[51] Oh Y, Lee S, Rioux R, Singh P, Jia MH, Jia Y, Mysore KS. Analysis of Differentially Expressed Rice Genes Reveals the Role of ATP-Binding Cassette (ABC) Transporters Against the Sheath Blight Pathogen, *Rhizoctonia solani*. Rice 2021; Submitted.

[52] Zheng A, Lin R, Zhang D, Qin P, Xu L, Ai P, et al. The evolution and pathogenic mechanisms of the rice sheath blight pathogen. Nature Communications. 2013;4:1424. doi: 10.1038/ncomms2427

[53] Johanson A, Turner H C, Mckay GJ, Brown AE. A PCR-based method to distinguish fungi of the rice sheath blight complex, *Rhizoctonia solani*, *R. oryzae* and *R. oryzae-sativae*. FEMS Microbiology Letters. 1998;162:289-294.

[54] Singh P, Jia Y, Cartwright R, Lee FN, Rothrock CS, Eizenga GC, Rutger JN. Characterization of isolate diversity of rice sheath blight pathogen *Rhizoctonia solani* by anastomosis, rDNA-ITS and pathogenicity assays. GenBank, National Center for Biotechnology Information Accession No AY185104 to AY185115 [Internet] 2002. Available from: http://www.ncbi.nlm.nih.gov/nuccore.

[55] Zhou J-M, and Zhang Y. Plant immunity: Danger perception and signaling. Cell. 2020;181:979-989.

[56] Boller T, Felix G. A renaissance of elicitors: perception of microbe-associated molecular patterns and danger signals by pattern-recognition receptors. Annual Review of Plant Biology. 2009; 60:379-406. DOI:10.1146/annurev.arplant.57.032905.105346

[57] Maekawa T, Kufer TA, Schulze-Lefert P. NLR functions in plant and animal immune systems: so far and yet so close. Nature Immunology. 2011;12:817-26. DOI:10.1038/ni. 2083

[58] Jacob F, Vernaldi S, Maekawa T. Evolution and Conservation of Plant NLR Functions. Frontiers in Immunology. 2013;4:297. Published 2013 Sep 25. doi:10.3389/fimmu.2013.00297

[59] Bryan GT, Wu KS, Farrall L, Jia Y, Hershey H, McAdams S, Tarchini R, Donaldson G, Faulk K, Valent B. A single amino acid difference distinguishes resistant and susceptible allele of the rice blast resistance gene *Pi-ta*. Plant Cell. 2000;12:2033-2045.

[60] Chen X, Shang J, Chen D, Lei C, Zou Y, Zhai W, Liu G, Xu J, Lin, Z, Cao G, Ma B, Wang Y, Zhao X, Li S, Zhu L. A B-lectin receptor kinase gene conferring rice blast resistance. Plant Journal. 2006;46:794-804.

[61] Liu X, Lin F, Wang L, Pan Q. The *in silico* map-based cloning of Pi36, a rice coiled-coil–nucleotide-binding site–leucine-rich repeat gene that confers race-specific resistance to the blast fungus. Genetics. 2007;176:2541-2549.

[62] Jia Y, Bryan GT, Farrall L, Valent B. Natural variation at the *Pi-ta* rice blast resistance locus. Phytopathology. 2003;93:1452-1459.

[63] Jia Y, Wang Z, Fjellstrom RG, Moldenhauer KAK, Azam MA, Correll J, Lee FN, Xia Y, Rutger JN. Rice *Pi-ta* gene confers resistance to the major pathotypes of the rice blast fungus in the US. Phytopathology. 2004;94:296-301.

[64] Wang Z, Jia Y, Rutge J, Xia Y. Rapid survey for presence of a blast resistance gene *Pi-ta* in rice cultivars using the dominant DNA markers derived from portions of the *Pi-ta* gene. Plant Breeding. 2007;126:36-42.

[65] Wang X, Jia, Y, Shu Q, Wu D. Haplotype diversity at the *Pi-ta* locus in cultivated rice and its wild relatives. Phytopathology. 2008;98:1305-1311.

[66] Wang X, Fjellstrom R, Jia Y, Yan WG, Jia MH, Scheffler BE, Wu D, Shu Q, McClung A. Characterization of *Pi-ta* blast resistance gene in an international rice core collection. Plant Breeding. 2010;129:491-501.

[67] Jia Y, McAdams SA, Bryan GT, Hershey H, Valent B. Direct interaction

of resistance gene and avirulence gene products confers rice blast resistance. EMBO Journal. 2000;19:4004-4014.

[68] Orbach M J, Farrall L, Sweigard J, Chumley FG, Valent B. A telomeric avirulence gene determines efficacy for the rice blast resistance gene *Pi-ta*. Plant Cell. 2000;12:2019-2032.

[69] Jia Y, Martin R. Identification of a new locus Ptr(t) required for rice blast resistance gene Pi-ta-mediated resistance. Molecular Plant-microbe Interactions. 2008;21:396-403.

[70] Zhao H, Wang X, Jia Y, Minkenberg B, Wheatley M, Fan J, Jia MH, Famoso A, Edwards JD, Wamishe Y, Valent B, Wang G-L, Yang Y. The Rice Blast Resistance Gene *Ptr* Encodes an Atypical Protein and Confers Broad Spectrum Disease Resistance. Nature Communications. 2018;9:2039 (DOI: 10.1038/s4147-018-04369-4).

[71] Meng X, Xiao G, Telebanco-Yanoria MJ, Siazon PM, Padilla J et al. The broad-spectrum rice blast resistance (*R*) gene *Pita2* encodes a novel R protein unique from *Pita*. Rice. 2020;13:19. https://doi.org/10.1186/s12284-020-00377-5

[72] Jia Y, Wang Z, Jia MH, Rutger JN, Moldenhauer K. Development and characterization of a large mutant population of a rice variety Katy for functional genomics studies and breeding. Crop Breeding, Genetics and Genomics. 2019;1(2): e190014. https://doi.org/10.20900/cbgg20190014.

[73] Sharma M, Singh A, Shankar A, Pandey A, Baranwal V, Kapoor S et al. Comprehensive expression analysis of rice Armadillo gene family during abiotic stress and development. DNA Research. 2014;21:267-283. doi: 10.1093/dnares/dst056

[74] Liu X, Yang Q, Lin F, Hua L, Wang C, Wang L, Pan Q. Identification and fine mapping of *Pi39(t)*, a major gene conferring the broad-spectrum resistance to *Magnaporthe oryzae*. Molecular Genetic Genomics. 2007;278:403-410.

[75] Kumar P, Pathania S, Katoch P, Sharma TR, Plaha P, Rathour R. Genetic and physical mapping of blast resistance gene *Pi42(t)* on the short arm of rice chromosome 12. Mol Breeding. 2010;25:217-228.

[76] Bhatta BP, Ponniah SK, Jia Y, Muthusamy M. CRISPR/Cas9-Mediated Gene Editing in Rice (*Oryza sativa L. japonica* cv. Katy) for Stable Resistance against Blast Fungus (*Magnaporthe oryzae*). In: Proceedings of Plant and Animal Genome, January 13-17, 2018; San Diego, California.

[77] Valent B, Farrall L, Chumley FG. *Magnaporthe grisea* genes for pathogenicity and virulence identified through a series of backcrosses. Genetics. 1991;127:87-101.

[78] Jia Y, Valent B, Lee FN. Determination of host responses to *Magnaporthe grisea* on detached rice leaves using a spot inoculation method. Plant Disease. 2003; 87:129-133.

[79] Jia Y, Correa-Victoria F, McClung A, Zhu L, Liu G, Wamishe Y, Xie J, Marchetti M A, Pinson SRM, Rutger JN, Correll JC. Rapid determination of rice cultivar responses to the sheath blight pathogen *Rhizoctonia solani* using a micro-chamber screening method. Plant Disease. 2007;91:485-491.

[80] Correa-Victora FJ, Zeigler RS, Levy M. Virulence characteristics of genetic families of *Pyricularia grisea* in Colombia. In: Zeigler RS, Leong SA, Teng PS, editors. Rice Blast Disease, Oxon (UK):CAB International; 1994. p.211-229.

[81] Jia Y. Marker assisted selection for the control of rice blast disease. Pesticide Outlook. 2003;14:150-152.

[82] Godwa M, Venu RC, Roopalakshmi K, Sreerekha MV, Kulkarni RS. Advances in rice breeding, genetics and genomics. Molecular Breeding. 2003;11:337-352.

[83] Molla KA, Karmakar S, Molla J, Bajaj P, Varshney RK, Datta S K. Understanding sheath blight resistance in rice: the road behind and the road ahead. Plant Biotechnology Journal. 2020;18: 895-915. DOI:10.1111/pbi.13312

[84] Kalia S, Rathour R. Current status on mapping of genes for resistance to leaf- and neck-blast disease in rice. 3 Biotechology. 2019;9:209 DOI.org/10.1007/s13205-019-1738-0

[85] Fjellstrom R, Conaway-Bormans CA, McClung, A., Marchetti, M. A., Shank, A. R. and Park, W. D. 2004. Development of DNA markers suitable for marker assisted selection of three *Pi-* genes conferring resistance to multiple *Pyricularia grisea* pathotypes. Crop Science. 44:1790-1798.

[86] Jia Y, Wang Z, Singh P. Development of dominant rice blast resistance *Pi-ta* gene markers. Crop Science. 2002;42:2145-2149.

[87] Jia Y, Redus M, Wang Z, Rutger JN. Development of a SNLP marker from the *Pi-ta* blast resistance gene by tri-Primer PCR. Euphytica. 2004;138:97-105.

[88] Fjellstrom RG, McClung, AM, Shank, AR. SSR markers closely linked to the *Piz* locus are useful for selection of blast resistance in a broad array of rice germplasm. Molecular Breeding 2006;17:149-157.

[89] Jia Y, Jia MH, Wang X, Liu *G. indica* and japonica crosses resulting in linkage block and recombination suppression on rice chromosome 12. PLoS ONE. 2012;7:10. e43066.

[90] Wang X, Jia MH, Ghai P, Lee FN, Jia Y. Genome-wide association of rice blast resistance and yield related components of rice. Molecular Plant Microbe Interactions. 2015;28:1383-1392.

[91] Christou P. Rice transformation: Bombardment. Plant Molecular Biology. 1997;35:193-203.

[92] Hiei Y, Ohta S, Komari T, Kumashiro T. Efficient transformation of rice mediated by *Agro-bacterium* and sequence analysis of the boundaries of the T-DNA. Plant Journal. 1994;6:271-282.

[93] Lin W, Anuratha CS, Datta K, Potrykus I, Muthukrishnan S, Datta SK. Genetic engineering of rice to resistance to sheath blight. Biotechnology. 1995;13:686-691.

[94] Pokhrel S, Ponniah SK, Jia Y, Yu O, Manoharan M. 2021. Transgenic Rice Transgenic rice expressing isoflavone synthase gene from soybean shows resistance against blast fungus (*Magnaporthe oryzae*). Plant Disease. https://doi.org/10.1094/PDIS-08-20-1777-RE.

[95] Telem RS, Wani SH, Singh NB, Nandini R, Sadhukhan R, Bhattacharya S, Mandal N. Cisgenics - A Sustainable Approach for Crop Improvement. Current Genomics. 2013;14:468-476.

[96] Goad DM, Jia Y, Gibbons A, Liu Y, Gealy D, Caicedo AL, Olsen KM. Identification of novel QTLs conferring sheath blight resistance in two weedy rice mapping populations. Rice. 2020;13:21.

[97] Tsiboe F, Nalley LL, Durand A, Thoma G, Shew A. The Economic and Environmental Benefits of Sheath Blight Resistance in Rice. Journal of

Agricultural and Resource Economics. 2017; 42:215-235.

[98] Zhou E, Jia Y, Singh P, Correll J, Lee FN. Instability of the *Magnaporthe oryzae* avirulence gene *AVR-Pita* alters virulence. Fungal Genetic Biology. 2007;44:1024-1034.

[99] Xie K, Yang Y. RNA-guided genome editing in plants using a CRISPR-Cas system. Molecular Plant. 2013;6:1975-83. doi: 10.1093/mp/sst119.

[100] Eizenga GC, Agama HA, Lee FN, Jia Y. Identifying novel resistance genes in newly introduced blast resistant rice germplasm. Crop Science. 2006;46:1870-1878.

[101] Jia Y, Gealy D. Weedy red rice has novel resistance resources to biotic stress. Crop Journal.2018; 6:443-450. https://doi.org/10.1016/j.cj.2018.07.001.

[102] Liu Y, Qi X, Gealy DR, Olsen KM, Caicedo AL, Jia Y. QTLs analysis for resistance to blast disease in US weedy rice. Molecular Plant-Microbe Interactions. 2015;28:834-844. doi.org/10.1094/MPMI-12-14-0386-R.

[103] Jia Y. Registration of lesion mimic mutant of Katy rice. Crop Science. 2005;45:1675.

[104] Jia Y, Wang Z, Jia MH, Rutger JN, Moldenhauer K. Development and characterization of a large mutant population of a rice variety Katy for functional genomics studies and breeding. Crop Breeding, Genetics and Genomics. 2019; 2019;1:e190014. https://doi.org/10.20900/cbgg20190014

[105] Zhu Y, Chen H, Fan J, Wang Y, Li Y, Chen J, Fan J, Yang S, Hu L, Leung H, Mew TW, Teng PS, Wang Z, Mundt CC. Genetic diversity and disease control in rice. Nature. 2006;406:718-722.

Chapter 6

# Enhancing Abiotic Stress Tolerance to Develop Climate-Smart Rice Using Holistic Breeding Approach

*M. Akhlasur Rahman, Hasina Khatun,*
*M. Ruhul Amin Sarker, Hosneara Hossain, M. Ruhul Quddus,*
*Khandakar M. Iftekharuddaula*
*and M. Shahjahan Kabir*

## Abstract

Agricultural land and resources reduced annually because of climate change thus it is necessary to further increase the productivity of the major staple food rice to sustain food security worldwide. However, rice productivity enhancement is one of the key challenges in abiotic stress-prone environments. The integration of cutting-edge breeding approaches and research management methods in the current varietal improvement pipelines can make a step-change towards varietal improvement for the abiotic stress-prone environments. Proper implementation of breeder's equations in the crop improvement pipeline can deliver a higher rate of genetic gain. Single Seed Descent based Rapid Generation Advance (RGA) technique in field and greenhouse is the most promising innovations and low-cost, high-throughput marker-assisted selection approaches are applied for rapid and efficient selection for abiotic stress-tolerances. Also improving efficiency, intensity, and accuracy of selection and reducing breeding cycle time through holistic rice breeding that can play an important role in developing climate-smart abiotic stress-tolerant rice for target environments. This information can use as the future direction for rice breeders and other researchers.

**Keywords:** abiotic-stress tolerance, high-night temperature (HNT), holistic rice breeding, QTL, gene, product profile, rice

## 1. Introduction

Rice is the major staple food for more than half of the world population that supplies 30–50% daily calorie intake. Rice security is synonymous with food security in Bangladesh. If rice production hampers because of different abiotic stresses then food security also become vulnerable. Abiotic stresses such as salinity, drought, flood, high-night temperature (HNT), and heat/high-temperature increase enormous challenges that limit agricultural production and food security.


IntechOpen

Coastal agriculture faces these challenges because salinity affects directly more than one million hectares of agricultural land in Bangladesh. Salinity causes due to chemical weathering of minerals that release salts in the soils as Na, Ca, Mg ions; coastal agricultural land is inundated by salt-water during flash flood/tidal upsurge; unplanned saltwater intrusion into the shrimp gher (shallow shrimp cultivation pond) in the southern coastal zone and lifting groundwater with EC >3.0 dS/m for irrigation.

Several studies revealed that the detrimental effects and suitable genetic [1, 2] and physiological mechanisms of salt-stress tolerance in various sensitive growth stages of seedling stages like early seedling stage during seedling establishment in the field after transplanting, and different susceptible stages of reproductive phase such as panicle initiation/emergence, booting, flowering/heading, spikelet fertility-sterility, and seed set, yield and other salt tolerance-related traits [3–10].

For maintaining food security and sustainability in rice production, both drought and heat/high-temperature tolerance is important in the respective stress-prone rice-growing areas for increasing rice production sustainably. However, steady growth in the rice sector is crucial during this pandemic situation to sustain self-sufficiency in different rice-growing countries in South Asia including Bangladesh.

Drought is also an important abiotic stress that threat for rainfed ecosystem. This stress adversely impacts on rice production. Drought tolerance is a complex polygenic trait with a complicated phenotype that affects various growth and developmental stages sensitive to drought-stress in rice. Different drought-responsive QTLs and genes regulate the degree of sensitivity or tolerance of rice through triggering signal transduction pathways to several drought conditions [11].

High-night temperature (HNT), different abiotic stress from heat stress, is emerging abiotic stress because of climate change. This stress (HNT) is drawing the attention of plant breeders and physiologists due to its detrimental effects on rice productivity. HNT varies 25-30°C that adversely affects yield and grain quality such as chalkiness in rice. This stress was reported in the Rajshahi region of Bangladesh (M. A. Rahman, unpublished data).

Fragile flood-prone environments belong to 18% of areas of Bangladesh that suffer from varying degrees of flooding causes due to flash flood submergence, monsoon flood, and irregular rainfall. The flood adversely influences the rice production of more than a million ha of land in Bangladesh. Submergence tolerant high-yielding rice varieties are grown by the farmers of the flash flood-prone areas. However, deepwater rice (DWR) is cultivated in areas where flood water-depth varied from 1.5 to 2.0 m and these areas are more vulnerable to rice production as there is no high-yielding DWR variety for this harsh ecosystem in Bangladesh. Thus, rice-growing areas under unfavorable environments need to enhance productivity by developing climate-smart rice to cope with the harmful effects of climate change.

In this review, we discuss the abiotic stresses and the development of climate resilient rice addressing adverse effect of climate change.

## 2. Abiotic stresses influencing rice production and food security

### 2.1 Heat tolerance

#### *2.1.1 Strategies to enhance heat tolerance*

Heat/high temperature tolerance is important in the heat-prone rice growing areas for increasing rice production sustainably. To address climate change, screening and breeding for higher level of heat tolerance is needed. Strategies such as agronomic intervention through sowing time adjustment, chemicals/plant hormones application, genetic and genomic approaches [12], breeding for heat resistant variety development, marker-assisted introgression of *qEMF3 QTL [13] for developing and selecting* cultivars with early morning flowering (EMF) before temperature rise are involved for improving heat tolerance.

QTLs associated with heat tolerance related traits using bulked segregant analysis in Rice to evaluate the genetic effect of QTLs controlling heat tolerance at flowering stage in ricc. A population comprising 279 $F_2$ individuals developed from 996 (heat tolerant)/4628 (heat-sensitive cultivar), was investigated for their segregation pattern of the difference in seed set rate under normal/optimum temperature condition and stress/high temperature condition that exhibited normal distribution, suggesting the polygenic control on the heat tolerance [14].

Eleven QTLs identified for heat tolerance using RIL population derived from IAPAR-9 (sensitive)/Liaoyan241 (heat tolerant) at the heading and flowering stage in rice. Four major QTLs such as *qNS1*, *qNS4*, *qNS6,* and *qRRS1* found stable in both seasons/years in various environments [15].

Jagadish [16] dissected QTLs for relative spikelet fertility during anthesis in rice *qtl_1.1* (38.35 Mb) and *qtl_11.1*, 24.16 Mb, QTL contributor is Azucena) and one (*qtl_10.1*, 20.14 Mb from Bala). Total 24.1% phenotypic variation was explained by these three QTLs.

### 2.2 Drought stress tolerance

Improving grain yield is the key and universal objective of any crop breeding programs including rice. Identification of yield and yield-related traits and their introgression into adapted varieties is one of the best strategies to increase grain yield under drought. A number of yield QTLs identified in different chromosomes of rice under drought stress. QTL *qDTY12.1* was the first identified major grain yield QTL on chromosome 12 in rice under drought at the reproductive stage [17]. Another large-effect QTL for grain yield, *qDTY1.1*, was identified on chromosome 1 [18]. Moreover, other QTLs with major effect such as *qDTY2.2*, *qDTY3.1*, *qDTY3.2*, *qDTY4.1*, *qDTY6.1*, *qDTY9.1* controlling drought tolerance in rice were reported by several investigators [19–22]. The identified QTLs should be consistent in multiple genetic backgrounds and various target environments [23, 24]. Efficient QTL stacking of the major effect QTLs in the adapted varietal background is necessary to achieve higher grain yield under drought [25].

QTLs related to drought tolerance in rice have been shown in **Table 1**. However, only a few QTL studies on grain yield under drought stress have been reported. Most of the QTLs detected for regulating drought stress in rice analyzed for different important drought-related traits such as osmotic adjustment [48, 49], drought avoidance [32], root and shoot responses [50], photosynthesis and whole plant response [51] to drought tolerance.

### 2.3 Flood/submergence tolerance

Flood-prone ecosystems are fragile characterized by varying level of flooding, erratic precipitation that affect the rice production of more than one million ha in Bangladesh. Deepwater rice is cultivated more than 100000 ha in Bangladesh and the typical deepwater rice with nodal tillering, kneeing ability to keep the top three leaves in the air (above the water level) to capture and use solar energy for photosynthesis, internode elongation ability to prevent drowning with high yielding potential comparing with local varieties like Hijoldigha, Laxmidigha, Kartiksail,

| Drought-tolerance traits/indices | Cross combination and mapping population | Molecular marker used | % Phenotypic variation | No. of identified QTLs for drought tolerance | References |
|---|---|---|---|---|---|
| Root penetration index, root and tiller number | CO39/Moroberekan (RILs) | RFLP | 8.0–14.0 | 39 | [26] |
| Drought traits related with osmotic adjustment and dehydration tolerance | CO39/Moroberekan (RILs) | RFLP | — | 1 | [27] |
| Root morphology and distribution | IR64/Azucena (Double-haploid) | RFLP | 6.0–22.0 | — | [28] |
| Root traits related with drought | IR58821–23-B-1–2-1/ IR52561-UBN-1–1-2 (RIL) | AFLP & RFLP | 6.0–27.0 | — | [29] |
| Cellular membrane stability | CT9993–5-10–1-M/ IR62266-42–6-2 (DH) | RFLP, AFLP & SSR | 11.8–54.3 | 9 | [30] |
| Drought resistance osmotic adjustment and root traits | CT9993/IR62266 (Doubled haploid lines) | RFLP, AFLP & SSR | 8.0–38.0 | 5 | [31] |
| Drought avoidance, leaf rolling and drying | Bala/Azucena (F5 population) | RFLP, AFLP & SSR | 7.4–25.6 | 17 | [32] |
| Root traits (Seminal and lateral root length; adventitious and lateral root number) | IR1552/Azucena (RILs) | SSR | — | 23 | [33] |
| Morphological and physiological traits | IR64/Azucena (Doubled haploid Lines) | RFLP | — | 15 | [34] |
| Root-penetration ability | Bala/Azucena (RILs) | RFLP & AFLP | — | 18 | [35] |
| Reproductive-stage drought tolerance | Vandana/Way Rarem ($F_3$–derived lines) | SSR | 33.0 | — | [17] |
| Grain yield under drought stress | CT9993/IR62266 (Doubled haploid lines) | AFLP | | 1 | [36] |
| Seedling stage drought tolerance | Indica/Japonica (Azucena) (RIL) | RFLP, AFLP & SSR | 10.0–27.0 | 7 | [37] |
| Morphological and physiological traits related to drought avoidance | Azucena/Bala (RIL) | RFLP, AFLP & SSR | — | — | [38] |
| Grain yield under lowland drought stress | Apo/2*Swarna (RILs) | SSR | 13.0–16.0 | 1 | [39] |
| Grain yield performance under aerobic condition | Three populations, Apo/(2)*Swarna, Apo/(2)*IR72, and Vandana/(2)*IR72 | SSR | 39.0–66.0 | 1 | [40] |

| Drought-tolerance traits/indices | Cross combination and mapping population | Molecular marker used | % Phenotypic variation | No. of identified QTLs for drought tolerance | References |
|---|---|---|---|---|---|
| Yield performance under drought stress | Two populations Basmati334/Swarna and N22/MTU1010 ($F_{3:4}$ population) | SSR | | — | [41] |
| Reproductive-stage drought stress | Aday Sel/IR77298-5-6-B-11 (backcross inbred lines (BILs)) | SSR | 19.0 | 9 | [20] |
| Yield under lowland drought in different environments | R77298/Sabitri, (BC1 derived) | SSR | — | 1 | [21] |
| Drought stress at reproductive stage | Two populations Kali Aus/IR64, Kali Aus/MTU1010 (RILs) | SSR | 6.0–9.0 | 2 | [42] |
| Grain yield and yield characters during reproductive stage | IR64/Cabacu (RILs) | SNP | — | 1 | [43] |
| Grain yield under stress at reproductive stage | Swarna/WAB (Backcross inbred lines) | SSR | — | 1 | [44] |
| Reproductive stage drought tolerance | Teqing/Lemont (Introgression lines) | SNP | | | [45] |
| Reproductive stage drought tolerance | IR55419-04/2*TDK1 ($BC_1F_{3:4}$ population) | SSR | 36.0 | 6 | [46] |
| Ratio of deep rooting (RDR) | 3 populations (RILs, mini-core collection and landraces) | SSR, SNP | — | 6 | [47] |

**Table 1.**
*Useful QTLs responsible for drought-stress tolerance in rice.*

Khoiyamtor, Lalmohan, and Shishumati. These local germplasm has the ideal ideotype for deepwater ecosystem but only limitation is low yielding ability. To address sustainable development goals (SDGs) and maintain food security, we need to increase the production in the abiotic stress prone environments such as salinity, flood/submergence, drought and heat-prone areas through horizontal expansion (expansion of arable land in abiotic stress prone areas which are not yet under cultivation) of abiotic stress tolerant rice varieties in these areas. However, Flood-prone ecosystems in Bangladesh are four types such as long time flooding zone (>35 days; 1.5–2.0 m water-depth), flash flood submergence zone (15–30 days; up to 1.5 m water depth), deepwater (> 2.0 m water-depth) zone and submergence during germination (10–12 cm depth) at relay Transplant Aman, direct seeded rice (DSR *Aus*) and broadcast Aman (B. Aman) Rice areas anaerobic germination In Asia, submergence affects rice yield adverselyin 20 million ha, causing food insecurity. The *SUB1*gene governing submergence tolerance cloned and introgressed into a number of rice varieties in South Asia, South East Asia and Africa. Yield advantages of Sub1 varieties ranged from 1.0 to >3.0 t $ha^{-t}$ after submergence comparing with non-Sub1 varieties. These submergence tolerant varieties reached more than 3.8 million farmers within 3 years of release in Asia [52].

Biosynthesis of growth regulator (Gibberellin) and signal transduction is important pathways for internode elongation of the deepwater rice [53]. Two large-effect QTLs located on chromosomes 3 (*qGTIL3,qGLEI3*, *qGNEI3* located between 38 and 87 cM) and chromosome 9 (*qGTIL9*, *qGLEI9*, *qGNEI9* positioned between 16 and 88 cM) are controlling traits such as total length of internode (TIL), lowest elongated internode (LEI) and number of elongated internode number (NEI). Three factors involved to characterize deepwater rice internode elongation ability: (1) total length of elongated internode (TLEI); (2) elongated internodes number (EIN); and (3) minimum elongated internode (MEI) [54–57]. Among these, MEI is the main parameter for initiating the internode elongation of deepwater rice [54] because MEI is leaf stage- dependent and first starts of internode elongation at the sixth leaf stage in deepwater rice.

Catling (1992) [58] described the genetic basis of internode elongation during submergence of deepwater rice that is regulated by several minor and two major genes. Suge [59] identified one gene with neither allele is found dominant (incomplete dominance) that responsible for elongation ability. Internode elongation depends on the increasing activity of cell division and cell elongation in specific areas of the internode.

### 2.4 Salt tolerance

Salt-stress entails changes in different physiological and metabolic pathways, based on severity and duration of the stress, and eventually decreases rice productivity [10, 60–62].

Genetic characterization of salt tolerance related traits is important to estimate phenotypic co-efficient of variation (PCV), genotypic co-efficient of variation (GCV), broad-sense heritability and genetic Advance (GA).

If sufficient variation with high heritability and genetic advance exists in the germplasm for salt tolerance related traits; consequently there is possibility to improve the traits related with salt tolerance in rice by exploiting salt tolerant landraces/germplasm in the breeding programs.

Genetic component analysis (GCA) study showed that both additive and dominance gene effects controlled low Na-K ratio [63]. The combining ability analysis shows that both general combing ability (GCA) and specific combining ability (SCA) effects are also important for deciphering the genetics of salt tolerance. They also revealed that selection may be made in later generation under controlled conditions for minimizing environmental effects for low heritable traits like Na-K ratio.

The additive effects could enhance fixation of the desirable combination of genes and also additive gene action is correlated to narrow sense heritability [64]. Thus, selection for salinity tolerance may be made in the early generation. Understanding the gene action for salt tolerance in rice will be useful in the future breeding program [65].

#### *2.4.1 Molecular basis of complex salt tolerance*

The molecular dissection of salt tolerance has considerably enhanced using the molecular platforms for identifying quantitative trait loci (QTL) controlling related genetic variation in crops including rice [1, 2, 8, 60, 66–72].

Moreover, several QTLs related with physiological, agronomic traits conferring salinity tolerance at seedling and reproductive stage have been reported [1, 8, 71, 73, 74] including major QTLs for salinity tolerance such as *SKC1* [75] (a sodium transporter *OsHKT1; 5* in the *SKC1* locus [76] and *Saltol* [71, 77] on chromosome 1

| Rice growth stage | QTL/Gene name | Mapping population and parentage | Population Size | Salinity tolerant trait/ index | Flanking marker | Positive allele and Position (cM or Mb) | PVE (% $R^2$) | References |
|---|---|---|---|---|---|---|---|---|
| Seedling | *qSaltol* | RIL (IR29/Pokkali) | 80 | $Na^+$ uptake; $Na^+/K^+$ | RM140 - C1733S | Pokkali; 51.6–65.9 cM/13.87 Mb | 39.2 | [71, 77] |
| | *OsHKT1;5* | Advanced backcross | 192 | Shoot $K^+$ concentration | K159 - K061 | Nona Bokra; 11.46 Mb | — | [76] |
| | *(qSKC-1)* | (Koshihikari*3/Nona Bokra) | | | | | | |
| | *qST1* | RIL (Milyang 23/Gihobyeo) | 164 | Young seedling stage | Est12 -RZ569A | Gihobyeo; 40 cM | 27.8 | [78] |
| | *qST3* | | | | RG179 - RZ596 | Milyang 23; 138 cM | 9.2 | |
| | *qSES1* | F2 (NERICA-L-19/Hasawi) | 153 | SES | RM8094-RM582 | NERICA-L-19 | 19.6 | [79] |
| | *qSES6* | | | | RM586-RM253 | NERICA-L-19 | 39.7 | |
| | *qSES10* | | | | RM228-RM333 | Hasawi | 30.7 | |
| | *qSES11* | | | | RM536-RM287 | Hasawi | 37.2 | |
| | *qSES2.1* | RIL (IR29/Hasawi) | 142 | SES | id2004774–id2007526 | Hasawi; 64.8 cM | 11.1 | [80] |
| | *qSES1.3* | | | SES | id1024972-id1023892 | Hasawi; 170.0 cM | 39.9 | [1] |
| | *qSES3.1* | F2 (Capsule/BRRI dhan29) | 94 | SES | RM5626- R3M53 | Capsule; 111.0 cM | 23.0 | [2] |
| | *qSES12.3* | | | | RM252B RM27615-RM27877 | Capsule; 31.0 cM | 17.0 | |
| | *qSL7* | RIL (93–11/PA64s) | 132 | Higher shoot length | SNP7–191–SNP7–226 | PA64s; 86.31 cM | 9.9 | [81] |
| Reproductive | *qNa1.7* | F2 (Cheriviruppu/Pusa Basmati 1) | 218 | Na (mmol g_1 dwt) | RM1349-RM7250 | Cheriviruppu;; 31.06 Mb | 13.5 | [8] |
| | *qNaKR1.8* | | | Na/K Ratio | RM1349-RM7250 | Cheriviruppu; 31.06 Mb | 11.0 | |
| | *qNFS2.1* | F2 (NSIC Rc222/BRRI dhan47) | 92 | Filled spikelet number | id2004774 | BRRI dhan47; 40.1 cM | 15.3 | [73] |
| | *qPFS2.1* | | | Filled spikelet (%) | id2013434 | BRRI dhan47; 122.1 cM | 18.4 | |
| | *qTS11.1* | | | Total spikelet (no) | id11000858 | BRRI dhan47; 14.0 cM | 15.8 | |

| Rice growth stage | QTL/Gene name | Mapping population and parentage | Population Size | Salinity tolerant trait/ index | Flanking marker | Positive allele and Position (cM or Mb) | PVE (% $R^2$) | References |
|---|---|---|---|---|---|---|---|---|
| | *qYLD2.1* | | | Yield | id2004774 | BRRI dhan47; 40.1 cM | 14.6 | |
| | *qDEG-S-2-2* | Backcross | | Spikelet degeneration | | 6.8 | 34.44 | [82] |
| | *qDEG-S-4-3* | Backcross | | Spikelet degeneration | | 4.19 | 17.43 | |

*SES: Overall phenotypic performance; SL: Shoot length; Na: $Na^+$: concentration; K: $K^+$: concentration; NaK-R: Na-K ratio, Sur: Survival; Chr: Chromosome number; PVE: Phenotypic variation explained.*

**Table 2.**
*Recently identified QTL for salt tolerance with controlling/responsible traits and phenotypic variation using different mapping populations at seedling and reproductive stages in rice.*

(**Table 2**). Recently unraveled molecular basis of various rice landraces such as Pokkali [71, 77], Nona Bokra [76], Hasawi [1], Capsule [2], Changmaogu [74] and Horkuch [72] can withstand different levels of salt-stress at various sensitive growth stages of rice.

#### *2.4.2 Enhancing genetic gain for salt-stress*

Widely used tools for quantitative genetics such as genomic estimated breeding value (GEBV) and best linear unbiased predictions (BLUPs) are applied to evaluate the performance to construct high throughput new breeding populations for selecting the superior breeding lines when combined with genetic relatedness or kinship matrix/information (i.e. coefficient of coancestry) using pedigree information to calculate estimated breeding values (EBVs). This is a key parameter for selecting complex traits like salt tolerance and yield through estimating parents' genetic potential to produce better descendants on the basis of parent's own performance, pedigree information and progeny data. EBVs play important role to select parent for higher rate of genetic gain [83].

Theory of genetic gain in breeding and classical biometrical genetics indicates the improvement of performance compared to a standard or baseline. It is generally evaluated after advancing one generation to the subsequent generation and artificial selection is done [84]. However, genetic gain per year is also known as genetic trend that measured varietal performance over year by comparing benchmark or dominant varieties [85].

To maintain the food security under this Covid-19 situation horizontal expansion (increase of rice growing areas in high saline zone) of rice cultivation is needed areas in the coastal zone. Thus large areas will be brought under cultivation in high saline prone areas.

### 2.5 Modern breeding

Genomic assisted breeding (GAB) is regularly practiced for the genetic improvement of salt-tolerant rice applying various innovative tools for genomic breeding such as forward breeding, rapid breeding, and haplotype-based breeding [86]; 5G breeding methods such as genome sequence availability (genome assembly), characterization of germplasm at genomic and morpho-agronomic level, gene detection and understanding function, genomic breeding, and genome editing for enhancing superior performance of genotypes [87] could be used for enhancing efficiency and accuracy of breeding for complex traits related with abiotic stress tolerance. A precise SNP-assisted introgression of the *hst1*(*hitomebore salt-tolerant 1*) gene improved salt tolerance in the high-yielding rice variety was achieved through SNP based speed breeding [88].

Modern breeding emphasizes data-based parent selection. Local and exotic germplasms are subjected to genetic distance measurement, trait characterization through diagnostic trait markers, genomic selection, and breeding value estimation. Sometimes trait of interest (ToI) like salinity and submergence tolerance, aroma, disease resistance is found in local germplasm with low yield potential. Then that ToI is first transferred to an elite background possessing high yield potential for developing pre-breeding materials. After that, the elite line with the desired traits is used in breeding purposes to fulfill the product profile. **Figure 1** shows the schematic illustration for optimizing breeding scheme to achieve genetic gain.

We evaluated 1436 breeding lines for trait assay using 20-trait specific single nucleotide polymorphism (SNP) markers. These lines characterized for important traits such as disease (blast, bacterial leaf blight: BLB) and insect (brown plant

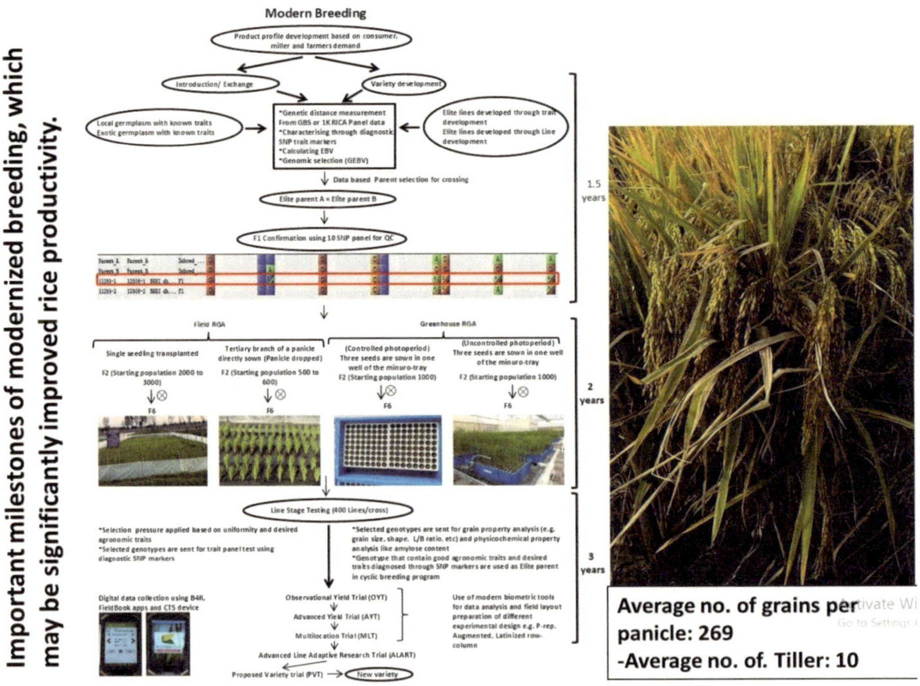


**Figure 1.**
*Varietal development through breeding modernization for rapid varietal turnover and replacement for rice growers of target regions.*

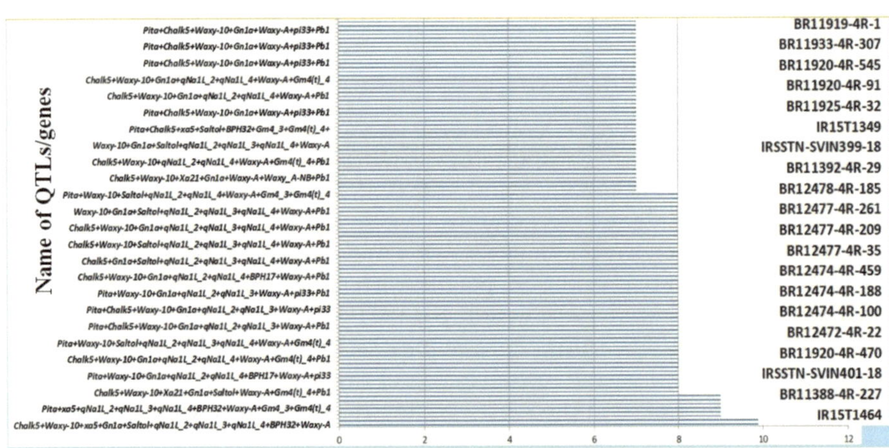


**Figure 2.**
*Genetically important lines (GILs) in the STR Breeding program, T. Aman, 2020-21.*

hopper: BPH; gall midge) resistance, grain quality (amylose content, chalkiness), grain number (*Gn1a*) and salinity tolerance at seedling stage (sodium exclusion, SES) traits. Out of 1436 breeding lines, 100 lines harbored the 7–10 QTLs/genes that regulating trait of interest (**Figure 2**) that designated as Genetically Important Lines (GILs). Each line assayed against QTLs and genes of interest to assess the presence or absence of useful traits. IRRI developed trait specific SNP markers used (https://gsl.irri.org/) and SNP genotyping assay was carried out by Intertek as an external service provider. The trait-based SNP markers associated with the traits for

| Trait category | Trait | Trait specific markers (favorable allele) | Chromosome | QTL/Trait contributor |
|---|---|---|---|---|
| Blast | *Pb1* | snpOS00478 (T) | 11 | Pokkali 26869 |
| Blast | *Pi9* | snpOS00451(C) | 6 | Pi9 (DQ285630) |
| Blast | *Pita* | snpOS00006 (C) | 12 | — |
| Blast | *qPi33* | snpOS00468 (T) | 8 | IR64 |
| BLB | *xa13* | snpOS00493 (C) | 8 | IRBB60, some aus |
| BLB | *Xa21* | snpOS00061(C) | 11 | IRBB60 |
| BLB | *xa5* | snpOS00054 | 5 | IRRI 154, FR13A |
| Insect (BPH) | *BPH17* | snpOS00430 (G) | 4 | Rathu Heenati |
| Insect (BPH) | *BPH32* | snpOS00442 (G) | 6 | Honderawala, Rc222 |
| Insect (Gall midge) | *Gm4(t)* | snpOS00466 (A) | 8 | Abhaya |
| Insect (Gall midge) | *Gm4(t)* | snpOS00467 (A) | 8 | Abhaya |
| Amylose | *Waxy* | snpOS00445 (C) | 6 | Wx(a) - except Basmati |
| Amylose | *Waxy* | snpOS00446 (T) | 6 | Wx(a)-Rc222, Exon10 |
| Chalkiness | *Chalk5* | snpOS00024 (G) | 5 | Minghui63 |
| Grain number | *Gn1a* | snpOS00396 (T) | 1 | Swarna (A8/AP9) allele |
| Salinity-seedling | *qNa1L* | snpOS00405 (A) | 1 | FL478 |
| Salinity-seedling | *qNa1L* | snpOS00409 (C) | 1 | FL478, Capsule |
| Salinity-seedling | *qNa1L* | snpOS00410 (A) | 1 | FL478, Capsule |
| Salinity-seedling | *qNa1L* | snpOS00411 (T) | 1 | FL478, Capsule |
| Salinity-seedling | *Saltol* | snpOS00397 (T) | 1 | FL478, IR 107321–1–141–3-120 |

**Table 3.**
*Useful traits, trait-based SNP markers and their contributors of favorable allele.*

instance, snpOS00478 (*Pb1*), snpOS00451 (*Pi9*), and snpOS0054 (*xa5*), snpOS0061 (*Xa21*) etc. were applied for genotyping (**Table 3**).

These ten SNP markers produced 40% polymorphism across the *indica* germplasm derived pairwise combinations where in 95% of crosses made at least 1 polymorphic SNP marker was found within the IRRI rice breeding pool or *indica* subspecies (gsl@irri.org[https://sites.google.com/a/irri.org/snp-genotyping-mmal/genotyping/quality-control-panel/indica-rice-qc-10-snp-panel).

Single seed descent method with the facility of rapid generation advance technique is expected to have better efficiency in the increment of genetic gain compared to pedigree and other methods of breeding [2]. From each cross, 200–400 fixed lines evaluated in line stage testing (LST) trial and selection performed using high selection pressure. Then the selected lines are evaluated in yield trials and include in new variety release system (**Figure 1**).

Modern biometric tools are used in data analysis and field layout preparation of several experiment design e.g. P-rep, Augmented RCB, Alpha-Lattice, Latinized row-column, RCBD. Observation trials are mainly conducted using P-rep, Augmented RCB, Alpha-Lattice design of experiment based on the entry number and land availability. Selected genotypes are subjected to grain property analysis (e.g. grain size, shape, L/B ratio, etc) and physicochemical property analysis (e.g. amylose content). Selected genotypes that hold good agronomic traits, grain quality characters and desired traits diagnosed through SNP markers are used as Elite parent in cyclic breeding program.

During this period phenotypic data collection is aided with data collection machine (Grain counter, Nondestructive moisture meter, destructive moisture meter, Phenoapp, CT5) and different software's like B4R (Breeding 4 Rice), FieldBook etc.

We need to explore a large number of cross combination derived fixed lines to experience a remarkable genetic gain with the shortest possible time. Line fixation can be accomplished within three years using the RGA [89, 90] facility which allows a breeder to contribute more in varietal development.

| **Salinity (T. Aman) Breeding Product Profiles (20% Breeding Program Focus)** | | | | | | |
|---|---|---|---|---|---|---|
| Country/Region: Salt affected coastal region and/ Whole Bangladesh | | %Resource allocation for trait improvement | Market Leading Variety #1: BRRI dhan41 (Medium slender, high head rice yield, medium growth duration) | | | |
| Market Slot: T. Aman Rice (Medium slender to long slender grain) | | | Market Important Variety #2: BR23 (Medium bold, high yield potential, photo period sensitive) | | | |
| **Trait Family** | **Key Economic Traits** | | **Trait value** | **Standard Variety Assessment** | **Trait Benchmarking Details** | **Trait availability in the Breeding Program** |
| Yield (Paddy) | Yield (>10% higher) | | 1-Must Have Trait | BRRI dhan41 | > BRRI dhan41 | Program is actively working with trait |
| Maturity | Intermediate /long range | 10 | 1-Must Have Trait | BRRI dhan23 | <= BRRI dhan23 | Program is actively working with trait |
| Abiotic Stress Tolerance | Salinity tolerance | 20 | 1-Must Have Trait | BRRI dhan67 | > = BRRI dhan67 | Program is actively working with trait |
| Biotic Resistance (Fungal) | Blast | 20 | 1-Must Have Trait | BRRI dhan67 | Standard Evaluation Scale =< 3 | Trait Limited or NOT available |
| Biotic Resistance (Bacterial) | BLB | 10 | 3-Trait Values | BRRI dhan67 | Standard Evaluation Scale =< 3 | Program has trait available |
| Biotic Resistance (Insect) | BPH | 10 | 3-Trait Values | BRRI dhan67 | = Rathuheenati (*bph32, bph17*) | Trait Limited or NOT available |
| Consumer Traits | Amylose content | 15 | 1-Must Have Trait | BRRI dhan67 | => 24% | Program is actively working with trait |
| | Zinc content | 5 | 3= Value added | BRRI dhan72 | BRRI dhan72; => 24 mg/kg Zn | Program is actively working with trait |
| Yield (Economic)- Head Rice | Head rice recovery | 10 | 1-Must Have Trait | BRRI dhan67 | > 50% | Program is actively working with trait |
| Total | | 100 | | | | |

*Legend: 1= must have traits, 2= nice to have, 3= value added or game changer*

**Table 4.**
*Product profile with market demand-led traits for developing salinity tolerant rice (STR) variety for target region under STR breeding program at BRRI.*

However, promising breeding lines selected in OYT will be recycled to initiate next cycle of breeding for population improvement. By applying this transformed/modernized breeding approach abiotic stress tolerant varieties with enhanced grain yield and quality traits will be developed.

### 2.6 Product profile (PP) for target region

Product profile (PP) is a realistic roadmap for varietal development which addresses requirements of plant breeder, farmers, millers/market and consumers demand. Example of a product profile for salinity tolerant variety development for Southern Coastal Zone in T. Aman season is shown in **Table 4**. Variety development based on formal PP will be able to replace market leading variety with the new one; thus farmer could get early turnover from the new variety. However, the market leading variety may be or may not be a mega variety in target region. The way of developing PP is:

- Designating the target region that use common PP or trait requirements
- Identifying the market leading variety and second important variety
- Placing key traits into appropriate trait family
- Estimating the amount of emphasis or breeding efforts need to give in the breeding program for selecting trait.

## 3. Varietal development

### 3.1 Germplasm collection and characterization for salt-stress tolerance, and utilization

A total of 107 landraces collected from southern coastal zone of Bangladesh.

The genetic base of salt-tolerant donors needs to broaden for developing climate-smart rice [91] varieties for salt-affected regions with higher level of adaptation. All germplasm [92] were used for diversity analysis using a genome-wide set of 376 single nucleotide polymorphism (SNP) markers to identify and characterize novel sources of salt tolerance.

Three major clusters -the indica, aus and aromatic subgroups were identified. The largest group was indica, with the salt-tolerant Pokkali accessions in one sub-cluster, while Bangladeshi landraces, including Akundi, Ashfal, Capsule, Chikirampatnai and Kutipatnai, were in a different sub-cluster. The salt-tolerant landrace Hasawi and Kalarata clustered into a distinct aus group near to indica. Allelic diversity study at the major QTL *Saltol* shows different alleles at the *Saltol* locus for Akundi, Ashfal, Capsule, Chikirampatnai and Kutipatnai.

Sixty-nine landraces were further screened for physiological traits associated with salt stress at the seedling stage. Seven landraces such as Akundi, Ashfal, Capsule, Chikirampatnai, Jatai Balam, Kalarata and Kutipatnai uptake less Na and comparatively more K and maintain lower Na/K ratio in leaves. They efficiently restrict sodium transport root to the shoot.

Newly identified salt-tolerant landraces are genetically and physiologically different from known donors (Pokkali and Nona Bokra). These landraces can be used to develop salt-tolerant varieties with higher tolerance [10].

These landraces may be harbored novel sources of QTL/alleles for salt tolerance that will be useful in molecular breeding.

### 3.2 Participatory varietal selection (PVS)

Participatory varietal selection (PVS) demonstrates the varietal/line selection on the basis of farmers' need/demand-based choice of varieties that differs from plant breeders' selection process. Plant Breeders evaluation of varietal performance – mostly following statistical designed and quantitative data based methods which is usually different from the farmers [93–95]. However, variety selection criteria may vary according to gender, environmental condition, market demand and economic/ social status [95–97]. Until now, breeding objectives in different countries have not been properly focused on the opinions of farmers, their preferences and needs for the adverse growing conditions of their regions [94, 98, 99]. To unravel this situation, participatory varietal selection is the important way of decentralized-bottom-up breeding or farmer breeding approach that integrates farmers and their complex criteria for variety selection into varietal development programs [99–103].

BRRI dhan47 (IR 63307-4B-4-3) was first selected through PVS and released as salt tolerant variety for *Boro* season in Bangladesh.

Different high-yielding rice varieties for salt (15 from BRRI and two from BINA), drought, submergence tolerance, upland rice varieties and other developed and released for Transplanted Aman (T. Aman-RLR) and Irrigated Ecosystem (Boro-dry season) (**Table 5**).

| Variety name | Cross | Salient features with growing season | Year of release |
|---|---|---|---|
| Salt-tolerant varieties | | | |
| BR23 (BR716–7–2-1-1) | DA29/BR4 | Moderately salt tolerant photosensitive *T. Aman* rice variety for Rainfed Lowland ecosystem | 1988 |
| BRRI dhan40 (BR5331–93–2-8-3) | IR4595-4-1-15/BR10 | Moderately salt tolerant *T. Aman* rice variety for Rainfed Lowland ecosystem | 2003 |
| BRRI dhan41 (BR5828–11–1-4) | BR23/BR1185-2B-16-1 | Moderately salt tolerant *T. Aman* rice variety for Rainfed Lowland ecosystem | 2003 |
| BRRI dhan47 (IR 63307-4B-4-3) | IR51511-B-B-34-B/ TCCP266–2-49-B-B-3 | Salt tolerant *Boro* rice variety for Irrigated ecosystem | 2007 |
| BRRI dhan53 (BR5778–156–1-3-HR14) | BR10/BR23//BR847–76–1-1 | Salt tolerant *T. Aman (RLR)* rice variety | 2010 |
| BRRI dhan54 (BR5999-82–3-2-HR1) | BR1185-2B-16-1/ BR548–128–1-3 | Salt tolerant *T. Aman* (RLR) rice variety | 2010 |
| BRRI dhan55 (IR 73678–6-9-B: AS996) | IR64/*Oryza rufipogon* | Moderately salt, cold and drought tolerant rice variety | 2011 |
| **BRRI dhan28-*Saltol* (IR 89573–84) | BRRI dhan28*3/ FL478 | Salt tolerant *Boro* rice line for Irrigated ecosystem | MABC product |
| **BR11-*Saltol* (IR 89574–7) | BR11*3/FL478 | Salt tolerant *T. Aman*(RLR) rice line | MABC product |
| BRRI dhan61 (BR7105-4R-2) | IR64419-3B-4-3/ BRRI dhan29 | Salt tolerant *Boro* rice variety/Irrigated rice | 2013 |

| Variety name | Cross | Salient features with growing season | Year of release |
|---|---|---|---|
| BRRI dhan67 (BR7100-2R-6-6) | IR61247–3B-8-2-1/ BRRI dhan36 | Salt tolerant *Boro*/Irrigated rice variety | 2014 |
| BRRI dhan73 (IR78767-B-SATB1–28–3-24) | BRRI dhan40/NSIC Rc106 (IR61920-3B-22–1-1) | Salt tolerant *T. Aman*(RLR) rice variety | 2015 |
| BRRI dhan78 (IR77092-B-2R-B-10) | IR84645/IR84649 | Dual tolerant (Salinity+*SUB1T. Aman* rice variety | 2016 |
| BRRI dhan97 (IR83484–3-B-7-1-1-1) | IRRI 113/BRRI dhan40 | Salt tolerant irrigated (Boro) Rice | 2020 |
| BRRI dhan99 HHZ5-DT20-DT2-DT1 (GSR IR1–5-D20-D2-D1) | Huang-Hua-Zhan/ OM1723 | Salt tolerant irrigated (Boro) Rice | 2020 |
| Binadhan-8 (IR66946-3R-149-1-1) | IR29/Pokkali | Salt tolerant irrigated (Boro) Rice | 2010 |
| Binadhan-10 (IR64197–3B-14-2) | IR42598-B-B-B-B-12/Nona Bokra | Salt tolerant irrigated (Boro) Rice | 2012 |
| Drought tolerant varieties | | | |
| BRRI dhan42 (BR6058-6-3-3) | BR14/IR25588–7–3-1 | Moderately drought-tolerant *Upland* (DSR) rice variety | 2004 |
| BRRI dhan43 (BR5543-5-1-2-4) | BR24/BR21 | Moderately drought-tolerant *Upland* (DSR) rice variety | 2004 |
| BRRI dhan56 (IR74371–70–1-1-B) | Way Rarem/ 2*IR5519–4 | Drought-tolerant *T. Aman* (RLR) rice variety | 2011 |
| BRRI dhan57 (BR7873-5NIL)-51-HR6 | BR11/5*CR146-7027-224 | Drought-tolerant *T. Aman* (RLR) rice variety | 2011 |
| BRRI dhan66 (IR82635-B-B-75-2) | IR78875–176-B-2/ IR78875–207-B-3 | Drought-tolerant *T. Aman* (RLR) rice variety | 2014 |
| BRRI dhan71 (IR82589-B-B-84–3) | IR55423–01 (NSIC Rc9)/IRRI148 | Drought-tolerant *T. Aman* (RLR) rice variety | 2015 |
| **Submergence Tolerant varieties** | | | |
| BRRI dhan51 (IR81213–246–237) | Swarna/IR49830–7–1-2-3 | Flood-tolerant *T. Aman* (RLR) rice variety | 2010 |
| BRRI dhan52 (IR85260–66–654-Gaz2) | BR11*3/IR40931-33-1-3-2 | Flood-tolerant *T. Aman* (RLR) rice variety | 2010 |
| BRRI dhan79 (BR9159-8-5-40-14-57) | BRRI dhan49*6/ BRRI dhan52 | Flood-tolerant *T. Aman* (RLR) rice variety | 2017 |
| **Deepwater Rice variety** | | | |
| BRRI dhan91 (BR10230–15-27-7B) | Tilokkachari/BRRI dhan41 | Suitable for shallow flooded ecosystem | 2019 |

**A large range of salt tolerant improve genotypes was grown in mother and baby trials of participatory varietal selection (PVS) in coastal areas of Bangladesh. BRRI dhan47 (IR 63307-4B-4-3) was finally selected and released as salt tolerant variety for* Boro *season in Bangladesh.*
***First introgression of* Saltol *locus into the mega varieties and developed two salt tolerant rice genotypes, IR89573–84 (BRRI dhan28-*Saltol*) and IR89574–7 (BR11-*Saltol*) through marker-assisted breeding at IRRI that are under advance stage of testing for release in Bangladesh and Myanmar.*

**Table 5.**
*List of abiotic stress tolerant varieties released in Bangladesh for Upland rice (Aus), T. Aman (RLR- wet) and Boro (dry) season.*

### 3.3 Marker-assisted selection

FL378 (IR66496-3R-78-1-1), a salinity tolerant recombinant inbred line derived from cross between IR29 and Pokkali was used as donor for *Saltol*. FL378 had the Pokkali introgression on the chromosome 1 from RM1287 (10.9 Mb) to RM493 (12.2 Mb) for about 1.3 Mb and its average tolerance score was around 4.7 [104–106]. The recurrent varieties were BR11, BRRI dhan28 and BRRI dhan29, three mega varieties of Bangladesh are widely grown in wet and dry season.

BRRI dhan28-*Saltol* seeds were developed at IRRI and FL478 as donor parents. The introgression lines of BR11-*Saltol* and BRRI dhan28-*Saltol* were evaluated in salt affected coastal district of Satkhira during dry season.

Moreover, *Saltol* QTL was introgressed into the genetic background of BRRI dhan49 [107] and Pusa44 and Sarjoo52 [92] through marker-assisted backcrossing. FL478 was used as a donor for *Saltol* QTL. A number of *Saltol* introgression lines (NILs: BRRI dhan49-*Saltol* lines) were developed [107]. Krishnamurthy [92] reported that the NILs PU99, PU176, PU200, PU215, PU229, PU240, PU241, PU244, PU252, PU263 of Pusa44 and SAR17, SAR23, SAR35, SAR39, SAR77, SAR87, SAR123, SAR136 NILs of Sarjoo52 exhibited salt tolerance with low salt injury score (SIS) of 3 or 5.

## 4. Conclusions

Effect of abiotic stresses increases due to worldwide climate change. The holistic breeding approach combines different cutting-edge/modern breeding strategies (data based parents selection for crossing, rapid breeding, genomics-assisted breeding and haplotype-based breeding) including efficient gene stacking facilitate the development of climate-resilient rice varieties. Genotypes could cope with the climatic threats, increase the varietal turnover of farmers, and contribute to meet challenges of abiotic stress-prone ecosystems through enhancing productivity and sustaining food security. Also, rice cultivation areas will be expanded under the high abiotic stress-prone areas where salt-stress is a key problem for rice production during both dry and wet seasons in different rice-growing countries including Bangladesh. Moreover, the stress related to HNT needs to be emphasized because this stress may also become the challenge for food security where the rice is a staple food.

## Acknowledgements

This research work was funded by Bill and Melinda Gates Foundation, and Government of Bangladesh.

## Conflict of interest

No conflict of interest.

## Author details

M. Akhlasur Rahman*, Hasina Khatun, M. Ruhul Amin Sarker, Hosneara Hossain, M. Ruhul Quddus, Khandakar M. Iftekharuddaula and M. Shahjahan Kabir
Bangladesh Rice Research Institute, Gazipur, Bangladesh

*Address all correspondence to: akhlas08@gmail.com

## References


[1] Rahman MA, Bimpong IK, Bizimana JB, Pascual ED, Arceta M, Swamy BM, Diaw F, Rahman MS, Singh RK. Mapping QTLs using a novel source of salinity tolerance from Hasawi and their interaction with environments in rice. Rice. 2017; 10(1):1-7

[2] Rahman MA, Thomson MJ, De Ocampo M, Egdane JA, Salam MA, Shah-E-Alam M, Ismail AM. Assessing trait contribution and mapping novel QTL for salinity tolerance using the Bangladeshi rice landrace Capsule. Rice. 2019;12(1):1-18

[3] Khatun S, Flowers TJ. Effects of salinity on seed set in rice. Plant, Cell & Environment. 1995;18(1):61-67

[4] Singh RK, Gregorio GB, Jain RK. QTL mapping for salinity tolerance in rice. Physiology and Molecular Biology of Plants. 2007;13(2):87

[5] Hasanuzzaman M, Fujita M, Islam MN, Ahamed KU, Nahar K. Performance of four irrigated rice varieties under different levels of salinity stress. International Journal of Integrative Biology. 2009; 6(2):85-90

[6] Singh RK, Flowers TJ. The Physiology and Molecular Biology of the Effects of Salinity on Rice. In: M. Pessarakli (ed.). Handbook of Plant and Crop Stress. 3rd edn. Taylor and Francis, Boca Raton, Florida. 2010:901-942

[7] Singh RK, Redona ED, Refuerzo L. 2010: Varietal improvement for abiotic stress tolerance in crop plants: special reference to salinity in rice. In: Pareek A, Sopory SK, Bohnert HJ, Govindjee (eds). Abiotic Stress Adaptation in Plants: Physiological, Molecular and Genomic Foundation. Springer, Dordrecht, the Netherlands. 2010:387-415

[8] Hossain H, Rahman MA, Alam M S, Singh RK. Mapping of quantitative trait loci associated with reproductive-stage salt tolerance in rice. Journal of Agronomy and Crop Science. 2015;201 (1):17-31 doi:10.1111/jac.12086

[9] Munns R, Gilliham M. Salinity tolerance of crops–what is the cost?. New phytologist. 2015; 208(3):668-673

[10] Rahman MA, Thomson MJ, Shah-E-Alam M, de Ocampo M, Egdane J, Ismail AM. Exploring novel genetic sources of salinity tolerance in rice through molecular and physiological characterization. Annals of Botany. 2016;117(6):1083-1097

[11] Oladosu Y, Rafii MY, Samuel C, Fatai A, Magaji U, Kareem I, Kamarudin ZS, Muhammad II, Kolapo K. Drought resistance in rice from conventional to molecular breeding: a review. International Journal of Molecular Sciences. 2019;20(14):3519

[12] Khan S, Anwar S, Ashraf MY, Khaliq B, Sun M, Hussain S, Gao ZQ, Noor H, Alam S. Mechanisms and adaptation strategies to improve heat tolerance in rice. A review. Plants. 2019; 8(11):508..doi:10.3390/plants8110508

[13] Hirabayashi H, Sasaki K, Kambe T, Gannaban RB, Miras MA, Mendioro MS, Simon EV, Lumanglas PD, Fujita D, Takemoto-Kun OY, Takeuc Y, Kaji R, Kondo M, Kobayashi N, Ogawa, T, Ando I, Jagadish KSV, Ishimaru T. qEMF3, a novel QTL for the early-morning flowering trait from wild rice, *Oryza officinalis*, to mitigate heat stress damage at flowering in rice, *O. sativa*. Journal of Experimental Botany. 2015;66 (5):1227-1236 doi:10.1093/jxb/eru474

[14] Zhang G L, Chen LY, Xiao GY, Xiao YH, Chen XB, Zhang ST. Bulked segregant analysis to detect QTL related to heat tolerance in rice (*Oryza sativa* L.) using SSR markers. Agricultural Sciences in China. 2009;8(4):482-487

[15] Li MM, Li X, Yu LQ, Wu JW, Li H, Liu J, Ma XD, Jo SM, Park DS, Song Y, Shin D. Identification of QTLs associated with heat tolerance at the heading and flowering stage in rice (*Oryza sativa* L.). Euphytica. 2018 Apr; 214(4):1-1

[16] Jagadish SVK, Cairns J, Lafitte R, Wheeler TR, Price AH, Craufurd PQ. Genetic analysis of heat tolerance at anthesis in rice. Crop Science. 2010 Dec; 50(5):1633-1641

[17] Bernier J, Kumar A, Ramaiah V, Spaner D, Atlin G. A large-effect QTL for grain yield under reproductive-stage drought stress in upland rice. Crop Science. 2007 Mar;47(2):507-16

[18] Ghimire KH, Quiatchon LA, Vikram P, Swamy BM, Dixit S, Ahmed H, Kumar A. Identification and mapping of a QTL (qDTY1.1) with a consistent effect on grain yield under drought. Field Crops Research, 2012;131: 88-96

[19] Venuprasad R, Bool ME, Quiatchon L, Atlin GN. A QTL for rice grain yield in aerobic environments with large effects in three genetic backgrounds. Theoretical and Applied Genetics. 2012b;124(2):323-332

[20] Swamy BM, Kumar A. Genomics-based precision breeding approaches to improve drought tolerance in rice. Biotechnology Advances. 2013 Dec 1; 31 (8):1308-1318

[21] Yadav RB, Dixit S, Raman A, Mishra KK, Vikram P, Swamy BM, Cruz MT, Maturan PT, Pandey M, Kumar A. A QTL for high grain yield under lowland drought in the background of popular rice variety Sabitri from Nepal. Field Crops Research. 2013 Mar 20;144:281-287

[22] Sandhu N, Singh A, Dixit S, Cruz MT, Maturan PC, Jain RK, Kumar A. Identification and mapping of stable QTL with main and epistasis effect on rice grain yield under upland drought stress. BMC Genetics. 2014 Dec;15(1):1-5

[23] Vikram P, Swamy BM, Dixit S, Ahmed HU, Cruz MT, Singh AK, Kumar A. qDTY 1.1, a major QTL for rice grain yield under reproductive-stage drought stress with a consistent effect in multiple elite genetic backgrounds. BMC Genetics. 2011 Dec; 12(1):1-5

[24] Prince SJ, Song L, Qiu D, Dos Santos JV, Chai C, Joshi T, Patil G, Valliyodan B, Vuong TD, Murphy M, Krampis K. Genetic variants in root architecture-related genes in a *Glycine soja* accession, a potential resource to improve cultivated soybean. BMC Genomics. 2015;16(1):1-20

[25] Shamsudin NA, Swamy BM, Ratnam W, Cruz MT, Sandhu N, Raman AK, Kumar A. Pyramiding of drought yield QTLs into a high quality Malaysian rice cultivar MRQ74 improves yield under reproductive stage drought. Rice. 2016;9(1):1-3

[26] Ray JD, Yu L, McCouch SR, Champoux MC, Wang G, Nguyen HT. Mapping quantitative trait loci associated with root penetration ability in rice (*Oryza sativa* L.). Theoretical and Applied Genetics. 1996 May 1; 92(6):627-36

[27] Lilley JM, Ludlow MM, McCouch SR, O'toole JC. Locating QTL for osmotic adjustment and dehydration tolerance in rice. Journal of Experimental Botany. 1996 Sep 1;47(9):1427-36

[28] Yadav R, Courtois B, Huang N, McLaren G. Mapping genes controlling root morphology and root distribution in a doubled-haploid population of rice. Theoretical and Applied Genetics. 1997 Apr 1;94(5):619-32

[29] Ali ML, Pathan MS, Zhang J, Bai G, Sarkarung S, Nguyen HT. Mapping

QTLs for root traits in a recombinant inbred population from two indica ecotypes in rice. Theoretical and Applied Genetics. 2000 Oct 1;101(5-6):756-66

[30] Tripathy JN, Zhang J, Robin S, Nguyen TT, Nguyen HT. QTLs for cell-membrane stability mapped in rice (*Oryza sativa* L.) under drought stress. Theoretical and Applied Genetics. 2000 Jun 1;100(8):1197-202

[31] Zhang J, Zheng HG, Aarti A, Pantuwan G, Nguyen TT, Tripathy JN, Sarial AK, Robin S, Babu RC, Nguyen BD, Sarkarung S. Locating genomic regions associated with components of drought resistance in rice: comparative mapping within and across species. Theoretical and Applied Genetics. 2001 Jul 1;103(1):19-29

[32] Price AH, Townend J, Jones MP, Audebert A, Courtois B. Mapping QTLs associated with drought avoidance in upland rice grown in the Philippines and West Africa. Plant Molecular Biology. 2002 Mar;48(5):683-95

[33] Zheng BS, Yang L, Zhang WP, Mao CZ, Wu YR, Yi KK, Liu FY, Wu P. Mapping QTLs and candidate genes for rice root traits under different water-supply conditions and comparative analysis across three populations. Theoretical and Applied Genetics. 2003 Nov 1;107(8):1505-15

[34] Hemamalini GS, Shashidhar HE, Hittalmani S. Molecular marker assisted tagging of morphological and physiological traits under two contrasting moisture regimes at peak vegetative stage in rice (*Oryza sativa* L.). Euphytica. 2000 Mar;112(1):69-78

[35] Price AH; Steele KA, Moore BJ, Barraclough PP, Clark LJ. A combined RFLP and AFLP linkage map of upland rice (*Oryza sativa* L.) used to identify QTLs for root-penetration ability. Theoretical Applied Genetics. 2000;100: 49-56

[36] Kumar R, Venuprasad R, Atlin GN. Genetic analysis of rainfed lowland rice drought tolerance under naturally-occurring stress in eastern India: heritability and QTL effects. Field Crops Research. 2007 Jul 25;103(1):42-52

[37] Zheng BS, Yang L, Mao CZ, Huang YJ, Wu P. Mapping QTLs for morphological traits under two water supply conditions at the young seedling stage in rice. Plant Science. 2008 Dec 1; 175(6):767-76

[38] Khowaja FS, Price AH. QTL mapping rolling, stomatal conductance and dimension traits of excised leaves in the Bala×Azucena recombinant inbred population of rice. Field Crops Research. 2008 Mar 20;106(3):248-57

[39] Venuprasad R, Dalid CO, Del Valle M, Zhao D, Espiritu M, Cruz MS, Amante M, Kumar A, Atlin GN. Identification and characterization of large-effect quantitative trait loci for grain yield under lowland drought stress in rice using bulk-segregant analysis. Theoretical and Applied Genetics. 2009 Dec;120(1):177-90

[40] Venuprasad R, Bool ME, Quiatchon L, Atlin GN. A QTL for rice grain yield in aerobic environments with large effects in three genetic backgrounds. Theoretical Applied Genetics. 2012;124:323-332.

[41] Vikram P, Swamy BM, Dixit S, Ahmed H, Cruz MS, Singh AK, Ye G, Kumar A. Bulk segregant analysis: "An effective approach for mapping consistent-effect drought grain yield QTLs in rice". Field Crops Research. 2012 Aug 12;134:185-92

[42] Palanog AD, Swamy BM, Shamsudin NA, Dixit S, Hernandez JE, Boromeo TH, Cruz PC, Kumar A. Grain yield QTLs with consistent-effect under reproductive-stage drought stress in rice. Field Crops Research. 2014 May 1; 161:46-54

[43] Trijatmiko KR, Prasetiyono J, Thomson MJ, Cruz CM, Moeljopawiro S, Pereira A. Meta-analysis of quantitative trait loci for grain yield and component traits under reproductive-stage drought stress in an upland rice population. Molecular Breeding. 2014 Aug;34(2):283-295

[44] Saikumar S, Gouda PK, Saiharini A, Varma CM, Vineesha O, Padmavathi G, Shenoy VV. Major QTL for enhancing rice grain yield under lowland reproductive drought stress identified using an *O. sativa*/*O. glaberrima* introgression line. Field Crops Research. 2014 Jul 1;163:119-131

[45] Wang Y, Zhang Q, Zheng T, Cui Y, Zhang W, Xu J, Li Z. Drought-tolerance QTLs commonly detected in two sets of reciprocal introgression lines in rice. Crop and Pasture Science. 2014 Mar 18; 65(2):171-184

[46] Dixit S, Singh A, Kumar A. Rice breeding for high grain yield under drought: a strategic solution to a complex problem. International Journal of Agronomy. 2014 Oct;2014

[47] Lou Q, Chen L, Mei H, Wei H, Feng F, Wang P, Xia H, Li T, Luo L. Quantitative trait locus mapping of deep rooting by linkage and association analysis in rice. Journal of Experimental Botany. 2015 Aug 1;66(15):4749-57

[48] Jongdee B, Fukai S, Cooper M. Leaf water potential and osmotic adjustment as physiological traits to improve drought tolerance in rice. Field Crops Research. 2002;76(2-3):153-163

[49] Blum A. (2017). Osmotic adjustment is a prime drought stress adaptive engine in support of plant production. Plant, Cell & Environment. 2017;40(1):4-10

[50] Kim Y, Chung YS, Lee E, Tripathi P, Heo S, Kim KH. Root response to drought stress in rice (*Oryza sativa* L.). International Journal of Molecular Sciences. 2020;21(4):1513

[51] Chaves MM, Flexas J, Pinheiro C. Photosynthesis under drought and salt stress: regulation mechanisms from whole plant to cell. Annals of Botany. 2009;103(4):551-560

[52] Ismail AM, Singh US, Singh S, Dar MH, Mackill DJ. The contribution of submergence-tolerant (Sub1) rice varieties to food security in flood-prone rainfed lowland areas in Asia. Field Crops Research. 2013;152:83-93

[53] Ayano M, Kani T, Kojima M, Sakakibara H, Kitaoka T, Kuroha T. ... Ashikari M. Gibberellin biosynthesis and signal transduction is essential for internode elongation in deepwater rice. Plant, Cell & Environment. 2014;37 (10):2313-2324

[54] Inouye J, mogami Y. On the position of the lowest elongated internode of floating rices originated in different countries. Japanese Journal of Tropical Agriculture. 1980 Mar 1;24(1):13-7

[55] Nemoto K, Ukai Y, Tang DQ, Kasai Y, Morita M. Inheritance of early elongation ability in floating rice revealed by diallel and QTL analyses. Theoretical and Applied Genetics. 2004; 109(1):42-7

[56] Hattori Y, Miura K, Asano K, Yamamoto E, Mori H, Kitano H, Matsuoka M, Ashikari M. A major QTL confers rapid internode elongation in response to water rise in deepwater rice. Breeding Science. 2007;57(4): 305-14

[57] Kawano R, Doi K, Yasui H, Mochizuki T, Yoshimura A. Mapping of QTLs for floating ability in rice. Breeding Science. 2008;58(1):47-53

[58] Catling D. Rice in deep water. Springer; 1993.

[59] Suge H. Physiological genetics of internodal elongation under submergence in floating rice. The Japanese Journal of Genetics. 1987;62 (1):69-80

[60] Munns R. Genes and salt tolerance: bringing them together. New Phytologist. 2005;167(3):645-663

[61] Rozema, J, Flowers T. Crops for a salinized world. Science. 2008; 1478-1480

[62] Rahnama A, James RA, Poustini K, Munns R. Stomatal conductance as a screen for osmotic stress tolerance in durum wheat growing in saline soil. Functional Plant Biology. 2010;37(3): 255-263

[63] Gregorio GB, Senadhira D. Genetic analysis of salinity tolerance in rice (*Oryza sativa* L.). Theoretical and Applied Genetics. 1993;86(2-3):333-338

[64] Mohammadi R, Mendioro MS, Diaz GQ, Gregorio GB, Singh RK. Genetic analysis of salt tolerance at seedling and reproductive stages in rice (*Oryza sativa* L). Plant Breeding. 2014; 133(5):548-559

[65] Souleymane O, Salifou M, Hamidou M, Manneh B, Danquah E, Ofori K. Genes action in salinity tolerance and the implication in rice breeding. Journal of Plant Breeding and Genetics. 2017;5(3):115-120

[66] Tanksley SD. Mapping polygenes. Annual review of genetics. 1993 Dec;27 (1):205-233

[67] Ribaut JM, Hoisington D. Marker-assisted selection: new tools and strategies. Trends in Plant Science. 1998 Jun 1;3(6):236-239

[68] Flowers TJ, Koyama ML, Flowers SA, Sudhakar C, Singh KP, Yeo AR. QTL: their place in engineering tolerance of rice to salinity. Journal of Experimental Botany. 2000 Jan 1;51 (342):99-106

[69] Koyama ML, Levesley A, Koebner RM, Flowers TJ, Yeo AR. Quantitative trait loci for component physiological traits determining salt tolerance in rice. Plant physiology. 2001 Jan 1;125(1):406-422

[70] Tuberosa R, Salvi S. Genomics-based approaches to improve drought tolerance of crops. Trends in Plant Science. 2006 Aug 1;11(8):405-412

[71] Thomson MJ, de Ocampo M, Egdane J, Rahman MA, Sajise AG, Adorada DL, Tumimbang-Raiz E, Blumwald E, Seraj ZI, Singh RK, Gregorio GB. Characterizing the Saltol quantitative trait locus for salinity tolerance in rice. Rice. 2010 Sep 1;3 (2-3):148-160

[72] Haque T, Elias SM, Razzaque S, Biswas S, Khan SF, Jewel GN, Rahman MS, Juenger TE, Seraj ZI. Natural variation in growth and physiology under salt stress in rice: QTL mapping in a Horkuch×IR29 mapping population at seedling and reproductive stages. bioRxiv. 2020 doi.org/10.1101/2020.03.01.971895

[73] Mondal S, Borromeo TH, Diaz MGQ, Amas J, Rahman M A, Thomson M J, Gregorio GB. Dissecting QTLs for reproductive stage salinity tolerance in rice from BRRI dhan 47. Plant Breeding and Biotechnology. 2019; 7(4):302-312

[74] Sun BR, Fu CY, Fan ZL, Chen Y, Chen WF, Zhang J, Jiang LQ, Lv S, Pan DJ, Li C. Genomic and transcriptomic analysis reveal molecular basis of salinity tolerance in a novel strong salt-tolerant rice landrace Changmaogu. Rice. 2019;12:99

[75] Lin HX, Zhu MZ, Yano M, Gao JP, Liang ZW, Su WA, Hu XH, Ren ZH, Chao DY. QTLs for $Na^+$ and $K^+$ uptake

of the shoots and roots controlling rice salt tolerance. Theoretical and Applied Genetics. 2004;108(2): 253-260

[76] Ren ZH, Gao JP, Li LG, Cai XL, Huang W, Chao DY, Zhu MZ, Wang ZY, Luan S, Lin HX. A rice quantitative trait locus for salt tolerance encodes a sodium transporter. Nature Genetics. 2005;37(10):1141-6

[77] Bonilla P, Dvorak J, Mackell D, Deal K, Gregorio G. RFLP and SSLP mapping of salinity tolerance genes in chromosome 1 of rice (*Oryza sativa* L.) using recombinant inbred lines. Philippine Agricultural Scientist (Philippines). 2002.

[78] Lee SY, Ahn JH, Cha YS, Yun DW, Lee MC, Ko JC, Lee KS, Eun MY. Mapping of quantitative trait loci for salt tolerance at the seedling stage in rice. Molecules & Cells (Springer Science & Business Media BV). 2006 Apr 1;21(2)

[79] Bimpong IK, Manneh B, Diop B, Ghislain K, Sow A, Amoah NK, Gregorio G, Singh RK, Ortiz R, Wopereis M. New quantitative trait loci for enhancing adaptation to salinity in rice from Hasawi, a Saudi landrace into three African cultivars at the reproductive stage. Euphytica. 2014; 200(1):45-60

[80] Bizimana JB, Luzi-Kihupi A, Murori RW, Singh RK. Identification of quantitative trait loci for salinity tolerance in rice (*Oryza sativa* L.) using IR29/Hasawi mapping population. Journal of Genetics. 2017 Sep;96(4): 571-582

[81] Jahan N, Zhang Y, Lv Y, Song M, Zhao C, Hu H, Cui Y, Wang Z, Yang S, Zhang A, Hu J. QTL analysis for rice salinity tolerance and fine mapping of a candidate locus qSL7 for shoot length under salt stress. Plant Growth Regulation. 2020;90(2):307-319

[82] O'Connor K, Hayes B, Hardner C, Nock C, Baten A, Alam M, Henry R, Topp B. Genome-wide association studies for yield component traits in a macadamia breeding population. BMC Genomics. 2020;21(1):1-2

[83] Quddus MR, Rahman MA, Jahan N, Debsharma SK, Disha RF, Hasan MM, Aditya TL, Iftekharuddaula KM, Collard BC. Estimating pedigree-based breeding values and stability parameters of elite rice breeding lines for yield under salt stress during the Boro season in Bangladesh. Plant Breeding and Biotechnology. 2019;7(3): 257-271

[84] Falconer DS. Introduction to quantitative genetics. 1996. 4th Edition. Addison Wesley Longman, Harlow.

[85] Peng S, Cassman KG, Virmani SS, Sheehy J, Khush GS. Yield potential trends of tropical rice since the release of IR8 and the challenge of increasing rice yield potential. Crop Science. 1999; 39(6):1552-9

[86] Sinha P, Singh VK, Saxena RK, Khan AW, Abbai R, Chitikineni A, Desai A, Molla J, Upadhyaya HD, Kumar A, Varshney RK. Superior haplotypes for haplotype-based breeding for drought tolerance in pigeonpea (*Cajanus cajan* L.). Plant Biotechnology Journal. 2020;18(12): 2482-2490

[87] Varshney RK, Sinha P, Singh VK, Kumar A, Zhang Q, Bennetzen JL. 5Gs for crop genetic improvement. Current Opinion in Plant Biology. 2020.

[88] Rana MM, Takamatsu T, Baslam M, Kaneko K, Itoh K, Harada N, Sugiyama T, Ohnishi T, Kinoshita T, Takagi H, Mitsui T. Salt tolerance improvement in rice through efficient SNP marker-assisted selection coupled with speed-breeding. International Journal of Molecular Sciences. 2019;20 (10):2585

[89] Collard BC, Beredo JC, Lenaerts B, Mendoza R, Santelices R, Lopena V, Verdeprado H, Raghavan C, Gregorio GB, Vial L, Demont M. Revisiting rice breeding methods-evaluating the use of rapid generation advance (RGA) for routine rice breeding. Plant Production Science. 2017 Oct 2;20(4):337-352

[90] Rahman MA, Quddus MR, Jahan N, Rahman MA, Sarker MRA, Hossain H, Iftekharuddaula KM. Field Rapid Generation Advance: An Effective Technique for Industrial Scale Rice Breeding Program. The Experiment. 2019;47(2):2659-2670

[91] Yadav S, Sandhu N, Dixit S, Singh VK, Catolos M, Mazumder RR, Rahman MA, Kumar A. Genomics-assisted breeding for successful development of multiple-stress-tolerant, climate-smart rice for southern and southeastern Asia. The plant genome, e20074. Advance online publication. 2021. https://doi.org/10.1002/tpg2.20074

[92] Krishnamurthy SL, Pundir P, Warraich AS, Rathor S, Lokeshkumar BM, Singh NK, Sharma PC. Introgressed *Saltol* QTL lines improves the salinity tolerance in rice at seedling stage. Frontiers in Plant Science. 2020;11:833

[93] Ceccarelli S, Grando S, Tutwiler R, Baha J, Martini AM, Salahieh H, Goodchild A, Michael M. A methodological study on participatory barley breeding I. Selection phase. Euphytica. 2000;111(2):91-104

[94] Mekbib F. Farmer and formal breeding of sorghum (*Sorghum bicolor* (L.) Moench) and the implications for integrated plant breeding. Euphytica. 2006;152(2):163-176

[95] Rahman MA, Thant AA, Win M, Tun MS, Moet P, Thu AM, Win KT, Myint T, Myint O, Tuntun Y, Labios RV. Participatory varietal selection (PVS): a" bottom-up" breeding approach helps rice farmers in the Ayeyarwady Delta, Myanmar. SABRAO Journal of Breeding & Genetics. 2015;47(3)

[96] Defoer T, Kamara A, De Groote H. Gender and variety selection: farmers' assessment of local maize varieties in southern Mali. African Crop Science Journal. 1997;5(1):65-76

[97] Weltzien, E and Whitaker, M L and Rattunde, H F W and Dhamotharan, M and Anders, M M .*Participatory Approaches in Pearl Millet Breeding.* In: Seeds of Choice: Making the most of new varieties for small farmers. Oxford & IBH Publishing, New Delhi, India, pp. 1998;143-170. ISBN 8120412400

[98] Witcombe JR, Gyawali S, Sunwar S, Sthapit BR, Joshi KD. Participatory plant breeding is better described as highly client-oriented plant breeding. II. Optional farmer collaboration in the segregating generations. Experimental Agriculture. 2006;42(1):79

[99] Islam MR, Salam MA, Bhuiyan MA, Rahman MA, Yasmeen R, Rahman MS, Uddin MK, Gregorio GB, Ismail AM. BRRI dhan47: A salt tolerant rice variety for Boro season isolated through participatory variety selection.2008a.

[100] Weltzien E, VomBrocke K, Rattunde F. Planning plant breeding activities with farmers. In: A. Christinck, E. Weltzien, V. Hoffmann, eds., Setting breeding objectives and developing seed systems with farmers. A handbook for practical use in participatory plant breeding projects. Margraf Publishers GmBH Scientific Books, Weikersheim, Germany. 2005; 123–152

[101] Ceccarelli S, Grando S. Decentralized-participatory plant breeding: an example of demand driven research. Euphytica. 2007;155(3): 349-360

[102] Thapa DB, Sharma RC, Mudwari A, Ortiz-Ferrara G, Sharma S, Basnet RK, Witcombe JR, Virk DS, Joshi KD. Identifying superior wheat cultivars in participatory research on resource poor farms. Field Crops Research. 2009 Jun 26;112(2-3):124-130

[103] Vom Brocke K, Trouche G, Weltzien E, Barro-Kondombo CP, Gozé E, Chantereau J. Participatory variety development for sorghum in Burkina Faso: Farmers' selection and farmers' criteria. Field Crops Research. 2010;119(1):183-194

[104] Thomson MJ, Septiningsih EM, Suwardjo F, Santoso TJ, Silitonga TS, McCouch SR. Genetic diversity analysis of traditional and improved Indonesian rice (*Oryza sativa* L.) germplasm using microsatellite markers. Theoretical and Applied Genetics. 2007;114(3):559-568

[105] Islam MR, Gregorio GB, Salam MA, Singh RK, Collard BC, Tumimbang-Raiz E, Adorada DL, Mendoza RD, Hassan L. Validation of molecular markers and haplotype diversity at the *SalTol* locus on chromosome 1 of rice. Molecular Plant Breeding. 2007.

[106] Islam MR, Gregorio GB, Salam MA, Collard BC, Singh RK, Hassan L. Validation of SalTol linked markers and haplotype diversity on chromosome 1 of rice. Molecular Plant Breeding. 2012 Jun 29;3(1)

[107] Hoque AB, Haque MA, Sarker MRA, Rahman MA. Marker-assisted introgression of saltol locus into genetic background of BRRI dhan49. Int. J. Biosci. 2015;6:71-80

Chapter 7

# How Sorghum Root Traits Can Contribute to Cereal Yield Increase

*Tobias Wojciechowski and Josefine Kant*


## Abstract

In recent decades the effects of climate change became more visible and the problems it causes for agricultural production and yield maintenance. Future crops need to be higher yielding than today, but at the same time more resilient to drought and increased temperatures, especially in drought-prone regions with erratic precipitation. Sorghum, more heat and drought tolerant than maize, presents an interesting candidate for potential genetic material to provide this increased resilience, containing traits and the underlying genetic loci conferring better performance. Compared to the above-ground tissues, root systems are less investigated, but an improvement in this "hidden half" also improves yield. Due to their close relationship, findings in sorghum may be easily incorporated into maize breeding programs. In this chapter we will review recent literature on sorghum and other cereal root system improvements and provide unpublished data on the natural variation of sorghum root development.


**Keywords:** Sorghum, root, root hair, natural variation, genomic diversity

## 1. Introduction

The domestication of Sorghum (*Sorghum bicolor* (L.) Moench) occurred in the region of present-day Sudan approximately 10,000 years ago. It diffused to diverse climates and regions across Africa, India, the Middle East, and Asia between 8,000–1,500 years ago [1, 2]. More recently, diffusion occurred to more temperate zones of northern China and the introduction to North America expanding the range of sorghum cultivation even further [1, 3]. *Sorghum bicolor*, the 5th most important cereal in the world behind maize, rice, wheat, and barley, is grown in both subsistence and commercial agriculture. It is a major crop in the semiarid regions and a dietary staple for more than 500 million people predominantly in Sub-Saharan Africa and South Africa [4–6]. Sorghum, a C4 grass, is cultivated for production of grain, forage, sugar / syrup, brewing, lingocellulosic biomass, and bioethanol [7–9]. Climate change threats the agricultural production and food security in semiarid regions increasing the importance of drought-tolerant crops. Although grain yield gains for maize have been higher than for sorghum, especially under rain-fed management in high water-holding capacity soils [10], sorghum has a higher water use efficiency compared with maize, when grown under optimal growing conditions [11]. Sub-Saharan Africa and South Asia are predicted to have the greatest decline in agricultural productivity due to a significant risk of rising temperatures [12]. These geographical regions overlap with areas with drought and erratic rainfall, where sorghum is already grown as a major staple food. There, sorghum is an important crop for subsistence farmers in these regions due to higher yields compared other cereals in drought and low perceptions regions, which make these areas unsuitable for maize and rice [13, 14]. However, grain yield of sorghum is likely to be affected by post-anthesis drought stress in rainfed farming systems of northeastern Australia [15], India's western-central monsoon region [16], the southern USA [17], and sub-Saharan Africa [18–20]. Selection for stay-green in elite sorghum hybrids should have the potential to increase yield, profitability, and sustainability for farmers in rainfed environments worldwide, without greater yield penalties during wetter years.

Sorghum evolved after splitting from the shared ancestors with rice in Africa 50–70 million years ago, but diffusion into other regions and the widespread cultivation led to high natural genetic diversity within sorghum nowadays [1], which has resulted in distinct phenotypic variance defined by their floral architecture and seed characteristics [6, 21]. Sorghum is a diploid C4 grass with 10 chromosomes and a genome of approximately 800 Mbp [22, 23]. A first reference genome was reported in 2009 [8]. The reference genome of sorghum is derived from the inbred 'BTx623,' a genotype with reduced height and early maturation, which is primarily used for production of grain. The phenotype of this reference genotype is very distinct from the tall, late maturing sorghums, which are usually grown for sugars or high biomass yield [1]. Commercial production systems in Argentina, Australia, Brazil, Mexico or USA utilize sorghum hybrids. However, subsistence agriculture mainly plants sorghum inbred lines for their livelihood. The preference of both consumers and regulators for non-GM sorghum has focused significantly on identifying and utilizing the natural genetic variation of sorghum to improve yield and quality. Currently, Sorghum breeding focusses on tackling abiotic and biotic stresses such as drought, acid soils, and insect and fungal pests [4]. The genetic resources that are largely created by public research are important to understand crop physiology to improve crop performance and production. It is aided by genome-wide map of SNP variation that will accelerate marker-assisted breeding. The adaptability and stress tolerance found in sorghum accessions allows to study the genotype–phenotype relationship as well as dissect genotype-by-environment (G x E x M) interactions for complex, quantitative traits [24] permitting future insights in drought tolerance and thereby mitigating the impacts of climate change. Especially, the exploration of the unknown and unexplored genetic potential taking advantage for the improvement of other cereals, especially maize.

The origin in Africa, distribution to other ecosystems, and agricultural practices is reflected in the phenotypic variation [21] ranging from traditional varieties across Africa and Asia to modern germplasm in China, Australia, and the Americas. This provides a wide variance of morphological and physiological traits for crop improvement [3, 7, 25]. Rainy periods are long and erratic in parts of West Africa, and subsequently, open panicle guinea types are preferred to reduce yield penalties such as grain mold and insect damage. In contrast, other parts of South and East Africa, where rainy seasons are relatively short and predictable, dense panicle kafir and durra types are preferred to increase grain yield per plant [4]. Further selection has occurred in the United States in the last 150 years as temperate and tropical sorghum from Africa and Asia has been bred for commercial agriculture [26].

While research on climate change impact on sorghum is limited, the importance of its root system has been highlighted. Modelling studies have shown that sorghum root systems have a relative adaptive advantage over maize in water-limited conditions [27]. The differences between maize and sorghum root system might facilitate adaptation to drought-prone regions with erratic precipitation. Maize and sorghum differed in root development at the seedling stage for both the number of seminal roots and the timing of nodal root appearance [28]. After germination, sorghum produced a single primary root and a coleoptile, by day 7 the two leaves stage was reached and the primary root had started to form lateral branches. In contrast to maize, no nodal or seminal roots had formed by day 7 (see also **Figure 1**). Sorghum produced only one primary root from seed and nodal roots emerged at the 4th–5th leaf stage, whereas maize produced 3–7 roots from the seed and nodal roots emerged at the 2nd leaf stage [28]. The differences in root development and the adaptation to different environmental and agricultural practices of sorghum root

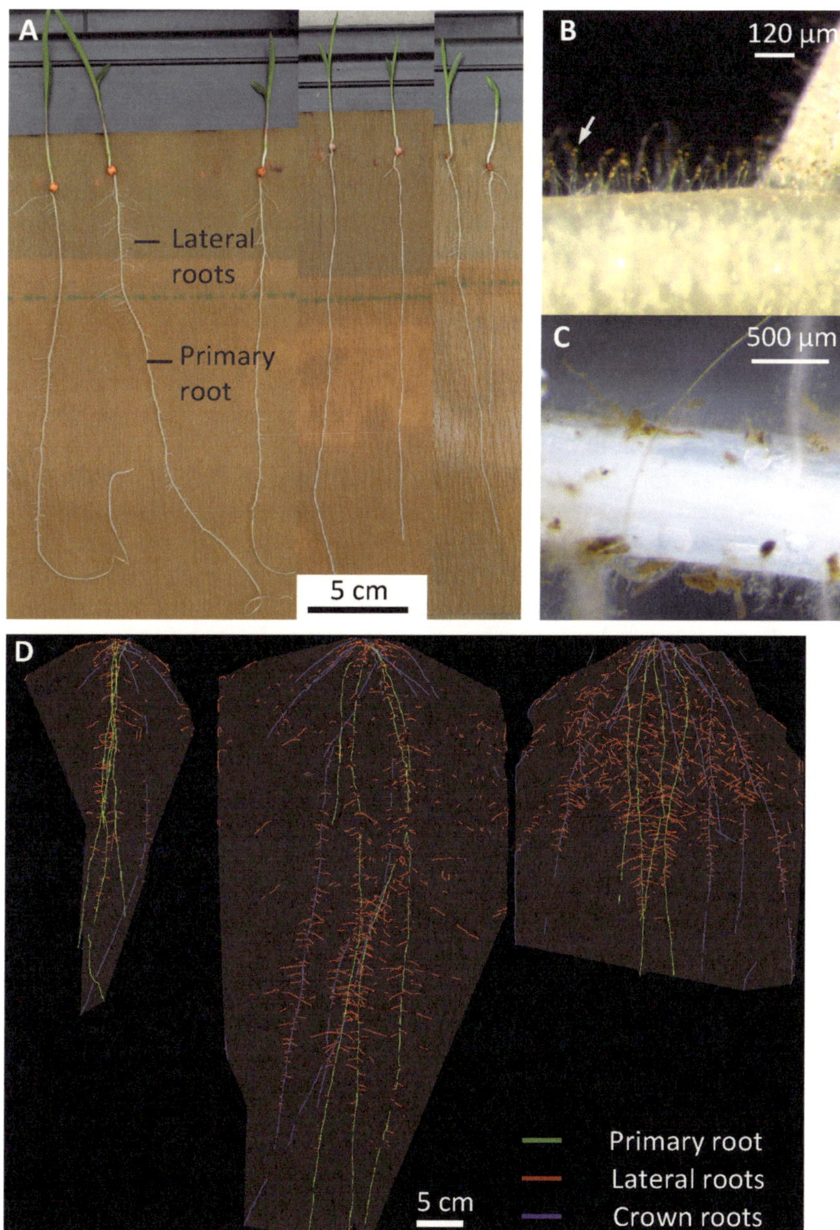


**Figure 1.**
*The root system of* Sorghum bicolor. *Depicted are three european sorghum genotypes either germinated and grown in filterpaper for 14 (A) or 10 days (B) and for 21 (C) or 14 days in soil (D). Root systems and root hair formation of the varieties 'WL08–713' (A,B,D) and 'Zerberus' (A,C,D), and 'SOR19'(A, D) are shown from left to right. The arrow in (B) highlights sorgoleone excretion at root hair tips. In (D) the Maximum Intensity Projections of traced roots grown in rhizotrons of 4 plants per genotype are visible.*

systems might explain the better performance of sorghum in drought-prone regions with erratic precipitation when compared with maize. Increased access to water can be achieved either by better water acquisition from the soil exploring an increased soil volume, which could be achieved by deeper rooting or greater lateral spread [29, 30]. A relationship between drought adaptation and nodal root angle was reported, which further supports the role of below-ground biomass traits in sorghum production under water stress [31]. Furthermore, QTLs were mapped for nodal root angle in sorghum at the 6-leaf stage and evaluated the relevance of the trait for improving drought adaptation via marker-assisted selection. All four nodal root angle QTLs in sorghum identified co-located with previously identified QTLs for stay-green loci [31]. The grain yield benefit of the stay-green phenotype under drought was found to be a result of reduced vegetative biomass and water uptake during the pre-flowering growth stages [32]. Under artificial conditions, sorghum root length during the seedling stage was found to be a major factor in drought tolerance [33].

The parasitic plants, *Striga asiatica* and *Striga hermonthica,* cause serious yield penalties in subsistence and commercial sorghum production. Striga is an obligate root parasite, which seeds will not germinate unless it receives a chemical signal from a potential host plant [34]. Chemicals identified in sorghum root exudates promote seed germination of Striga, the most potent are the strigolactones, a class of related compounds used by most terrestrial plants as hormones to regulate shoot and lateral root development [35, 36], and symbiotic colonization by arbuscular mycorrhizal fungi (AMF) [37]. Sorghum produces several strigolactones and exudes them from its root hairs, particularly under conditions of phosphorous and nitrogen limitations, promoting mycorrhizal association [38]. Colonization with AM fungi greatly improve the performance of sorghum in low-nutrient and drought environments [39]. Striga seems to utilize the signaling to detect its proximity to sorghum roots, so germination at the right time and place will increase the chances of infestation and completing its life cycle. The breeding strategy of Striga-tolerant lines included the introgression of lines that exude less of the Striga-promoting strigolactones, reducing yield penalties through Striga.

Root growth is impeded by aluminum, the third most abundant element in the Earth's crust. A major physiological mechanism facilitating plant aluminum tolerance is aluminum exclusion from root apices based on organic acid release forming stable, nontoxic $Al^{3+}$-complexes in the rhizosphere. Quantitative RT-PCR analysis showed that the responsible gene (*AltSB*) was expressed only in roots of the aluminum-tolerant near isogenic lines and was induced by aluminum. Its expression was highest in the first centimeter of the root [40]. The aluminum and Striga-tolerance are rare examples of adaptive root traits being part of breeding programs. Sorgoleone has similar potential due to its allopathic properties and facilitation of arbuscular mycorrhiza. Sorghum performance on tropical soils is the result of adaptations to multiple stress conditions, including tolerance to aluminum toxicity, efficient acquisition of drought conditions increase the impedance of soils. Sorgoleone is a major component of sorghum root exudates (see also **Figure 1**). It composes from 76 to 99% of the total exudates from sorghum root hairs [41], and is one of the most studied allopathic chemicals [42]. Phosphorus (P) is immobilized in aluminum and iron complexes due to low pH in tropical soils (reviewed in [42–44]). Several root system properties can facilitate P uptake efficiency by responses of root system morphology and architecture [45], modulation of P transporters, organic acid exudation, phosphatase secretion, and association with arbuscular mycorrhizal fungi (AMF) (e.g. [46]). The utilization of sorgoleone in breeding programs could facilitate crop production in drought-prone regions and mitigate the effects of climate change. Sorgoleone synthesis is constitutive and compartmentalized within root hairs, which can accumulate up to 20 μg of exudate per milligram of root dry weight [47, 48]. Attracting AMF for an increased P uptake efficiency is one opportunity, root hairs offer another opportunity to mitigate the effects of climate change such as drought and low precipitation as at low P concentration in tropical soils, root exudation and subsequent mycorrhizal colonization will increase the phosphorous uptake efficiency [44, 48–51]. Furthermore, root hairs play an important role in the uptake of soil phosphorous and water [52–55] as they facilitate acquisition of immobile nutrients such as phosphorous and potassium through increased soil exploration. Acquiring nutrients and water from tropical soils requires a root system that explores the soil volume to deliver these water and nutrients. Soil exploration is often impeded by increased soil strength [43, 56, 57] but a study suggests that root hairs can provide anchorage force required to penetrate tropical soils [56] concluding that root hairs provide anchorage for individual maize root tips and that could provide anchorage for root penetration. Nevertheless, the degree of anchorage provided by root hairs will depend substantially on root hairs and mucilage production. The phenotypic and underlying genotypic potential of sorghum, especially the less studied root traits such as root hairs, has a great potential for breeding as breeding is a necessity in production of new, ideally improved varieties. It requires traits-of-interest with proven effects and phenotypic variation, ideally based on genetic diversity in a known population [57, 58]. To what degree diversity in root phenotypes can be expected will be presented and discussed in the following sections.

## 2. Results and discussion

To exemplify what degree of variation can be expected by sorghum varieties, a set of European sorghum lines was grown under sterile conditions in filter paper and in soil-filled rhizotrons. Grown in filter paper, the primary root and its lateral roots were identified easily, while no seminal roots were observed (**Figure 1**). In agreement with [28] 14 and 21 days after sowing (DAS), no seminal roots were observed in any of the varieties grown, while a varying number of crown roots was found. All root types of all the tested sorghum varieties did have root hairs and all those hairs were excreting sorgoleone, visible as droplets on each root hair tip. On soil-grown roots no sorgoleone was observed, but that might have been absorbed by the surrounding soil or washed away during the washing procedures at harvest.

### 2.1 Diverse set of sorghum genotypes, but little aboveground diversity

30 diverse sorghum genotypes, selected for variation in origin and breeding status, including physiological traits such as drought tolerance, and flowering time, (summarized in **Table 1**) were grown in soil-filled rhizotrons and their roots and shoots non-invasively phenotyped over three weeks. Although genotypic variation was large for most traits, the mean shoot height over all genotypes followed a linear increase (**Figure 2**) and its variation was stable in the last week of growth. Both, shoot height as well as shoot dry weight had a variation of ~2x and ~ 4x, meaning the largest genotype had a dry weight or shoot height twice or four times as large as the smallest genotype. At harvest, 21 DAS, the most contrasting lines had 35 cm compared to 60 cm high shoots and 0.23 g compared to 0.82 g shoot dry matter. Among the varieties with largest shoot height were 'Mace Da Kunya', 'SC35', and 'Mota Maradi', landraces described either as drought tolerant, post-flowering or pre-flowering drought tolerant, respectively. The shortest three varieties were 'Tx430', 'Tx631', and 'Tx436', all American feed-grade hybrids. Genotypes with large shoot height tended to also have higher shoot dry mass compared to genotypes with shorter

| ID | Pedigree | Description | Origin |
|---|---|---|---|
| 1 | T × 430 | Feed-grade hybrid pollinator-parent | USA |
| 2 | T × 2752 | Feed-grade hybrid seed-parent | USA |
| 3 | T × 631 | Food-grade hybrid seed-parent | USA |
| 4 | T × ARG1 | Food-grade hybrid seed-parent | USA |
| 5 | T × 436 | Food-grade hybrid pollinator-parent | USA |
| 6 | B N223 | Food-grade hybrid seed-parent | Niger |
| 7 | SC599 | Post-flowering drought tolerant accession | USA |
| 8 | SC35 | Post-flowering drought tolerant accession | USA |
| 9 | Kuyuma | Improved, open pollinated variety | Zambia |
| 10 | Sepon82 | Improved, open pollinated variety | Niger |
| 11 | SK 5912 Short Kaura | Improved, open pollinated variety | Nigeria |
| 12 | Ajabsido | Drought tolerant landrace | Sudan |
| 13 | CE-151-262-A1 | Improved, open pollinated variety | Senegal |
| 14 | CSM-63 | Drought tolerant landrace | Mali |
| 15 | Mota Maradi | Pre-flowering drought tolerant landrace | Niger |
| 16 | Koro Kollo | Pre-flowering drought tolerant landrace | Sudan |
| 17 | Feterita Gishesh | Pre-flowering drought tolerant landrace | Sudan |
| 18 | Segeolane | Pre-flowering drought tolerant landrace | Botswana |
| 19 | PI609567 | Post-flowering drought tolerant accession | Mali |
| 20 | MR732 | Elite, food-grade, hybrid pollinator-parent | Niger |
| 21 | Wassa | Improved, open pollinated variety | Mali |
| 22 | Seguetana | Improved, open pollinated variety | Mali |
| 23 | El Mota - S241 | Pre-flowering drought tolerant landrace | Niger |
| 24 | Honey Drip | Sweet-stem sorghum | USA |
| 25 | Theis | Sweet-stem sorghum | USA |
| 26 | Framida | Improved, Striga-resistant variety | Burkina Faso |
| 27 | ICSV1049 | Improved, Striga-resistant variety | Burkina Faso |
| 28 | Sariaso 14 | Improved, Striga-resistant variety | Burkina Faso |
| 29 | Grinkan | Improved, open pollinated variety | Mali |
| 30 | Mace Da Kunya | Drought tolerant landrace | Niger |

**Table 1.**
*Commercial sorghum parent lines and accessions.* Sorghum bicolor *lines selected for whole genome sequencing including diverse varieties from Africa, Striga-resistant lines from West Africa, and elite sorghum parent lines.*

shoots. A higher variation in shoot biomass compared to shoot height implicated additional factors influencing the first independent of the latter, such as leaf number, width and thickness. Given the highly diverse origin of these selected genotypes (**Table 1**) a high phenotypic variation above- and below ground was expected.

## 2.2 Diverse set of sorghum genotypes with much higher belowground diversity

In contrast to the relatively small above ground variation in the rhizotron-grown sorghum lines, root dry matter varied much more after three weeks of

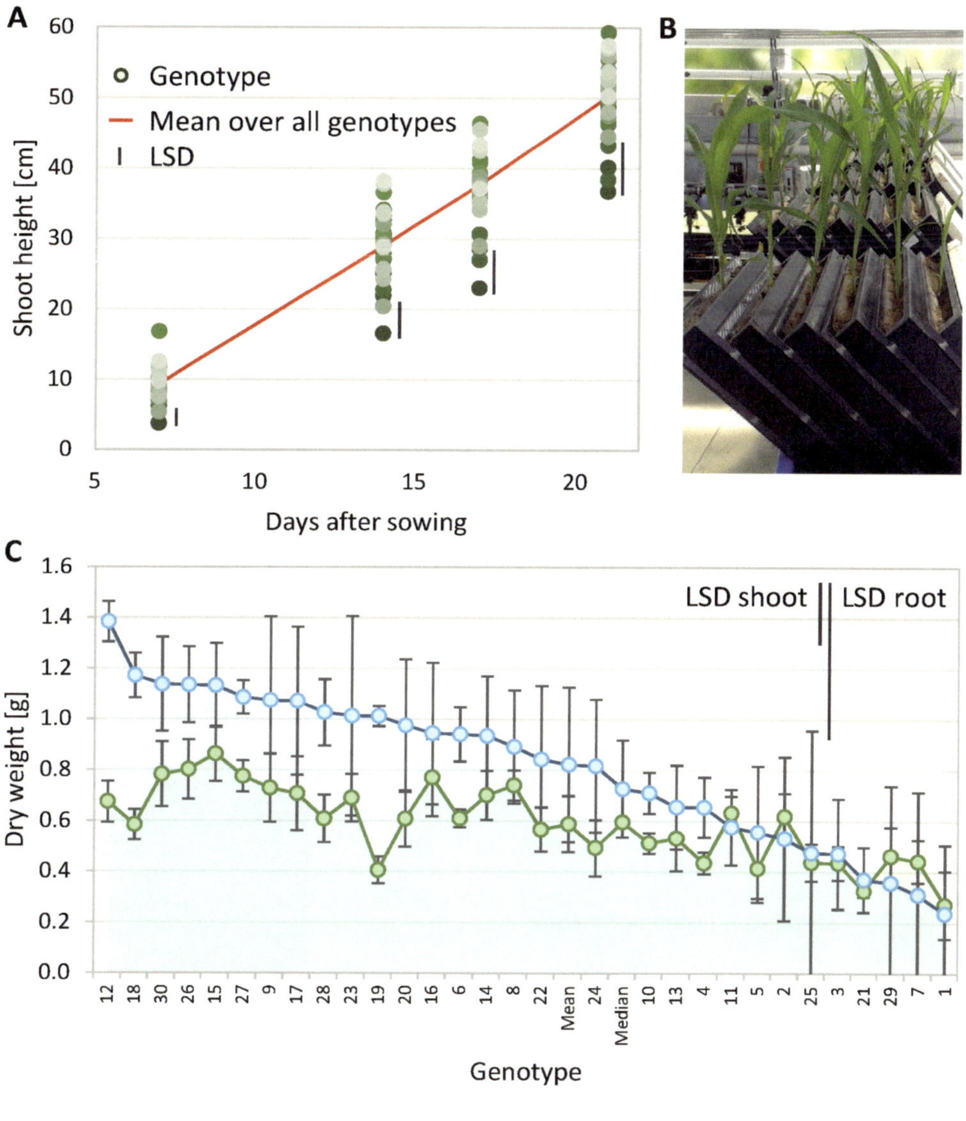


**Figure 2.**
*Variation in growth of 30* Sorghum bicolor *genotypes. Shoot height of 30 sorghum genotypes over 3 weeks grown in soil-filled rhizotrones are shown (A). Single genotype values are means over 4 replicates. At harvest, 21 DAS (example image in B), root and shoot dry weight was determined (C) and shown per genotypes as mean (n = 4) +/− SE. Per timepoint a one-way ANOVA Least Significant Difference (LSD) is depicted. Detailed information on genotypes can be found in* ***Table 1****.*

growth- almost 7x between the most extreme genotypes (**Figure 2**). At harvest the root dry weight varied between 0.22 g and 1.4 g. Again, the three largest root biomass varieties were observed as drought tolerant landraces ('Ajabsido', 'Segeolane', 'Mace Da Kunya'), while improved and hybrid varieties had lower root biomass ('Tx430', 'SC599', 'Wassa'). A similar wide range of variation (6-7x) was found for root length of all separated types- the primary root, nodal roots, and lateral roots (**Figure 3**), but it changed over time. One week after sowing the first emerging primary root showed the highest length and variation while crown and lateral roots were almost not detected. Primary root length reached a plateau between 14 and 17 DAS, both due to the physical rhizotron constraints and the

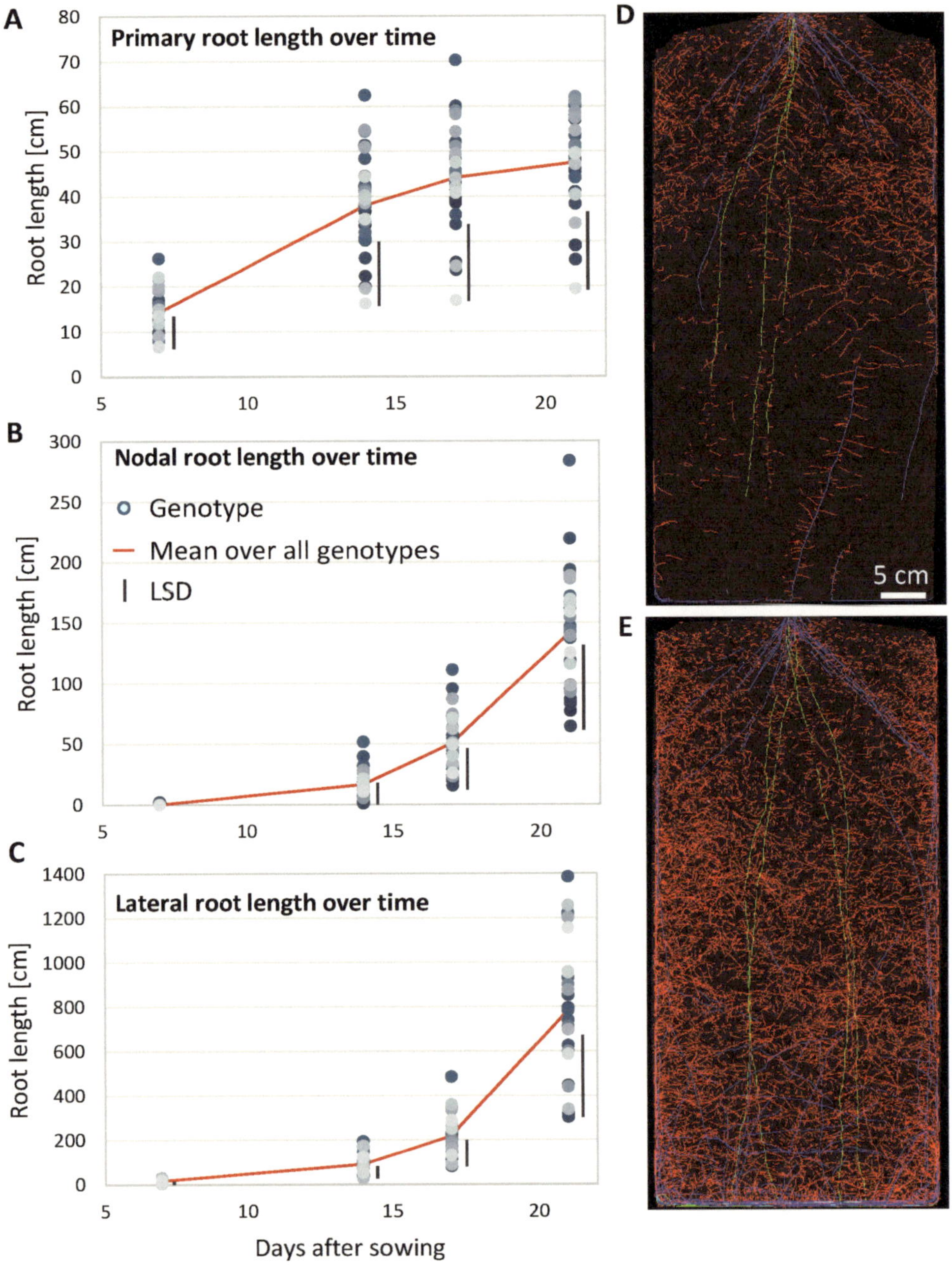


**Figure 3.**
*Variation in root growth of 30* Sorghum bicolor *genotypes. Root length of 30 sorghum genotypes grown for 3 weeks in soil-filled rhizotrones are shown. Single genotype values are means over 4 replicates. Visible and traced primary root (PR) length (A), nodal root (NR) length (B), and lateral root (LR) length (C) are depicted. Maximum intensity projections of all replicates per genotype are shown for the smalles LR and NR length (D, genotype 'SC599') and the largest LR and NR length (E, genotype 'Mota Maradi'). Coloured lines represent traced PR (green), NR (blue), and LRs (red). Per time point a one-way ANOVA Least Significant Difference (LSD) is depicted. Detailed information on genotypes can be found in* ***Table 1****.*

limited number of one primary root per plant. Two weeks after sowing, nodal root length varied from not detected to close to primary root length (50 cm), and just three days later their length doubled, and more than doubled again at harvest, 21 DAS. Lateral root length showed an even stronger increase in length over time, the genotype with the longest LRs had 1,400 cm LR length at harvest, while the most contrasting genotype on the other end had only 250 cm LR length. With increase in NR and LR length over time, their variation among the tested genotypes also increased both in absolute and relative values. Although the genotypic ranking per investigated root type varied slightly, also over time, a general trend of stable ranking became visible. Since these plants were grown without nutritional, water, light, or biological stress this expresses their genetic potential to either form rather small or large root systems, often also with higher numbers of main axis. In all three root types, 'Tx436', a food-grade hybrid pollinator parent, and 'SC599' (**Figure 3D**), a post-flowering drought tolerant accession, were among the lowest ranking genotypes. Among the largest root systems were 'Mota Maradi' (**Figure 3E**), a pre-flowering drought tolerant landrace, and 'SK5912' and short 'Kaura', an improved open pollinated variety. Thus, previously drought tolerant described varieties did not show comparable root system developments in contrast to their early shoot development.

To gain more detailed knowledge about root morphology of these 30 sorghum genotypes, microscopic analyses were performed. Per root type (PR, NR, LRs) root diameter, root hair length, and root hair density were measured (**Figure 4**). When root hair density was plotted against root hair length per root type, a dependency became visible: roughly the more root hairs the longer they were (**Figure 4A**). All root types except for nodal roots showed significant correlations of root hair length and density. On all root types the genotypes 'Tx430', 'Tx631', 'Tx436', and 'Mace Da Kunya' formed the shortest and fewest root hairs. As root hairs are known to be instrumental for water und nutrient uptake [59] it is surprising to find 'Mace Da Kunya' in this list as it was also producing high root and shoot biomass. It should be noted that without nutrient and water limitation short and fewer root hairs were shown to be sufficient for plant growth [60, 61]. The longest and most root hairs were formed on roots of 'Segeolane', 'MR732', and 'Mota Maradi'. Since the latter, a pre-flowering drought tolerant landrace, also had the largest root system, it overall has the highest root surface area leading to the most soil contact for water and nutrient uptake. Like 'Mota Maradi', the genotypes with most root hairs also have the potential to excrete more sorgoleone into their soil environment compared to varieties with smaller root systems and fewer root hairs. Nodal roots had longer root hairs compared to all other root types, followed by primary roots, but their lateral roots did not differ from each other. Overall these soil-grown roots did produce many, but short root hairs of ~150 μm length; similar ranges of root hair formation have been reported for soil-grown rice varieties [60]. On the other hand, field-grown barley genotypes were reported to form longer root hairs from 400 μm [62] up to 700 μm [61]. Root hair formation in these studies varied with environmental conditions, be it nutrient or water supply, or other soil properties, therefore it is likely that sorghum root hairs could be longer in less optimal conditions then the one they were grown in here. Studies on rice root type-dependent root hair formation also showed a high dependency on the growth media used [63, 64].

The rhizotron-grown 30 genotypes showed root type-specific separation of root diameters (**Figure 4B**). For every genotype, nodal roots were not only thicker than primary roots, they did separate clearly with PRs ranging from ~300–550 μm and NRs from ~700–1,050 μm thickness. In contrast, their lateral roots had similar diameters, and both main roots (NR & PR) had 'thin' and 'thick' lateral roots, the first with ~100 μm diameter and little variation, the latter with higher variation from ~150–300 μm. In rice, distinct classes of lateral roots, S-type (thin) and L-type (thick) have been identified that are distinguishable by their diameter, but also branching ability [65, 66]. Recently those LR types have also been indicated to have

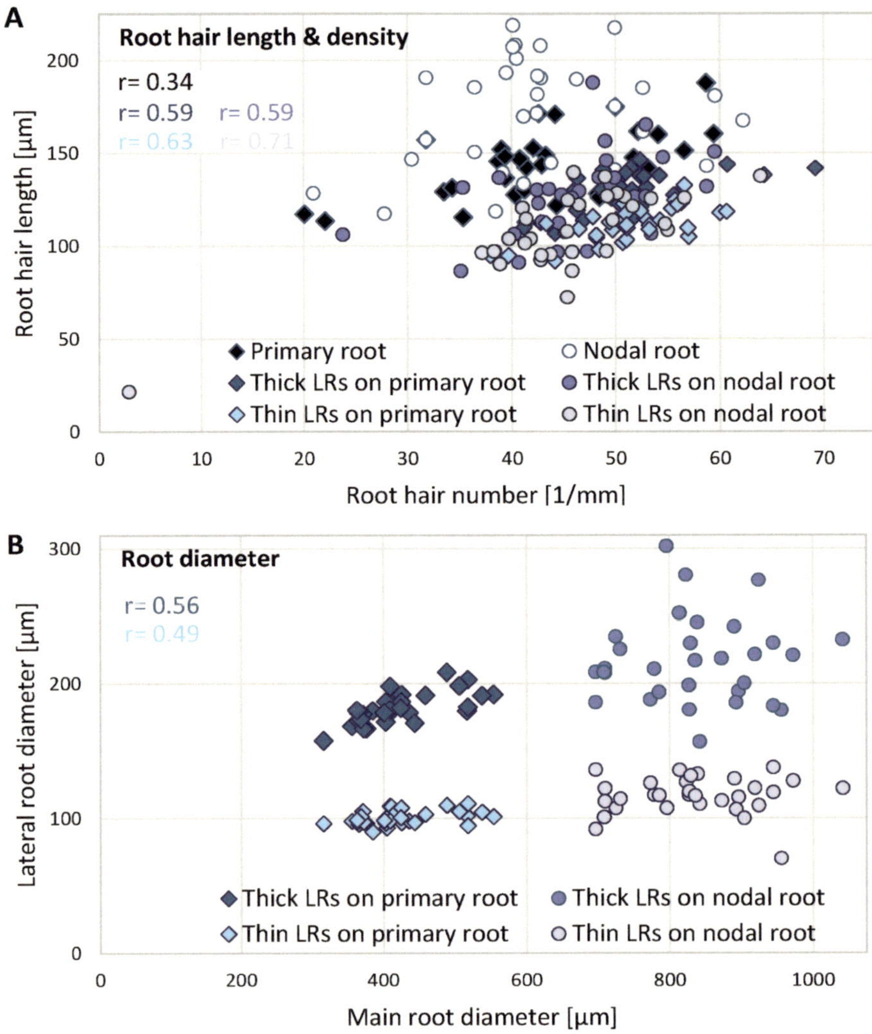


**Figure 4.**
*Root morphology of 30 Sorghum bicolor genotypes. Root morphological traits of 30 sorghum genotypes grown over 3 weeks in soil-filled rhizotrones were determined using a stereomicroscope and analysed with the software Image J. Each point represents a single genotype value which is a mean over 4 replicates. Shown are root hair length depending on root hair density per root type (A) and lateral root (LR) diameter depending on their main root diameter (B). Significant linear correlations are depicted. Detailed information on the genotypes can be found in* ***Table 1****.*

different functions in water and nutrient uptake and transport [67, 68]. If these different diameters do also indicate different LR functions in Sorghum would be interesting to investigate in future experiments, especially with resource limited conditions. Interestingly, while the primary root diameter did significantly correlate with the diameter of its lateral roots, this was not found for nodal roots and nodal root lateral roots. This may be due to the higher variation in lateral root diameter on nodal roots. The genotypes 'Tx436', 'Koro Kollo', and 'Tx631' had thin roots, while 'CSM-63', 'Feterita Gishesh', and 'Framida' were among the biggest root types. Interestingly, 'Kuyuma' had very thick PR and NRs, but very thin lateral roots, especially on NRs, while 'Ajabsido' behaved contrastingly. Overall, the thicker the root, the longer root hairs were measured (**Figure 4**), a trend that has also been observed in maize [69] and in rice [63].

## 3. Conclusion

Already, sorghum is grown in regions where maize production might fail due to adverse drought conditions. One difference between maize and sorghum lies in their root systems. The phenotypic and genotypic variation within *Sorghum bicolor*, especially for root traits such as root morphology, root hairs, and biomass, has great potential for breeding programs to mitigate climate change and to contribute to yield stability in drought prone regions with erratic precipitation. To explore the potential of the sorghum root system, we studied a subset of 30 selected genotypes, which varied in origin and breeding status, and found variation of above ground traits, but a much wider variance of root morphology, root biomass, and root hair density and length. Future experiments with resource limiting conditions will help to understand the underlying physiology of root hairs and their exudates facilitating water and nutrient acquisition, while impacting neighboring weeds and arbuscular mycorrhizal fungi.

## 4. Material and methods

### 4.1 Plant material and growth conditions

#### *4.1.1 Plant material*

For demonstration images (**Figure 1**) three different European sorghum genotypes were grown, WL08–713 and Zerberus from Germany, and SOR19 from Portugal. In the larger rhizotron screening 30 genotypes were grown, selected for variation in several physiological traits, including drought tolerance, flowering time, and origin (summarized in **Table 1**).

#### *4.1.2 Growth in filter paper*

Fungicide-coated seeds were sown either in between sheets of moist (DI water) white filter paper, placed in square petri dishes wrapped with parafilm or placed in moist brown filter paper which was rolled and placed standing upright in a 5 l container with 1 l DI water. After 10 (white) and 14 (brown) days in a greenhouse chamber with 22°C day (16 h) and 18°C night (8 h) sorghum roots were observed for root system structure, sorgoleone production and root hair formation using a stereomicroscope (MX12.5, Leica).

#### *4.1.3 Growth in rhizotrons*

Fungicide-coated seeds of 30 selected genotypes (**Table 1**) were sown in sheets of moist white filter paper, placed in a square petri dish and wrapped with parafilm. After two days a single germinated seed was placed into a rhizotron, four rhizotrons per genotype. The rhizotrons were 30x60 cm large and filled with 3 kg soil substrate (dried & sieved field soil 50:50 (v/v) mixed with dried organic soil low in nutrients "Nullerde"). Groups of six rhizotrons were grouped into a larger container, inclined to ~45° and covered to reduce light falling onto the transparent plexiglas side used for root imaging. The 30 genotypes in each replicate were randomized and grown in a greenhouse compartment at 24°C during the day (16 h, 70% humidity) and 20°C during nights (8 h, 90% humidity). After 7, 14, 18, and 21 DAS non-invasive phenotyping, and after 21 DAS invasive measurements were performed.

### 4.2 Non-invasive root and shoot measurements

Seedlings grown in filter paper were unwrapped gently without removing them from the paper. Roots were photographed with a digital camera and primary and their branched lateral roots imaged under a stereomicroscope (MX12.5, Leica) to evaluate sorgoleone production, root diameter, and root hair formation (length and density).

For rhizotron-grown plants, at the given time points, roots and shoots of the 30 genotypes were non-invasively measured. Shoot length was measured until the tip of the youngest elongated leaf. The root systems of rhizotron-grown plants were imaged using a photo-station equipped with a digital camera. The PaintRhizo software (FZJ) was used to mark and track primary, lateral, and nodal roots separately over time.

### 4.3 Invasive root and shoot measurements

#### *4.3.1 Shoot and root dry weight*

At harvest, 21 DAS and following the last non-invasive measurements, shoots were cut off, dried for seven days in a 60°C oven and then weighted to determine total shoot dry weight. After shoot removal, rhizotrons were opened and the soil was gently removed by washing using running tap water. The primary root as well as the longest crown root were gently separated from the remaining root system. From these roots several 1 cm segments (at 5, 10, 20, 30, 40 cm from the root tip) were cut and transferred to 50% ethanol (p.a.) for subsequent root morphological analyses. The remaining root system was dried for seven days in 60°C and then weighted for root dry weight determination.

#### *4.3.2 Root morphological analyses*

All root segments were imaged using a stereomicroscope (MX12.5, Leica) followed by analysis using the image J software (Fiji). Per root segment four images were taken; per image root diameter and ten root hairs were measured in length, while sorgoleone production was noted as presence or absence. Root hair density was scored following the procedure described in [63]. Distinguished were the primary root and the longest nodal root as well as their daughter roots, separated as 'thick' and 'thin' lateral roots.

### 4.4 Statistical analyses

The experimental data were analyzed with Excel (version 2019, Microsoft) and R (Rstudio, version 4.0.3). Genotypic variation per time point was analyzed by a one-way ANOVA followed by Tukey's Honest Significant Difference (HSD) and the LSD (Least Significant Difference). Linear correlations over all plants was calculated as Pearson's correlation.

## Acknowledgements

We thank Andreas Neuwohner for technical support. We thank Professor Dr. Mitchell R. Tuinstra for selection and generous provision of seeds of the 30 diverse sorghum genotypes. JK was funded by a Feodor-Lynen-Return Fellowship of the Alexander-von-Humboldt Foundation (DEU/1151987). The authors acknowledge institutional funding by the Helmholtz Association (POF III Program – Key Technologies for the Bioeconomy).

## Conflict of interest

The authors declare no conflict of interest.

## Author details

Tobias Wojciechowski and Josefine Kant*
Institute for Bio-and Geosciences, Plant Sciences (IBG-2), Forschungszentrum Juelich, Juelich, Germany

*Address all correspondence to: j.kant@juelich.de

## References


[1] Kimber C. Sorghum: origin, History, Technology and Production. John Wiley and Sons, 2000. ISBN: 978-0-471-24237-6

[2] Wendorf F, Close AE, Schild R, Wasylikowa K, Housley RA, Harlan JR, Królik, H. Saharan exploitation of plants 8000 years b.p. Nature. 1992;359. DOI:10.1038/359721a0.

[3] Deu M, Rattunde F, Chantereau J. A global view of genetic diversity in cultivated sorghums using a core collection. Genome. 2006;49:168-180. DOI: 10.1139/g05-092

[4] National ResearchCouncil Lost Crops of Africa. Lost Crops of Africa I. 1996. The National Academies Press. DOI: 10.17226/2305

[5] Mace ES, Tai S, Gilding EK, Li Y, Prentis PJ, Bian L, Campbell BC, Hu W, Innes DJ, Han X, Cruickshank A, Dai C, Frère C, Zhang H, Hunt CH, Wang X, Shatte T, Wang M, Su Z, Li J, Lin X, Godwin ID, Jordan DR, Wang J. Whole-genome sequencing reveals untapped genetic potential in Africa's indigenous cereal crop sorghum. Nature Communication. 2013;4. DOI: 10.1038/ncomms3320

[6] Morris GP, Ramu P, Deshpande SP, Hash CT, Shah T, Upadhyaya HD, Riera-Lizarazu O, Brown PJ, Acharya CB, Mitchell SE, Harriman J, Glaubitz JC, Buckler ES, Kresovich S. Population genomic and genome-wide association studies of agroclimatic traits in sorghum. Proceedings of the National Academy of Sciences U.S.A. 2013;110:453-458. DOI: 10.1073/pnas.1215985110

[7] Casa AM, Pressoir G, Brown PJ, Mitchell SE, Rooney WL, Tuinstra MR, Franks CD, Kresovich S. Community resources and strategies for association mapping in Sorghum. Crop Science. 2008;48:30-40. DOI: 10.2135/cropsci2007.02.0080

[8] Paterson AH, Bowers JE, Bruggmann R, Dubchak I, Grimwood J, Gundlach H, Haberer G, Hellsten U, Mitros T, Poliakov A, Schmutz J, Spannagl M, Tang H, Wang X, Wicker T, Bharti AK, Chapman J, Feltus FA, Gowik U, Grigoriev IV, Lyons E, Maher CA, Martis M, Narechania A, Otillar RP, Penning BW, Salamov AA, Wang Y, Zhang L, Carpita NC, Freeling M, Gingle AR, Hash CT, Keller B, Klein P, Kresovich S, McCann MC, Ming R, Peterson DG, Mehboob-Ur-Rahman WD, Westhoff P, Mayer KFX, Messing J, Rokhsar DS. The *Sorghum bicolor* genome and the diversification of grasses. Nature. 2009;457:551-556. DOI: 10.1038/nature07723

[9] Brenton ZW, Cooper EA, Myers MT, Boyles RE, Shakoor N, Zielinski KJ, Rauh BL, Bridges WC, Morris GP, Kresovich S. A genomic resource for the development, improvement, and exploitation of sorghum for bioenergy. Genetics. 2016;204:21-33. DOI: 10.1534/genetics.115.183947

[10] Mason SC, Kathol D, Eskridge KM, Galusha TD. Yield increase has been more rapid for maize than for grain sorghum. Crop Science. 2008;48:1560-1568. DOI: 10.2135/cropsci2007.09.0529

[11] Marsalis MA, Angadi SV, Contreras-Govea FE. Dry matter yield and nutritive value of corn, forage sorghum, and BMR forage sorghum at different plant populations and nitrogen rates. Functional Crop Research. 2010.116:52-57. DOI: 10.2135/cropsci2007.09.0529

[12] Battisti DS, Naylor RL. Historical warnings of future food insecurity with unprecedented seasonal heat. Science. 2009;323,240-244. DOI: 10.1126/science.1164363.

[13] Monk R, Franks C, Dahlberg J. Yield Gains in Major US Field Crops. Sorghum. Crop Science Society of America. 2014. DOI: 10.2135/cssaspecpub33.c11

[14] Staggenborg SA, Dhuyvetter KC, Gordon WB. Grain sorghum and corn comparisons: Yield, economic, and environmental responses. Agronomy Journal. 2008;100:1600-1604. DOI: 10.2134/agronj2008.0129

[15] Chapman SC, Cooper M, Hammer GL. Using crop simulation to generate genotype by environment interaction effects for sorghum in water-limited environments. Australian Journal of Agricultural Research. 2002;53:379-389. DOI: 10.1071/AR01070

[16] DeLacy IH, Kaul S, Rana BS, Cooper M. Genotypic variation for grain and stover yield of dryland (rabi) sorghum in India 2. A characterisation of genotype×environment interactions. Functional Crop Research. 2010;118:236-242. DOI: 10.1016/j.fcr.2010.05.014

[17] Bandaru, V., Stewart, B. A., Baumhardt, R. L., Ambati, S., Robinson, C. A., Schlegel, A. Growing dryland grain sorghum in clumps to reduce vegetative growth and increase yield. Agronomy Journal. 2006;98:1109-1120. DOI: 10.2134/agronj2005.0166

[18] Dingkuhn M, Singh BB, Clerget B, Chantereau J, Sultan B. Past, present and future criteria to breed crops for water-limited environments in West Africa. Agricultural Water Management. 2006;80:241-261. DOI: 10.1016/j.agwat.2005.07.016

[19] Mamoutou K, Dingkuhn M, Vaksmann M, Heinemann A. Adaptation to diverse semi-arid environments of sorghum genotypes having different plant type and sensitivity to photoperiod. Agriculture for Meteorology. 2008;148:357-371. DOI: 10.1016/j.agwat.2005.07.016

[20] Abdulai AL, Kouressy M, Vaksmann M, Asch F, Giese M, Holger B. Latitude and Date of Sowing Influences Phenology of Photoperiod-Sensitive Sorghums. Journal of Agronomical Crop Science. 2012;198:340-348. DOI: 10.1111/j.1439-037X.2012.00523.x

[21] Harlan JR, de Wet JMJ. A Simplified Classification of Cultivated Sorghum1. Crop Science. 1972;12:. DOI: cropsci1972.0011183X001200020005x

[22] Price HJ, Hodnett GL, Burson BL, Dillon SL, Rooney WL. A *Sorghum bicolor S. macrospermum* hybrid recovered by embryo rescue and culture. Australian Journal of Botany. 2005;53:579-582. DOI: 10.1071/BT04213

[23] Price HJ, Dillon SL, Hodnett G, Rooney WL, Ross L, Johnston JS. Genome evolution in the genus Sorghum (Poaceae). Annals of Botany. 2005;95:219-227. DOI: 10.1093/aob/mci015

[24] Lasky JR, Upadhyaya HD, Ramu P, Deshpande S, Hash CT, Bonnette J, Juenger TE, Hyma K, Acharya C, Mitchell SE, Buckler ES, Brenton Z, Kresovich S, Morris G. P Genome-environment associations in sorghum landraces predict adaptive traits. Science Advances. 2015;1:e1400218. DOI: 10.1126/sciadv.1400218

[25] Upadhyaya HD, Pundir RPS, Dwivedi SL, Gowda CLL, Reddy VG, Singh S. Developing a Mini Core Collection of Sorghum for Diversified Utilization of Germplasm. Crop Science. 2009;49:1769-1780. DOI: 10.2135/cropsci2009.01.0014.

[26] Quinby JR. The genetics of sorghum improvement. Journal of Heredety. 1975;66:56-62. DOI: 10.1016/B978-0-08-101879-8.00010-3

[27] Sinclair TR, Muchow RC. System analysis of plant traits to increase grain

yield on limited water supplies. Agronomy Journal. 2001;93:263-270. DOI: 10.2134/agronj2001.932263x

[28] Singh, V., Oosterom, E. J., Jordan, D. R., Messina, C. D., Cooper, M., Hammer, Gr., L. Morphological and architectural development of root systems in sorghum and maize. Plant and Soil. 2010;333:287-299. DOI: 10.1007/s11104-010-0343-0

[29] Manschadi AM, Christopher J, Devoil P, Hammer GL. The role of root architectural traits in adaptation of wheat to water-limited environments. Functional Plant Biology. 2006;33:823-837. DOI: 10.1071/FP06055

[30] Manschadi AM, Christopher JT, Hammer GL, Devoil P. Experimental and modelling studies of drought-adaptive root architectural traits in wheat (*Triticum aestivum* L.). Plant Biosystems. 2010;144:458-462. DOI: 10.1080/11263501003731805

[31] Mace ES, Singh V, van Oosterom EJ, Hammer GL, Hunt CH, Jordan DR. QTL for nodal root angle in sorghum (*Sorghum bicolor* L. Moench) co-locate with QTL for traits associated with drought adaptation. Theoretical and Applied Genetics. 2012;124:97-109. DOI: 10.1080/11263501003731805

[32] Borrell AK, Mullet JE, George-Jaeggli B, van Oosterom EJ, Hammer GL, Klein PE, Jordan D. R. Drought adaptation of stay-green sorghum is associated with canopy development, leaf anatomy, root growth, and water uptake. Journal of Experimental Botany. 2014;65:6251-6263. DOI: 10.1093/jxb/eru232

[33] Bibi A, Sadaqat HA, Tahir MHN, Akram H. M. Screening of sorghum (*sorghum bicolor* var moench) for drought tolerance at seedling stage in polyethylene glycol. Journal of Animal and Plant Sciences. 2012;22:671-678. ISSN: 1018-7081

[34] Rich PJ, Ejeta G. Biology of host-parasite interactions in Striga species. Integrating New Technologies for Striga Control: Towards Ending the Witch-Hunt. World Scientific, Singapore. 2007. DOI: 10.1142/9789812771506_0002

[35] Gomez-Roldan V, Fermas S, Brewer PB, Puech-Pagès V, Dun EA, Pillot JP, Letisse F, Matusova R, Danoun S, Portais JC, Bouwmeester H, Bécard G, Beveridge CA, Rameau C, Rochange SF. Strigolactone inhibition of shoot branching. Nature. 2008;455:189-194. DOI: 10.1038/nature07271

[36] Rasmussen A, Mason MG, de Cuyper C, Brewer PB, Herold S, Agusti J, Geelen D, Greb T, Goormachtig S, Beeckman T, Beveridge CA. Strigolactones suppress adventitious rooting in *Arabidopsis* and pea. Plant Physiology. 2012;158:1976-1987. DOI: 10.1104/pp.111.187104

[37] Akiyama K, Matsuzaki K, Hayashi H. Plant sesquiterpenes induce hyphal branching in arbuscular mycorrhizal fungi. Nature. 2005;435:824-827. DOI: 10.1038/nature03608

[38] Jamil M, Van Mourik TA, Charnikhova T, Bouwmeester HJ. Effect of diammonium phosphate application on strigolactone production and *Striga hermonthica* infection in three sorghum cultivars. Weed Research. 2013;53:121-130. DOI: 10.1111/wre.12003

[39] Sun XG, Tang M. Effect of arbuscular mycorrhizal fungi inoculation on root traits and root volatile organic compound emissions of *Sorghum bicolor*. South African Journal of Botany. 2013;88:373-379. DOI: 10.1016/j.sajb.2013.09.007

[40] Magalhaes JV, Liu J, Guimarães CT, Ubiraci LGP, Alves VMC, Wang YH, Schaffert RE, Hoekenga OA, Piñeros MA, Shaff JE, Klein PE, Carneiro NP, Coelho CM, Trick HN,

Kochian LV. A gene in the multidrug and toxic compound extrusion (MATE) family confers aluminum tolerance in sorghum. Nature Genetics. 2007;39:1156-1161. DOI: 10.1038/ng2074

[41] Jesudas AP, Kingsley JS. Sorgoleone from *Sorghum bicolor* as a Potent Bioherbicide. Research Journal of Recent Sciences. 2014;3:32-36.

[42] Tibugari H, Chiduza C, Mashingaidze AB, Mabasa S. Quantification of sorgoleone in sorghum accessions from eight southern African countries. South African Journal of Plant Soil. 2019;36:41-50. DOI: 10.1080/02571862.2018.1469794

[43] Lynch JP, Wojciechowski T. Opportunities and challenges in the subsoil: pathways to deeper rooted crops. Journal of Experimental Botany. 2015;66:2199-2210. DOI: 10.1093/jxb/eru508

[44] Schlemper TR, Leite MFA, Lucheta AR, Shimels M, Bouwmeester HJ, van Veen JA, Kuramae EE. Rhizobacterial community structure differences among sorghum cultivars in different growth stages and soils. FEMS Microbiology Ecology. 2017;93:1-11. DOI: 10.1093/femsec/fix096

[45] Correa J, Postma JA, Watt M, Wojciechowski T. Soil compaction and the architectural plasticity of root systems. Journal of Experimental Botany. 2019;70:6019-6034. DOI: 10.1093/jxb/erz383

[46] Bernardino KC, Pastina MM, Menezes CB, De Sousa SM, MacIel LS, Geraldo CGC, Guimarães CT, Barros BA, Da Costa EL, Carneiro PCS, Schaffert RE, Kochian LV, Magalhaes JV. The genetic architecture of phosphorus efficiency in sorghum involves pleiotropic QTL for root morphology and grain yield under low phosphorus availability in the soil. BMC Plant Biol. 2019;19:1-15. DOI: 10.1186/s12870-019-1689-y

[47] Dayan FE, Cantrell CL, Duke SO. Natural products in crop protection. Bioorganic & Medical Chemistry. 2009;17:4022-4034. DOI: 10.1016/j.bmc.2009.01.046

[48] Figueiredo de Oliveira I, Ferreira Simeone ML, Carvalho de Guimarães C, Garcia NS, Schaffert RE, Morais de Sousa S. Sorgoleone concentration influences mycorrhizal colonization in sorghum. Mycorrhiza. 2020;2-7. DOI: 10.1007/s00572-020-01006-1

[49] Schlemper TR, van Veen JA, Kuramae EE. Co-Variation of Bacterial and Fungal Communities in Different Sorghum Cultivars and Growth Stages is Soil Dependent. Microbial Ecology. 2018;76:205-214. DOI: 10.1007/s00248-017-1108-6

[50] Chiu CH, Paszkowski U. Mechanisms and Impact of Symbiotic Phosphate Acquisition. Cold Spring Harbor Perspectives in Biology. 2019;11:. DOI: 10.1101/cshperspect.a034603

[51] Kobae Y. Dynamic phosphate uptake in arbuscular mycorrhizal roots under field conditions. Frontiers in Environmental Science. 2019;6:1-12. DOI: 10.3389/fenvs.2018.00159

[52] Bates TR, Lynch JP. Root hairs confer a competitive advantage under low phosphorus availability. Plant & Soil. 2001;236:243-250.

[53] Gahoonia TS, Nielsen NE. Phosphorus (P) uptake and growth of a root hairless barley mutant (*bald root barley*, *brb*) and wild type in low- and high-P soils. Plant, Cell & Environment. 2003;26:1759-1766. DOI: 10.1046/j.1365-3040.2003.01093.x

[54] Haling RE, Brown LK, Bengough AG, Young IM, Hallett PD, White PJ, George TS. Root hairs improve

root penetration, root-soil contact, and phosphorus acquisition in soils of different strength. Journal of Experimental Botany. 2013;64:3711-3721. DOI: 10.1093/jxb/ert200

[55] Carminati A, Passioura JB, Zarebanadkouki M, Mutez AA, Ryan PR, Watt M, Delhaize El. Root hairs enable high transpiration rates in drying soils. New Phytologist. 2017;216:771-781. DOI: 10.1111/nph.14715

[56] Bengough AG, Mckenzie BM, Hallett PD, Valentine TA. Root elongation, water stress, and mechanical impedance: a review of limiting stresses and beneficial root tip traits. Journal of Experimental Botany. 2011;62:59-68. DOI: 10.1093/jxb/erq350

[57] Valentine T, Hallett PD, Binnie K, Young MW, Squire GR, Hawes C, Bengough AG. Soil strength and macropore volume limit root elongation rates in many UK agricultural soils. Annals of Botany. 2012;110:259-70. DOI: 10.1093/aob/mcs118

[58] Wissuwa M, Kant J. Does half a millimetre matter? Root hairs for yield stability. A commentary on 'Significance of root hairs for plant performance under contrasting field conditions and water deficit'. Annals of Botany. 2021:XX:i–iii. DOI:10.1093/aob/mcab027

[59] Gahoonia TS, Nielsen NE. Direct evidence on participation of root hairs in phosphorus ($^{32}$P) uptake from soil. Plant and Soil. 1998;198:147-152. DOI: 10.1023/A:1004346412006

[60] Nestler J, Wissuwa *M. Superior* root hair formation confers root efficiency in some, but not all, rice genotypes upon P deficiency. Frontiers in Plant Science. 2016;7:1935. DOI: 10.3389/fpls.2016.01935

[61] Marin M, Feeney DS, Brown LK, Naveed M, Ruiz S, Koebernick N, Bengough AG, Hallett PD, Roose T, Puértolas J, Dodd IC, George TS. Significance of root hairs for plant performance under contrasting field conditions and water deficit. Annals of Botany. 2020;,mcaa181. DOI: 10.1093/aob/mcaa181

[62] Haling RE, Brown LK, Bengough AG, Valentine TA, White PJ, Young IM, George TS. Root hair length and rhizosheath mass depend on soil porosity, strength and water content in barley genotypes. Planta. 2014;239:643-651. DOI: 10.1007/s00425-013-2002-1

[63] Nestler J, Keyes SD, Wissuwa M. Root hair formation in rice (*Oryza sativa* L.) differs between root types and is altered in artificial growth conditions. Journal of Experimental Botany. 2016;67:3699-3708. DOI: 10.1093/jxb/erw115

[64] Kant J, Ishizaki T, Pariasca-Tanaka J, Rose T, Wissuwa M, Watt M. Phosphorus efficient phenotype of rice. In: Rice. Intech Open. 2018. DOI: 10.5772/intechopen.75642

[65] Kono Y, Igeta M, Yamada N. Studies on the developmental physiology of the lateral roots in the rice seminal root. Japanese Journal of Crop Science. 1972;41:192-204. DOI: 10.1626/jcs.41.192

[66] Yamauchi A, Kono Y, Tatsumi J. Comparison of root system structures of 13 species of cereals. Japanese Journal of Crop Science. 1987;56:618-631. DOI: 10.1626/jcs.56.618

[67] De Bauw P, Mai TH, Schnepf A, Merckx R, Smolders E, Vanderborght J. A functional-structural model of upland rice root systems reveals the importance of laterals and growing root tips for phosphate uptake from wet and dry soils. Annals of Botany. 2020;XX:1-18. DOI: 10.1093/aob/mcaa120

[68] Watanabe Y, Kabuki T, Kakehashi T, Kano-Nakata M, Mitsuya S,

Yamauchi A. Morphological and histological differences among three types of component roots and their differential contribution to water uptake in the rice root system. Plant Production Science. 2020;23:191-201. DOI: 10.1080/1343943X.2020.1730701

[69] Nestler J, Liu S, Wen TJ, Paschold A, Marcon C, Tang HM, Li D, Li L, Meeley RB, Sakai H, Bruce W, Schnable PS, Hochholdinger F. *Roothairless5*, which functions in maize (*Zea mays* L.) root hair initiation and elongation encodes a monocot-specific NADPH oxidase. The Plant Journal. 2014;79:729-740. DOI: 10.1111/tpj.12578

Chapter 8

# Irrigated and Rain-Fed Lowland Rice Breeding in Uganda: A Review

*Jimmy Lamo, David Ochan, Desta Abebe, Zelalem Zewdu Ayalew, Anna Mlaki and Cyprien Ndikuryayo*

## Abstract

Since introduction of rice into Uganda in 1904, improvement of the irrigated and rain-fed lowland types was undertaken to address a number of production and quality constraints in three consecutive and overlapping phases. The initial phase was achieved through evaluation of introduction, selection of promising lines and subsequent release of the selected lines for production by the farmers. In the second phase, genetic potential of traits and characteristics of interest were analyzed and used to guide selection of suitable parents for hybridization and the third phase employed genotyping approach in screening and selection of the parental lines and the segregating populations to enhance the breeding efficiency for the traits of importance. Simultaneously, the key production constraints addressed included resistance to rice yellow mottle virus (RYMV), rice blast, bacterial leaf blight and narrow leaf spot diseases as well as submergence tolerance and cold tolerance. The quality traits considered for the improvement alongside the grain yield parameters were the grain aroma, amylose content, shape and size. These interventions have resulted into release and wide adoption of seven rice varieties in Uganda besides several breeding lines which have informally diffused into different major rice production agro-ecology. Subsequently, it can be concluded that a substantially strong and functional breeding platform for rice in Uganda has been established.

**Keywords:** evaluation, genotyping, grain quality and yield, hybridization, inbred lines and variety release

## 1. Introduction

Uganda is a tropical country that lies astride equator, but with modified climatic conditions due to large water bodies and high peaked mountains. The altitude varies from 614 to 5,111 metres above sea level (masl) with much of the rice production areas falling within an altitude ranging from 1,000 to 1,400 masl. The least and the highest recorded temperature within the rice production agro-ecology is 8°C and 38°C, respectively, whereas on the basis of the average for the entire country, lowest temperatures range between 10°C and 17°C and highest from 23°C-25°C. The annual rainfall intensity in Uganda varies from 600 mm to 2,500 mm with much of the country receiving between 900 mm and 1,800 mm of rainfall. Owing to the diversity in the climatic conditions, rice production ecologies in Uganda are classified into three broad categories of rain-fed upland, irrigated and rain-fed lowland production areas. The total land area suitable for rain-fed upland rice production constitutes an estimated 70% of the arable land in Uganda and are mostly located within Upper Nile basin in the Northern region and Albert basin in Western Uganda [1]. The area is classified into four zones on the basis of availability of water for production and proximity to critical services for rice production, namely Upper Nile basin, Albert basin, Victoria basin and Kyoga basin. The upper Nile basin drains an area of 48,911 sq. $km^2$ comprising Albert Nile (21,234 sq. $km^2$) and Aswa catchment. The Albert basin covers 21,875 sq. $km^2$ with 4.5% of the area being permanent wetlands and 3.5% seasonal/ temporary wetlands while Victoria zone covers 30, 880 sq. $km^2$ and Kyoga zone covering 137,500 sq. $km^2$. Overall, a total of 239,166 sq. $km^2$ catchment basin suitable for rice production is located in Uganda. Of the Uganda's 241,500 sq. $km^2$ catchment area, 15% is open water and 3% represent permanent wetland area while 9.4% comprise the seasonal wetland area. In essence, therefore, much of the Uganda's land surface is collectively suitable for rain-fed and irrigated rice production.

Rice was introduced to Uganda in the early 1900's, but the exact year remains contradicting as the different years of 1904 [2] and 1910 [3] were reported. In addition, other authors believed that rice was already introduced into the country by end of the 1870's, the time when the Arab community grew rice for their consumption [4]. However, it was believed that rice was first introduced in milled form for consumption by European administrators and Indian businessmen as well as Indian rail construction workers, the 'coolies' who built the railway line then referred to as the 'Iron Snake', from Mombasa to Uganda. Subsequently, small observation and minimal fields were latter established during the period from 1904 to 1940 by Swahili and Indian staff and Church Missionary Society (CMS) staff [2]. In the year 1921, rice was already recognized and reported as one of the food crops produced and promoted in the country [5]. Rice production was promoted further during the World War II in 1939–1945 to provide food to soldiers. Progressively, a successful irrigated rice trial was launched in central Uganda [2] and by 1940's, commercial production of rain-fed lowland rice in the country had increased [2, 6]. Later, during the 1950's, the Uganda government developed further interest in rice and potential for irrigated rice farming. Subsequently, two irrigation schemes of Kibimba and Doho were developed in the 1970's in Eastern Uganda and later on a third rice irrigation scheme was constructed at Olweny swamp in Northern Uganda.

Given the relatively many decades of rice cultivation of the introduced rice crop and with expanding acreage under production, a long-standing draw back to irrigated and lowland rice production of pests and diseases were reported [2, 3, 6–9]. In most cases, some of the then existing varieties were dropped due to susceptibility to insect pests and diseases [7]. With this background constraints threatening rice yield, the rice breeding program was inevitably established to identify and incorporate broad-spectrum and durable resistance (BSDR). The initial stage was to identify the genetic donors for the targeted traits for the improvement through characterization and conservation of germplasm collections. This was followed by development of segregating populations and selection of genotypes possessing the desirable traits and identify candidate rice genes contributing towards BSDR through co-localization with resistance to different stress genes. The new lines generated were then advanced through anther culture technique and modified Rapid Generation Advance (RGA) technique, for which the anther culture technique has proved useful in improvement of selection efficiency for yield and other traits of low heritability. In addition, the doubled haploid lines developed from the RGA have been associated more with additive genetic variance component compared to the conventional $F_2$ and $F_3$ generations as the dominance variance component is eliminated in double haploid technology, implying higher irritability of these economic traits in rice [10]. It was reported that in the case of $F_3$ and $F_4$, both additive and dominance gene effects contribute to phenotypic differences between individuals, which tends to mask the expression of the desired traits [11] whereas variation in doubled haploid progeny is only due to some environmental effects.

The elite breeding lines were then evaluated in replicated yield trials and multilocation testing to identify promising lines for onwards evaluations on-farm and testing in National Trials following the National Variety Release guidelines and subsequently the promising lines basing on the key preferred production and quality attributes are nominated to the National Variety Release Committee of Uganda for approval and release as new rice varieties.

Key attributes of preferred rice for production are highlighted below:

1. Agronomic traits: This target breeding for high yielding and preferred plant type such as high yielding rice exhibiting 20% yield increase compared to the existing rain-fed lowland and irrigated rice varieties, but with plant height ranging from 90 to 110 cm and maturity period of between 90 and 135 days after planting.

2. Biotic stress: The focus in regards to the abiotic stress has been development of new climate resilient varieties tolerant to abiotic stresses such as drought, iron toxicity, low nitrogen (high nitrogen use efficiency), submergence and cold stresses.

3. Biotic stress: Among the biotic stresses, efforts were directed to develop varieties resistant to currently important and emerging insect pests and diseases. Major pests include stem borers (*Scripophaga* sp., *Chilo* sp., *Diopsis* sp), African rice gall midge (*Orseolia oryzae*), leaf hopper (*Cnaphalocrosis medinalis*), grain sucking insects such as rice bugs (*Nezara viridula*, *Leptocorsia sp* and *Nephotettix spp*), root feeding insects including Rood weevil (*Lissorhoptrus oryzophilus*) and whorl maggot (*Hydellia* sp); while major diseases include rice leaf blast (*Magnaporthe oryzae*), panicle blast, bacterial leaf blight (*Xanthomonas oryzae* pv. *oryzae*). bacterial leaf streak, rice yellow mottle virus, sheath blight, sheath rot, narrow leaf spot and brown leaf spot diseases.

4. Grain quality: Development of rice with quality attributes preferably with low amylose content for diabetic and elderly persons of less than 20%, noodles (>29%), high milling recovery (>65%) and preferred cooking quality of non-sticky and aroma characteristics. Preference for aromatic, non-sticky, whole grain and white rice grains traits are also commonly preferred by most rice consumers in Uganda [7, 12]. In terms of size, medium and long grains are preferred in the country as well as rice with intermediate amylose and intermediate gelatinization temperature desired.

5. Physiological characteristics: This involves breeding for moderate threshability (non-shattering and tight grain attachment) as detailed in an earlier study conducted in Uganda [reference].

6. Produce quality breeder and foundation seed. This is one of the prescribed roles of a breeder to ensure availability of quality seeds in the right quantity either directly or through the agro-input dealers to the farming communities.

## 2. Evaluation and deployment of rice varieties

### 2.1 Introduction of rice varieties to Uganda

Irrigated and rain-fed lowland rice varieties were introduced into Uganda in two phases namely, first from 1921 to 1970 and second during 1971–2020. In the first phase, a total of eight irrigated and rain-fed lowland rice varieties, specifically Jaggery, Cakala, Matama, Kawemba, Kigaire, Seena, SUPA LOCAL and Bungala were introduced and grown (**Table 1**). On the basis of aroma, all the eight rice cultivars except Bungala were aromatic types.

In the second phase covering the period from 1971 to 2020, six released varieties and up to 70 unreleased but informally released rice varieties were commonly being grown by the farming communities. The 70 informally released varieties are classified into two different groups according to their generations. The first are called the K-series of rice introduced from China under technical assistance program, while the second are introductions from major breeding centers detailed in **Table 1**. Both groups are modern varieties with high-yielding capacity and tolerance for various biotic and abiotic field stress conditions. The K-series is an acronym of Kibimba lines, named so probably due to the series being grown then in the Kibimba government rice irrigation scheme. Fortunately, most of the K- series had a desirable combination of intermediate amylose content and intermediate gelatinization temperature and a notable variety is IR64, which has been widely accepted as a high-quality rice. Later, however, there were several devastating stresses that undermined the importance of these varieties for which reasons more introductions from other centres were requested and received. The major breeding centers that provided rice germplasm to Uganda include Africa Ricer Center (AfricaRice), International Rice Research Institute (IRRI), International Centre for Tropical Agriculture (CIAT), Colombia, Tanzania Agricultural Research Organization (TARO), Nigeria, Crop Research Institute (CRI) Ghana, Yunnan Agricultural University, Kunming, China and Chinese Academy of Agricultural Sciences (CAAS), China. Overall, the germplasm received include the following specie of *Oryza sativa*, generations involving crosses between *O. sativa* and *O. glaberrima*, *Oryza barthi* x *O. sativa* crosses, and *O. sativa* x *O. longistaminata* crosses.

### 2.2 Screening introduction, genetic studies and hybridization

#### *2.2.1 Rice diseases*

Rice yellow mottle virus (RYMV) disease is apparently the most serious disease of rice under irrigated and rain-fed lowland conditions in Uganda. In order to identify lines with resistance to RYMV disease, 934 rice lines were screened in the years 2015, 2016, 17 and 2018. These list excluded IR 64, K 34, K 38, K 85 and KOMBOKA (susceptible control), Gigante from AfricaRice (resistant check), Namche 2, NERICA 4 (MET P71), NERICA 8 (MET P72) and WITA 9 (local resistant check) that were included in each set [20, 21]. Each line was sown in the field at high plant population of 10 grams in a land area measuring 20 cm x 20 cm in size. Mechanical inoculations were carried out on seedlings (10 plants test line) at 3 weeks post germination as described by Thouvenel and Fauquet [22]. Symptom appearance was monitored on daily basis to assess the stage of disease initiation and thereafter, disease severity was scored using a scale of 1 (no symptoms) to 9 (severe symptoms) [23, 24] at 45 days post-inoculation (DPI). Of the 934 lines evaluated, a total of 54 either highly resistant or just resistant were identified (**Table 2**). Majority of the RYMV resistant lines were crosses with Tongil

| S/N | Variety | Designation | Origin | Time started growing | Key Preferred trait | Reference |
|---|---|---|---|---|---|---|
| 1 | Jaggery | Unknown | Tanzania | 1921 | Aromatic | [13] |
| 2 | Cakala, | Unknown | Tanzania | 1921 | Aromatic | [2] |
| 3 | Matama, | Unknown | Tanzania | 1921 | Aromatic | [2] |
| 4 | Kawemba, | Unknown | Tanzania | 1921 | Aromatic | [2] |
| 5 | Kigaire | Unknown | Tanzania | 1921 | Aromatic | [2] |
| 6 | Seena | Unknown | Tanzania | 1921 | Aromatic | [2] |
| 7 | SUPA LOCAL | Supa v 88 | Tanzania | 1970 | Aromatic | [13] |
| 8 | Bungala | Unknown | Tanzania | 1970 | Non- aromatic | [13] |
| 9 | Congo | Unknown | Tanzania | 1972 | Aromatic | [13] |
| 10 | Kaiso | Unknown | Tanzania | 1972 | Aromatic | [13] |
| 11 | K-12 | Unknown | Tanzania | 1972 | Non- aromatic | [13] |
| 12 | K-23 | Unknown | Tanzania | 1972 | Non- aromatic | [13] |
| 13 | K-34 | Unknown | Tanzania | 1972 | Non- aromatic | [13] |
| 14 | K-38 | Unknown | Tanzania | 1972 | Non- aromatic | [13] |
| 15 | K-85 | Unknown | Tanzania | 1972 | Non- aromatic | [13] |
| 16 | K-98 | Unknown | Tanzania | 1972 | Non- aromatic | [13] |
| 17 | K-264 | Unknown | Tanzania | 1972 | Non- aromatic | [13] |
| 18 | Benenego | Unknown | Tanzania | 1972 | Aromatic | [14] |
| 19 | Supa America | Unknown | Tanzania | 1972 | Aromatic | [14] |
| 20 | Bulemeezi | Unknown | Tanzania | 1972 | Aromatic | [15, 16] |
| 21 | Kyabukooli | Unknown | Tanzania | 1972 | Non- aromatic | [15, 16] |
| 22 | Pakistan | Unknown | Tanzania | 1972 | Aromatic | [15, 16] |

| S/N | Variety | Designation | Origin | Time started growing | Key Preferred trait | Reference |
|---|---|---|---|---|---|---|
| 23 | Maisombira | Unknown | Tanzania | 1972 | Aromatic | [15, 16] |
| 24 | Abenego | Unknown | Tanzania | 1972 | Aromatic | [15, 16] |
| 25 | Namahumbo | Unknown | Tanzania | 1972 | Aromatic | [15, 16] |
| 26 | Sebagala | Unknown | Tanzania | 1972 | Aromatic | [15, 16] |
| 27 | Vietnam | Unknown | Tanzania | 1972 | Non- aromatic | [15, 16] |
| 28 | Supa china | Unknown | Tanzania | 1972 | Aromatic | [15, 16] |
| 29 | Gabon | Unknown | Tanzania | 1972 | Aromatic | [15, 16] |
| 30 | Namala | Unknown | Tanzania | 1972 | Aromatic | [15, 16] |
| 31 | Kaki | Unknown | Tanzania | 1972 | Non- aromatic | [15, 16] |
| 32 | Kibimba | Unknown | Tanzania | 1972 | Non- aromatic | [15, 16] |
| 33 | Kabonge | Unknown | Tanzania | 1972 | Non- aromatic | [15, 16] |
| 34 | Kibuyu | Unknown | Tanzania | 1972 | Non- aromatic | [15, 16] |
| 35 | NASSAN | Unknown | Tanzania | 1972 | Non- aromatic | [15, 16] |
| 36 | Nylon | Unknown | Tanzania | 1972 | Aromatic | [15, 16] |
| 37 | Basmati 370 | Unknown | Tanzania | 1972 | Aromatic | [15, 16] |
| 38 | SIENNA | Unknown | Tanzania | 1972 | Aromatic | [13, 17] |
| 39 | TXD 306 | Unknown | Tanzania | 1972 | Aromatic | [13, 17] |
| 40 | Pishori | Unknown | Tanzania | 1972 | Aromatic | [17, 18] |
| 41 | ITA 335 | Unknown | Tanzania | 1972 | Aromatic | [15, 16] |
| 42 | TOX 6 | Unknown | Tanzania | 1972 | Aromatic | [13] |
| 43 | TOX 4 | Unknown | Tanzania | 2001 | Non- aromatic | [13] |
| 44 | TOX 5 | Unknown | Tanzania | 2001 | Non- aromatic | [13] |

| S/N | Variety | Designation | Origin | Time started growing | Key Preferred trait | Reference |
|---|---|---|---|---|---|---|
| 45 | TOX 9 | Unknown | Tanzania | 2001 | Non- aromatic | [13] |
| 46 | WITA 6 | Unknown | Tanzania | 2001 | Non- aromatic | [13] |
| 47 | WITA 7 | Unknown | Tanzania | 2001 | Non- aromatic | [13] |
| 48 | WAB 450 | Unknown | Tanzania | 2002 | Non- aromatic | [13] |
| 49 | WAB 189 | Unknown | Tanzania | 2002 | Aromatic | [13] |
| 50 | Kilombero | Unknown | Tanzania | 2005 | Aromatic | [13] |
| 51 | Kenya | Unknown | Tanzania | 1980 | Aromatic | [13] |
| 52 | MET 3 | ART35-114-1-6N-2 | AfricaRice | 2019 | Aromatic | [19] |
| 53 | MET 4 | ART34-146-1-8N-1 | AfricaRice | 2019 | Aromatic | [19] |
| 54 | MET 6 | ART35-49-1-4N-1 | AfricaRice | 2019 | Aromatic | [19] |
| 55 | MET 12 | ART34-88-1-2-B-1 | AfricaRice | 2019 | Aromatic | [19] |
| 56 | MET 13 | ART34-113-3-2-B-1 | AfricaRice | 2019 | Aromatic | [19] |
| 57 | MET 14 | ART34-256-3-1-B-2 | AfricaRice | 2019 | Aromatic | [19] |
| 58 | MET 16 | ART35-272-1-2-B-1 | AfricaRice | 2019 | Aromatic | [19] |
| 59 | MET 40 | ART27-190-1-4-2-1-1-3 | AfricaRice | 2019 | Aromatic | [19] |
| 60 | SUPA 1 | Unknown | IRRI | 2017 | Aromatic | [19] |
| 61 | SUPA 2 | Unknown | IRRI | 2017 | Aromatic | [19] |
| 62 | SUPA 3 | Unknown | IRRI | 2017 | Aromatic | [19] |
| 63 | SUPA 4 | Unknown | IRRI | 2017 | Aromatic | [19] |
| 64 | SUPA 5 | Unknown | IRRI | 2017 | Aromatic | [19] |
| 65 | SUPA 6 | Unknown | IRRI | 2017 | Aromatic | [19] |
| 66 | SUPA 1052 | Unknown | IRRI | 2017 | Aromatic | [19] |

| S/N | Variety | Designation | Origin | Time started growing | Key Preferred trait | Reference |
|---|---|---|---|---|---|---|
| 67 | SUPA 1024 | Unknown | IRRI | 2017 | Aromatic | [19] |
| 68 | PR 26 | Unknown | China | 2018 | Non- aromatic | [19] |
| 69 | PR 27 | Unknown | China | 2018 | Aromatic | [19] |
| 70 | PR 101 | Unknown | China | 2018 | Aromatic | [19] |
| 71 | ARU 1189 | Unknown | AfricaRice | 2017 | Aromatic | [19] |
| 72 | ARU 1190 | Unknown | AfricaRice | 2017 | Aromatic | [19] |
| 73 | ARU 1191 | Unknown | AfricaRice | 2017 | Aromatic | [19] |
| 74 | AGRA 41 | AGRA-CRI-UPL-3-4 | Ghana | 2017 | Aromatic | [19] |
| 75 | AGRA 55 | AGRA-CRI-UPL-4-4 | Ghana | 2017 | Aromatic | [19] |
| 76 | AGRA 60 | AGRA-CRI-UPL-4-13 | Ghana | 2017 | Aromatic | [19] |
| 77 | AGRA 78 | AGRA-CRI-UPL-2-1 | Ghana | 2017 | Aromatic | [19] |
| 78 | Yasmin aromatic | Unknown | Egypt | 2017 | Aromatic | [19] |

**Table 1.**
*Rice varieties grown in Uganda.*

| No | Genotype | Designation | Reaction | Breeding centre | Reference |
|---|---|---|---|---|---|
| 1 | ARC-1 | ARC36-2-1-2 | HR | AfricaRice | [25] |
| 2 | ARC-2 | ARC36-4-EP-2 | HR | AfricaRice | [25] |
| 3 | ARC-3 | ARC39-145-P-3 | HR | AfricaRice | [25] |
| 4 | ARC-4 | ARC39-145-P-2 | HR | AfricaRice | [25] |
| 5 | ARC-5 | ARS126-3-B-1-2 | HR | AfricaRice | [25] |
| 6 | ARS126-3-B-1-2 | ARS126-3-B-1-2 | HR | AfricaRice | [25] |
| 7 | IRL 2 (GP 54) | IRL 2 (GP 54) | R | AfricaRice | [25] |
| 8 | IRL 4 (69 GP 54)] | IRL 4 (69 GP 54)] | R | AfricaRice | [25] |
| 9 | Gigante | Gigante | HR | AfricaRice | [25] |
| 10 | ARICA-5/NamChe1 | WAB95-B-B-40-HB (ARICA 5) | HR | AfricaRice | [25] |
| 11 | Tog5672 | Tog5672 | HR | AfricaRice | [25] |
| 12 | Tog5674 | Tog5674 | HR | AfricaRice | [25] |
| 13 | Tog5681 | Tog5681 | HR | AfricaRice | [25] |
| 14 | MET14 | ART34-256-3-1-B-2 | HR | AfricaRice | [19]; [25] |
| 15 | NamChe2 | NM7-8-2-B-P-2-1 | HR | AfricaRice | [25, 26] |
| 16 | MET3 | ART35-114-1-6N-2 | HR | AfricaRice | [25, 26] |
| 17 | MET8 | ART35-100-1-7D-1 | HR | AfricaRice | [25, 26] |
| 18 | MET12 | ART34-88-1-2-B-1 | HR | AfricaRice | [25, 26] |
| 19 | MET13 | ART34-113-3-2-B-1 | HR | AfricaRice | [25, 26] |
| 20 | MET16 | ART35-272-1-2-B-1 | HR | AfricaRice | [25, 26] |
| 21 | MET35 | ART27-58-3-2-2-1 | HR | AfricaRice | [25, 26] |
| 22 | MET44 | PCT-11\0\0\2,Bo\2 \1>404-1-1-1-1-M | R | AfricaRice | [25, 26] |
| 23 | MET50 | PCT-11\0\0\2,Bo\2 \1>82-3-1-1-3-2-M | R | AfricaRice | [25, 26] |
| 24 | MET60 | PCT-4\0\0 \1>295-2-3-1-3-3-M | R | AfricaRice | [25, 26] |
| 25 | MET66 | PCT-4\SA\1\1,SA\2 \1>746-1-1-4-1-3-M | R | AfricaRice | [25, 26] |
| 26 | MET70 | PCT-4\SA\5 \1>1754-5-1-5-3-1-M | R | AfricaRice | [25, 26] |
| 27 | Nerica8 | | R | AfricaRice | [25, 26] |
| 28 | IR61979-138-1-3-2-3 | | R | AfricaRice | [26] |
| 29 | SR33859-HB3324-133 | SR33859-HB3324-133 | HR | AfricaRice | [26] |
| 30 | NM 15-1 | R33701-HB3330-78 X SR33859-HB3324-133 | R | AfricaRice | [26] |
| 31 | NM 15-2 | FAROX 521-357-H1 X SR33701-HB3330-78 | HR | AfricaRice | [26] |

| No | Genotype | Designation | Reaction | Breeding centre | Reference |
|---|---|---|---|---|---|
| 32 | NM 15-3 | NamChe-2/ SR33705- HB3381-62 | HR | AfricaRice | [26] |
| 33 | NM 15-4 | NamChe-5/ SR33705- HB3381-62 | HR | AfricaRice | [26] |
| 34 | NM 15-5 | ARC36-2-P-2/ SR33705- HB3381-62 | HR | AfricaRice | [26] |
| 35 | NM 15-6 | 32610A X NERICA-4 | HR | AfricaRice | [26] |
| 36 | NM 15-7 | 326104 X NamChe-5 | HR | AfricaRice | [26] |
| 37 | NM 15-8 | 326104 X NERICA-1 | HR | AfricaRice | [26] |
| 38 | SR34034-HB3471-10 | HB4052/Ungwang | HR | AfricaRice | [26] |
| 39 | SR34034-HB3471-23 | HB4052/Ungwang | HR | AfricaRice | [26] |
| 40 | SR34034-HB3471-24 | HB4052/Ungwang | HR | AfricaRice | [26] |
| 41 | SR34555-HB3472-18 | HB4055/Hopum | HR | AfricaRice | [26] |
| 42 | SR34039-HB3366-64 | HB4055/MS11 | HR | AfricaRice | [26] |
| 43 | SR34042-HB3475-6 | HB4057/Japonica 1 | HR | AfricaRice | [26] |
| 44 | SR34035-HB3477-61 | HB4057/Ungwang | HR | AfricaRice | [26] |
| 45 | SR34035-HB3477-96 | HB4057/Ungwang | HR | AfricaRice | [26] |
| 46 | SR33705-HB3381-62 | Hwaseong/ IR84421-4-47-B-1-3 | HR | AfricaRice | [26] |
| 47 | SR33705-HB3381-113 | Hwaseong/ IR84421-4-47-B-1-3 | HR | AfricaRice | [26] |
| 48 | SR33703-HB3482-8 | Ilpum/IR84421-4-47-B-1-3 | HR | AfricaRice | [26] |
| 49 | SR34040-HB3367-55 | Japonica 1/HB4052 | HR | AfricaRice | [26] |
| 50 | SR34038-HB3420-35 | MS11/HB4052 | HR | AfricaRice | [26] |
| 51 | SR34038-HB3420-38 | MS11/HB4052 | HR | AfricaRice | [26] |
| 52 | SR34038-HB3420-56 | MS11/HB4052 | HR | AfricaRice | [26] |
| 53 | SR33859-HB3324-93 | Samgwang/ Yeongdeok53 | HR | AfricaRice | [26] |
| 54 | SR33859-HB3324-133 | Samgwang/ Yeongdeok53 | HR | AfricaRice | [26] |
| 55 | SR33701-HB3330-78 | Odae/Borami | HR | AfricaRice | [26] |
| 56 | CN127 | D20-ARS-22-B | HR | AfricaRice | [26] |
| 57 | AR38 | ARS887-9-1-4-4 | HR | AfricaRice | [26] |
| 58 | LW21 | CSR36 | HR | AfricaRice | [26] |
| 59 | RF96 | IR 97071-24-1-1-2 | HR | AfricaRice | [26] |

**Table 2.**
*Rice genotypes resistant to RYMV.*

rice types in their background developed under Korea-Africa Food and Agriculture Cooperation Initiative (KAFACI) rice collaborative research.

In 2018, another set of 112 germplasm was screened for rice yellow mottle virus disease (RYMVD) resistance. The germplasm comprised of *O. barthi* interspecific lines generated from crosses between *Oryza sativa* L. and *O. barthi.* These breeding lines were selected for their high yield potential, resistance to diseases and other desirable culinary qualities. Seventeen promising genotypes were identified, among which six comprising of [ARS126–3-B-1-2 (11), ARC36–2-P-2 (2), ARC39–145-P-2 (5), Gigante, IRL 2 (GP 54) and IRL 4 (69 GP 54)] were highly resistant to RYMV disease. List of the 11 resistant lines for RYM derived from *O. barthi* crosses are shown in **Table 2**.

In 2019, a total of 307 lines including the 247 KAFACI lines were introduced into East African Regional Rice Research and Training centre at NaCRRI, Namulonge-Uganda and each of the introduced seed samples were divided into three parts. The first part was planted using a single row plots in the screen house at NaCRRI. Each row had 10 cm spacing with 15 hills and the rows were. Spaced 15 cm apart. A prepared RYMV inoculate was used to inoculate 5 plants in each row following procedure described by [22]. Symptom appearance was monitored on daily basis to assess the stage of disease initiation. Also, symptom expression post inoculation (DPI), severity on weekly intervals from 21 DPI, through 28 DPI, 35 DPI and 42 DPI were collected. We also took record of plant height for inoculated and non-inoculated plants and the percentage reduction for each of the lines were calculated The results revealed that 3 lines ofIR64/rymv1-2, IR64/rymv1-3 and IR64/RYMV3 showed severity score of 1 on score 1-9; 14 lines showed severity score of 3. These 7 lines are; SR34574-HB3565-284, ARS1958-1, SR34574-HB3565-285, SR34574-HB3565-290, HR32068F1-4-20-1-6-3-2, HR32068F1-4-20-1-6-4-2 and HR32068F1-4-20-1-6-4-3, which were also found to exhibit resistance to all other diseases assessed, namely rice blast, BB, sheath rot and narrow leaf brown spot. A total of 39 lines showed severity score of 5 while up to 57 lines showed reduction in plant height by at most 30%. Further analysis based on a rating scale for Susceptible (7-9), Medium resistance (4-6) and Resistance (0-3) indicated line, namely, HB4057, rymv1-2, rymv1-3, RYMV3, Hannam, NamChe-2 and Ungwang as the resistant donors.

In respect to bacterial leaf blight (BLB) disease, a total of 18 isogenic lines developed for BLB disease were screened in three hotspot areas of Namulonge-Wakiso, Olweny-Lira and Kibimba- Bugiri districts in Uganda. The results revealed IRBB21 (Xa 21) as the most resistant line in all the three locations followed by IRBB8 [27], implying that lines with these genes could be used to pyramid for multiple stress resistance in our breeding programme.

In another study, 32 lines were screened for resistance to BLB and six other lines, namely, CT 12, WITA 132 x NERICA 14, NERICA 10, NERICA 4 and NERICA 1 were reported resistant to BLB [27]. These four lines could be donor parents in breeding for BB resistance as currently, several rice lines have shown resistance to leaf bacterial blight (**Table 3**).

Rice blast disease has been mentioned in several articles as a major constraint in rice production in Uganda [2, 3, 5, 6]. Accordingly, 450 lines were progressively screened in 2014, 2015, 2016, 2017, 2018 and 2020 for sources of resistance to the rice blast disease. In 2014, a total of 50 lines introduced from South Korea though the KAFACI were screened alongside a resistant (IR-64) and a susceptible (NERICA 1) checks. These were the breeding population from a cross of an African cultivated rice, *Oryza glaberrima* of Niger Delta origin and Milyang 23, a Tongil-type Korean rice variety and a total of 29 lines were resistant to rice blast [26] (**Table 4**).

| No | Genotype | Designation | Reaction | Source |
|---|---|---|---|---|
| 1 | IRBB21 | Xa 21 | HR | NaCRRI |
| 2 | CT 12 | Unknown | HR | NaCRRI |
| 3 | NERICA 1 | WAB 450-1-B-P-38-HB | HR | NaCRRI |
| 4 | NERICA 4 | WAB 450 IBP91HB. | HR | NaCRRI |
| 5 | NERICA 10 | WAB 450-11-1-1P41-HB | HR | NaCRRI |
| 6 | WITA 132 x NERICA 14 | Unknown | HR | NaCRRI |

*Key: HR = high resistance; NaCRRI = National Crops Resources research Institute, Namulonge - Uganda. Source; Lussewa et al. [27].*

**Table 3.**
*Lines resistant to bacterial leaf blight disease.*

Another set of 46 rice genotypes introduced from South Korea though the KAFACI were screened for resistance to rice blast under screen house condition and in the field at NaCRRI Uganda in the year 2015 [28]. The screening exercise involved infecting and selecting the infected rice plants and by observing the symptoms on the leaves based on the rice blast identification guide. Data on leaf blast severity, lesion size, area under disease progress curve (AUDPC) for leaf blast severity and lesion size, panicle blast and yield were collected on five randomly selected plants in the field and on three plants in the screenhouse from each plot according to the standard evaluation system of rice [24]. Results revealed that genotypes SR33859-HB3324-133, SR33859-HB3324-93 and SR33701-HB3330-78 were highly resistance to rice blast and had good performance for yield [28] (**Table 4**).

In a related study aimed at understanding transmission of genes for resistance to rice blast, it was found that both additive and non-additive effects contributed to transmission of resistance genes for rice blast disease to the progenies. The inheritance of rice blast resistance has been indicated to be mainly controlled by additive gene effects, besides a small influence of a non-additive effects [28].

In an effort to combat brown spot disease caused by *Biplaris oryzae* pathogen, a total of 100 lines were screened for resistance to rice brown spot in 2017 and 2018 at Namulonge. The results showed that 18 lines were rated as highly resistant, 52 resistant, 27 moderately resistant and three lines including the checks were susceptible [29]. The list of the highly resistant lines recommended for further breeding work is presented in **Table 5**.

Further studies where $F_2$ progenies from the crosses involving parents with distinct phenotypic classes of brown spot revealed information of segregation ratios for the different crosses. In particular, crosses TXD 306 × NERICA 4, NERICA 1 × NERICA 4, NERICA 4 × NERICA 1 and E 22 × PAK conformed to the 3:1 ratio, suggesting the presence of at least one gene showing dominance [30].

Another study conducted to identify lines resistant to bacterial leaf streak (BLS) identified three lines of NERICA 1, NERICA 6 and IURON 2015-1 as highly resistant to the pathogen causing bacterial leaf streak [31].

### *2.2.2 Insect pests*

A study was also conducted to identify rice lines resistant to African rice gall midge (AfRGM) and 20 rice genotypes with diverse breeding background were evaluated for the resistance to AfRGM under controlled conditions in a screen house and under the field conditions at NaCRRI, Namulonge - Uganda [32]. Infestation was done in accordance with the method use by Ogah [33] where 3 females and 2

| No | Genotype | Designation | Reaction | Breeding centre | Reference |
|---|---|---|---|---|---|
| 1 | MET3 | ART35-114-1-6N-2 | R | AfricaRice | [26] |
| 2 | MET8 | ART35-100-1-7D-1 | R | AfricaRice | [26] |
| 3 | MET12 | ART34-88-1-2-B-1 | R | AfricaRice | [26] |
| 4 | MET13 | ART34-113-3-2-B-1 | R | AfricaRice | [26] |
| 5 | MET16 | ART35-272-1-2-B-1 | R | AfricaRice | [26] |
| 6 | MET35 | ART27-58-3-2-2-1 | R | AfricaRice | [26] |
| 7 | MET44 | PCT-11\0\0\2,Bo\2\1>404-1-1-1-1-1-M | R | AfricaRice | [26] |
| 8 | MET50 | PCT-11\0\0\2,Bo\2\1>82-3-1-1-3-2-M | R | AfricaRice | [26] |
| 9 | MET60 | PCT-4\0\0\1>295-2-3-1-3-3-M | R | AfricaRice | [26] |
| 10 | MET66 | PCT-4\SA\1\1,SA\2\1>746-1-1-4-1-3-M | R | AfricaRice | [26] |
| 11 | MET70 | PCT-4\SA\5\1>1754-5-1-5-3-1-M | R | AfricaRice | [26] |
| 12 | Nerica8 | | R | AfricaRice | [26] |
| 13 | SR34034-HB3471-10 | HB4052/Ungwang | R | KAFACI | [26] |
| 14 | SR34034-HB3471-23 | HB4052/Ungwang | R | KAFACI | [26] |
| 15 | SR34034-HB3471-24 | HB4052/Ungwang | R | KAFACI | [26] |
| 16 | SR34555-HB3472-18 | HB4055/Hopum | R | KAFACI | [26] |
| 17 | SR34039-HB3366-64 | HB4055/MS11 | R | KAFACI | [26] |
| 18 | SR34042-HB3475-6 | HB4057/Japonica 1 | R | KAFACI | [26] |
| 19 | SR34035-HB3477-61 | HB4057/Ungwang | R | KAFACI | [26] |
| 20 | SR34035-HB3477-96 | HB4057/Ungwang | R | KAFACI | [26] |
| 21 | SR33705-HB3381-62 | Hwaseong/IR84421-4-47-B-1-3 | R | KAFACI | [26] |
| 22 | SR33705-HB3381-113 | Hwaseong/IR84421-4-47-B-1-3 | R | KAFACI | [26] |
| 23 | SR33703-HB3482-8 | Ilpum/IR84421-4-47-B-1-3 | R | KAFACI | [26] |
| 24 | SR34040-HB3367-55 | Japonica 1/HB4052 | R | KAFACI | [26] |
| 25 | SR34038-HB3420-35 | MS11/HB4052 | R | KAFACI | [26] |
| 26 | SR34038-HB3420-38 | MS11/HB4052 | R | KAFACI | [26] |
| 27 | SR34038-HB3420-56 | MS11/HB4052 | R | KAFACI | [26] |
| 28 | SR33859-HB3324-93 | Samgwang/Yeongdeok53 | R | KAFACI | [26] |
| 29 | SR33859-HB3324-133 | Samgwang/Yeongdeok53 | R | KAFACI | [26] |
| 30 | SR-133, | SR33859-HB3324-133, | | KAFACI | [28] |
| 31 | SR-93 | SR33859-HB3324-93 | | KAFACI | [28] |
| 32 | SR-80 | SR33701-HB3330-80 | | KAFACI | [28] |

**Table 4.**
*Lines resistant to rice blast.*

male of AfRGM were released in each cage. A total of 180 females and 120 males of gall midge was used. Genotypes MET P-7 and NERICA 6, NERICA 4 and NERICA 1 consistently exhibited high tolerance to AfRGM. The most desirable (high negative)

| No | Genotype | Designation | Reaction | Source |
|---|---|---|---|---|
| 1 | NM-22-1 | NamChe-2/SR33705- HB3381-62 | HR | NaCRRI |
| 2 | NM-22-2 | NamChe-5/SR33705- HB3381-62 | HR | NaCRRI |
| 3 | NM-22-3 | ARC36-2-P-2/SR33705- HB3381-62 | HR | NaCRRI |
| 4 | E22 | NM7-22- 11- B-P-1-1 (WAB 450-1-BL1-136-HB /WAB 450-B-136-HB) | HR | NaCRRI |
| 5 | NERICA 10 | | HR | AfricaRice |
| 6 | NERICA 4 | WAB450-1-B-P-91-HB | HR | AfricaRice |
| 7 | E1 | NM7-22- 14- B-P-1-5 | HR | NaCRRI |
| 8 | E11 | NM7-1- 9- B-P-1-1 | HR | NaCRRI |
| 9 | P27H4 | NM7-27- 5- B-P-1-2 | HR | NaCRRI |
| 10 | E186 | NM7-186- 8- B-P-1-3 | HR | NaCRRI |
| 11 | E51 | NM7-51- 9- B-P-1-4 | HR | NaCRRI |
| 12 | P8H13 | NM7-8- 9- B-P-1-5 | HR | NaCRRI |
| 13 | E123 | NM7-123- 10- B-P-1-6 | HR | NaCRRI |
| 14 | E3 | NM7-3- 9- B-P-1-7 | HR | NaCRRI |
| 15 | E104 | NM7-104- 16- B-P-1-8 | HR | NaCRRI |
| 16 | E99 | NM7-99- 14- B-P-1-9 | HR | NaCRRI |
| 17 | E16 | NM7-16- 12- B-P-1-7 | HR | NaCRRI |
| 18 | E135 | NM7-135- 11- B-P-1-9 | HR | NaCRRI |
| 19 | E8 | NM7-8- 11- B-P-1-12 | HR | NaCRRI |
| 20 | P26H6 | NM7-26- 11- B-P-1-9 | HR | NaCRRI |
| 21 | P27H3 | NM7-27- 11- B-P-1-12 | HR | NaCRRI |
| 22 | P3R1 | NM7-3- 11- B-P-1-4 | HR | NaCRRI |
| 23 | P55H7 | NM7-55- 11- B-P-1-6 | HR | NaCRRI |

*Source: Marco et al. [29].*

**Table 5.**
*List of rice lines resistant to brown spot.*

SCA effects were observed in the crosses from NERICA 6 X E 22, NERICA 1 X K 85, NERICA 1 X KOMBOKA and NERICA 4 X KOMBOKA. Low (desirable) GCA values were witnessed in the case of genotypes of NERICA 6, NERICA 1, NERICA 4 and MET P-7, indicating the importance of the parents in contributing resistance towards AfRGM in rice.

In another study to identify sources of resistance to stalk-eyed fly pest (*D. longicornis*) in rice plants, four out of eight lines namely, NERICA 4, TXD 306, K 85 and NM7-22-11-B-P-1-1, showed high resistance to stalk eyed fly upon screening on-station at Namulonge in 2015 [34]. These four-high resistant (HR) lines were crossed with moderately susceptible lines NERICA 1, NERICA 6, Namche 2 and PAKISTAN in a North Carolina II mating design to determine their combing abilities for the insect pest resistance. The results showed that NERICA 4 and K 85 were good general combiners for resistance to the pest. The crosses Pakistan × TXD 306 and NERICA 1× NM7-22-11-B-P-1-1 were identified as promising lines for advancement. Further analysis revealed that the stalk-eyed fly in rice seems to be controlled both by additive and non-additive genes, thus selection at early generations ($F_1$ and $F_2$) would not be effective. Therefore, selection can be appropriately delayed to a later generations, between $F_4$ and $F_6$. Advancement of selected breeding lines (Pakistan × TXD 306 and NERICA 1× NM7-22-11-B-P-1-1) is, therefore, recommended for further evaluation for resistance to the stalk-eyed fly in later generations [34]. The parental lines NERICA 4 and K 85 are recommended as good general combiners and could be used as the donor parents in the breeding programme for the pest.

#### *2.2.3 Abiotic stresses*

Further, an investigation aimed at identifying rice lines with tolerance to cold was conducted and 41 lines at panicle initiation growth stage were subjected to controlled environmental condition of 17°C for 30 days prior to screening and the results revealed three lines, namely, Yunertian, Yunkeng and Zhongeng to exhibit tolerance to cold stress. Furthermore, a study was carried out to determine mode of transmission of genes resistance to cold stress and the crosses from Agoro x Zhongeng, TXD x Zhongeng, 1189 x Zhongeng, 1052 x Zhongeng, TXD x Yunertian and 1189 x Rumbuka presented high and positive specific combining ability (SCA) effect to cold stress at seedling stage indicating that the crosses could be used in selection of cold tolerant lines.

Submergence is a salient yield decimating factor in rice partly because water control and field level under irrigated and rain-fed conditions is weak. In order to address this challenge, a total of 29 rice genotypes were screened for submergence tolerance. Of these, six genotypes namely, *O. barthi* interspecific lines which were obtained from a cross of *O. barthi* and *O. sativa*, where *O. glaberrima* is a monocarpic annual derived from *O. barthi*. Two are released varieties (Namche 5, Namche 2) and six are candidate lines for release (ARS 126-3-B-1-2, ARU1189, ARU1190, ARU1191, E20 and E22) are potential candidate varieties for rain-fed lowland condition [35]. Evaluation of submergence tolerant rice genotypes following the IRRI standard protocol revealed a significant difference in seedling height assessed immediately after submergence stress. The genotypes were screened by submerging 14 days old seedling at 100 cm water depth for 14 days and another set at 45 cm water depth for 14 days following IRRI standard protocols. The study revealed four rice genotypes of Swarna, IRRI SUPA 3, KOMBOKA and SUPA 5 to be tolerant to submergence at 45 cm water depth for 10 days, with at least 85% survival rates. While varieties Swarna, SUPA 5, IRRI SUPA 3, KOMBOKA Mahsuri and IR 64 showed stable survival rate at both water depths with ≥75% survival rates.

#### *2.2.4 Grain quality*

Aroma in rice is a trait of high economic importance, thus are highly valued by consumers and thus commanding higher prices compared to the non-aromatic genotypes. However, the popular global varieties, namely, Basmati of India and Pakistan origins could not be adopted in Uganda because they succumbed to multiple biotic stresses rendering them not suitable for production in Uganda. In response to this challenge, the rice breeding program in Uganda, therefore, have employed two intervention strategies. The first was to screen all rice germplasm for aroma and initiate breeding program for improvement of aroma characteristics and through cooking and leaf sample testing, 39 aromatic rice varieties were identified (**Table 6**). In the second study, screening was conducted based on 2-acetyl-1-pyrroline (2-AP) concentration and the genotypes MET 3, SUPA 1052, Namche 1, ART-4 and BASMATI 370 exhibited not only high, but also stable 2-AP levels. These

| S/N | Genotype | Designation | Reaction | Breeding centre | Reference |
|---|---|---|---|---|---|
| 1 | AGRA 41 | AGRA-CRI-UPL-3-4 | Aromatic | CRI, Ghana | [19] |
| 2 | AGRA 55 | AGRA-CRI-UPL-4-4 | Aromatic | CRI, Ghana | [19] |
| 3 | AGRA 60 | AGRA-CRI-UPL-4-13 | Aromatic | CRI, Ghana | [19] |
| 4 | AGRA 78 | AGRA-CRI-UPL-2-1 | Aromatic | CRI, Ghana | [19] |
| 5 | ART 4 | ART15-22-10-8-1-B-2-2 | Aromatic | AfricaRice, Nigeria | [19] |
| 6 | ART 7 | ART15-17-7-8-1-1-1-B-1-1 | Aromatic | AfricaRice, Nigeria | [19] |
| 7 | ART 10 | ART15-21-2-4-1-B-1-B-1-1 | Aromatic | AfricaRice, Nigeria | [19] |
| 8 | Basmati 370 | Unknown | Aromatic | India | [19] |
| 9 | Komboka | IR 05N 221 | Aromatic | IRRI, Philippines | [19, 36] |
| 10 | MET 3 | ART35-114-1-6N-2 | Aromatic | AfricaRice, Nigeria | [19] |
| 11 | MET 4 | ART34-146-1-8N-1 | Aromatic | AfricaRice, Nigeria | [19] |
| 12 | MET 6 | ART35-49-1-4N-1 | Aromatic | AfricaRice, Nigeria | [19] |
| 13 | MET 12 | ART34-88-1-2-B-1 | Aromatic | AfricaRice, Nigeria | [19] |
| 14 | MET 13 | ART34-113-3-2-B-1 | Aromatic | AfricaRice, Nigeria | [19] |
| 15 | MET 14 | ART34-256-3-1-B-2 | Aromatic | AfricaRice, Nigeria | [19] |
| 16 | MET 16 | ART35-272-1-2-B-1 | Aromatic | AfricaRice, Nigeria | [19] |
| 17 | MET 40 | ART27-190-1-4-2-1-1-3 | Aromatic | AfricaRice, Nigeria | [19] |
| 18 | NERICA 10 | WAB 450-11-1-1-P41-HB | Aromatic | WARDA/Africa Rice | [19] |
| 19 | Sande | O. barthi interspecific lines | Aromatic | AfricaRice, Nigeria | [19] |
| 20 | Supa 3 | IR 97011-7-7-3-1-B | Aromatic | IRRI, Philippines | [19] |
| 21 | Supa 5 | IR 97011-16-2-4-B | Aromatic | IRRI, Philippines | [19] |
| 22 | Supa 6 | IR 9712-4-1-2-1-1 | Aromatic | IRRI, Philippines | [19] |
| 23 | Supa 1052 | SUPA V88*2/ IR09F154 | Aromatic | AfricaRice, Nigeria | [19] |
| 24 | Yasmin aromatic | Unknown | Aromatic | Egypt | [19] |
| 25 | GSR-1 | GSR IR1- 5-S14-S2-Y1 | Aromatic | IRRI, Philippines | [19] |
| 26 | GSR-2 | GSR IR1- 4-D3-Y1-Y1 | Aromatic | IRRI, Philippines | [19] |
| 27 | GSR-3 | GSR IR1- 4-D6-LI2-LI1 | Aromatic | IRRI, Philippines | [19] |
| 28 | GSR-4 | GSR IR1- 3-D7-LI1-S2 | Aromatic | IRRI, Philippines | [19] |
| 29 | GSR-5 | GSR IR1- 4-D3-LI1-LI1 | Aromatic | IRRI, Philippines | [19] |
| 30 | GSR-5 | GSR IR1- 3-S13-Y1-S1 | Aromatic | IRRI, Philippines | [19] |
| 31 | NM 19-12-13-1 | ARC36-2-P-2/Komboka | Aromatic | NARO, Uganda | [19] |
| 32 | NM 19-12-13-1 | ARS126-3-B-1- 2/Komboka | Aromatic | NARO, Uganda | [19] |
| 33 | NM 19-12-13-1 | SR33705-HB3381- 62/ Komboka | Aromatic | NARO, Uganda | [19] |
| 34 | NM 19-12-13-1 | NamChe-2/Komboka | Aromatic | NARO, Uganda | [19] |
| 35 | NM 19-12-13-1 | NamChe-5/Komboka | Aromatic | NARO, Uganda | [19] |
| 36 | Supa local | SUPA V88 | Aromatic | Tanzania | [37] |
| 38 | TXD-306 | TXD-306 | Aromatic | Tanzania | [37] |
| 39 | Pishori | Pishori | Aromatic | Tanzania | [18] |
| 40 | GSR1 | GSR IR1- 5-S14-S2-Y1 | Aromatic | IRRI/CAAS | |
| 41 | GSR2 | GSR IR1- 4-D3-Y1-Y1 | Aromatic | IRRI/CAAS | |

| S/N | Genotype | Designation | Reaction | Breeding centre | Reference |
|---|---|---|---|---|---|
| 42 | GSR3 | GSR IR1- 4-D6-LI2-LI1 | Aromatic | IRRI/CAAS | |
| 43 | GSR4 | GSR IR1- 3-D7-LI1-S2 | Aromatic | IRRI/CAAS | |
| 44 | GSR5 | GSR IR1- 4-D3-LI1-LI1 | Aromatic | IRRI/CAAS | |
| 45 | GSR6 | GSR IR1- 3-S13-Y1-S1 | Aromatic | IRRI/CAAS | |
| 46 | GSR7 | NM7-8-2-B-P-11-6 | Aromatic | IRRI/CAAS | |
| 47 | GSR8 | IR 83683-63-3-1-2 1 | Aromatic | IRRI/CAAS | |

**Table 6.**
*Rice lines showing aroma characteristics.*

parents identified with strong aroma characteristics could be used in the subsequent breeding program for aroma characteristics.

Amylose level influences grain cooking quality and therefore rice with high amylose content are not preferred, for example in Uganda, rice with intermediate amylose level ranging from 15–22% are the commonly grown varieties. Therefore, in a bid to maintain preferred amylose content within the Uganda's rice collections, a study was under taken to screen 60 lines for amylose content for two seasons in 2018 [38]. Of the 60 lines screened, seven lines consistently were of intermediate amylose content (AC), namely Namche 1 (21.84%), P62H17 (20.86%), Namche 1 (21.84%), Namche 2 (16.74%), Namche 3 (14.64%), Namche 5 (22.77%) and ARU 1190 (27.86%) in both environments. A study to understand the mode of transmission of genes for amylose content revealed that six crosses, namely, 1052 x Suparica, 326104 x NERICA 4, 1052 x Namche 2, Namche 2 x Namche 3, Namche 1 x NERICA 4 and Namche 3 x NERICA 4 with significant ($P \leq 0.05$) negative SCA effects indicating that there was reduced AC% in the crosses whereas the remaining seven crosses with positive significant SCA effects for amylose content indicated increase in the AC% of the progenies in these crosses.

## 2.3 Variety release

In light of the challenges of rice production and increasing demand for climate smart agriculture, varieties that are tolerant to known biotic and abiotic stresses were developed and released in Uganda (**Table** 7). Overall, 7 rainfed lowland rice varieties were released in addition to existing local rice varieties. Of the 7 varieties 2 were aromatic and 5 non-aromatic varieties [14, 39].

## 2.4 Current focus

This information will guide selection of parents to use in rice improvement in the country. In the development of improved rice varieties, core sets of population are critical. Identified SNP markers are accelerating this process. Currently, rice germplasm available are being genotypes for presence of known genes of importance in rice breeding. Over 50 SNP markers covering major biotic and biotic, grains quality, yield and physiological traits are the current target are being used on the Uganda germplasm. With SNP markers being developed already aiding the process of selecting core populations for breeding and accelerating selection of promising lines, we believe that efforts to identify more SNPs in populations that show presence of genes through morphological but not positive under current SNP panel be given urgent attention. This will provide broad accumulations of preferred genes at genome level for *O sativa*. This is critical considering that core populations may

| Variety details | Exceptional characteristics |
|---|---|
| Name: Chiga-1<br>Designation: DU 363-2<br>Cross: NA<br>Source: China<br>Year of release: 2019<br>Potential yield: 9.600 Kgs/ha<br>Maturity period: 129 days<br>Type: Hybrid | Bold and big grains like SUPA; Leaf blade has distinctive purple edge that warrants purity management; Strong stem, moderately resistant to RYMV |
| Name: ARIZE-1<br>Designation: ARIZE GOLD 6444<br>Cross: NA<br>Source: Bayer Crop Science<br>Year of release: 20149<br>Potential yield: 7,900 Kgs/ha<br>Maturity period: 108 days<br>Type: Hybrid | Has distinctive erect flag leaf; moderately tolerant to RYMV, tolerant to rice blast and BLS; Flaffy when cooked |
| Name: Komboka<br>Designation: IR 79253-55-1-4-6<br>Cross: IR 74052-297-2-1/IR 71700-247-1-1-2<br>Source: IRRI<br>Year of release: 2014<br>Potential yield: 6,900 Kgs/ha<br>Maturity period: 118 days<br>Type: Inbred | Has distinctive erect flag leaf; moderately tolerant to RYMV, tolerant to rice blast and BLS; Flaffy when cooked, Aromatic |
| Name: Agoro<br>Designation: IR 09 A 136<br>Cross: IR 75000-69-2-1-2 / IR 71684-36-3-3-2<br>Source: IRRI<br>Year of release: 2014<br>Potential yield: 6,100 Kgs/ha<br>Maturity period: 124 days<br>Type: Inbred | Has light green and erect flag leaf; moderately tolerant to RYMV, Rice blast and BLS; Flaffy when cooked |
| Name: Okile<br>Designation: GSR-I-0057<br>Cross: ZGY 1<br>Source: IRRI/CAAS<br>Year of release: 2014<br>Potential yield: 6,800 Kgs/ha<br>Maturity period: 140 days<br>Type: Inbred | Has large erect flag leaf; moderately tolerant to RYMV, Rice blast and BLS; Flaffy when cooked |
| Name: WITA-9<br>Designation: TOX 3058-28-1-1-1<br>Cross: IR 2042-178-1/CT19<br>Source: AfricaRice/IITA<br>Year of release: 2014<br>Potential yield:<br>Maturity period:<br>Type: Inbred | Has short erect flag leaf; tolerant to RYMV, resistant to Rice blast and BLS; Flaffy when cooked, purple stipes in the leaf sheath and leaf margin |
| Name: NERICA-6<br>Designation: WAB450-1-B-P-160-HB<br>Cross: CG 14/WAB56-104<br>Source: AfricaRice<br>Year of release: 2014<br>Potential yield: 6100<br>Maturity period:<br>Type: Inbred | Has long flag leaf; smooth leaf, Tolerant to RYMV, resistant to Rice blast and BLS; Flaffy when cooked |

**Table 7.**
*Irrigated and rainfed rice varieties released in Uganda.*

solve current challenges but not necessarily emerging constraints. This is important considering that several breeding programs Uganda rice breeding developing varieties with relatives of *O.sativa* especially *O. glaberrima* and *O. barthi* in their background that may open new causative genes one by one sources of traits of importance. The Uganda rice breeding program urgently needs the global rice community to urgently work towards identification of numerous causative genes and the phenotypic performances for the current populations being developed with *O. glaberrima* and *O. barthi* in their background and understand biochemical pathways associated with them.

## 3. Conclusion

This paper provides information on the trends in the development of irrigated and lowland rice in Uganda. It reveals that there are more unreleased rice genotypes under cultivation than the released varieties. This observation points to the fact that rice improvement and variety release is a recent development in the country. Also, that more effort has been in screening introductions and conducting adaptation trials. These efforts contributed to selection of widely adapted genotypes that became accepted in the rice breeding program in Uganda. The panel of adapted lines have diverse parental background that includes *Oryza barthii*, *Oryza glaberrima* and the parent *O. sativa*. These lines will form basic germplasm for use in integrating the classical breeding to molecular biology led breeding.

## Acknowledgements

The authors would like to acknowledge persons who were involved in various research on rice in Uganda and coordinating rice breeding initiatives in the country. We appreciate International Rice Research Institute (IRRI)-through STRASA and GSR projects, AfricaRice-through the AfricaWide Rice Breeding Task Force, CIAT Colombia, Crops Research Institute (Kumasi, Ghana) for sharing germplasm. We thank Prof Tongoona Pangirayi and Prof Patrick Okori who guided development of rice breeding program in Uganda through training and mentorship with support Alliance for Green Revolution AGRA.

## Conflict of interest

The authors declare no competing interests.

## Statement

Rice breeding is Uganda is still new. There is need to strengthen collaborative research with International and National partners in research and capacity (human and physical) development.

## Author details

Jimmy Lamo[1]*, David Ochan[2], Desta Abebe[3], Zelalem Zewdu Ayalew[3], Anna Mlaki[4] and Cyprien Ndikuryayo[5]

1 National Agricultural Research Organization (NARO), Uganda

2 Gulu University, Gulu, Uganda

3 Ethiopian Institute of Agricultural Research, Addis Ababa, Ethiopia

4 International Institute of Tropical Agriculture, Arusha, Tanzania

5 Institut des Sciences Agronomiques du Burundi (ISABU), Burundi

*Address all correspondence to: jlamoayo@gmail.com

## References

[1] Ojara M, A., Olivier Ann, L Aribo, B. A Ogwang B.A, and P Wasswa, *Predicting Suitability of Upland Rice for Adoption as Food Security and Poverty Alleviation Crop in Uganda* Journal of Geography and Earth Sciences, 2017(5: 1): p. 26-40.

[2] Biggs, C.E.J., *Agriculture in Uganda*. Agriculture in Uganda, ed. T. JD. 1940, London: Oxford University Press. 158-161.

[3] Tiley, G.E.D., *Rice – Oryza sativa L.* , in *Agriculture in Uganda*, J. JD, Editor. 1970, Oxford University Press: London. .

[4] Reid, R., *Political power in pre-colonial Buganda: Economy, Society and warfare in the nineteenth century*. 2002, Kampala Fountain Publisher

[5] McMaster, D.N., *A subsistence crop geography of Uganda*. 1962, Cornwall, England: Geographical Publications.

[6] Imanywaha, J.B., *Rice*, in *Agriculture in Uganda*, Mukiibi, Editor. 2001, Fountain Publishers: Kampala, Uganda. p. 70.

[7] Lamo, J., *Genetic studies on drought tolerance and grain shattering* in *Farmers preferred rice varieties and their implications for breeding in Uganda* 2010, Researchspace: University of Kwazulu Natal. p. 39-55.

[8] Alibu S, O.M., Adur-Okello SE, Lamo J, Ekobu M and Asea G. , *Farmer's Knowledge and Perceptions on Rice Insect Pests and Their Management in Uganda Agriculture* Agriculture, 2016. **6**(3):38 DOI: 10.3390/agriculture6030038

[9] Adur-Okello EE, A.S., Lamo J, Ekobu M and Otim MH, *Farmers' Knowledge and Management of Rice Diseases in Uganda.* Journal of Agricultural Sciences. , 2020. **1**(12:12).

[10] Winnyfred Amongi, S.T.N., Paul Gibson, Richard Edema and Mikdred Ochwo-Ssemakula. , *Genetics of drought tolerance in common beans genotypes adapted to Ugandan conditions* J.o.P.B.a. C. Science, Editor. 2015. p. 18-27. DOI: 10.5897/JPBS 2014.0462.

[11] Parviz Fasahat, A.R., Javad Mohseni Rad, John Derera, *Principles and utilization of combining ability in plant breeding*, B.a.B.I. Journal, Editor. 2016. p. 1–22. DOI: 10.15406/bbij.2016.04.00085.

[12] Kilimo-Trust., *Development of Inclusive Markets in Agriculture and Trade (DIMAT): The Nature and Markets of Rice Value Chains in Uganda.* . 2012.

[13] Lamo, J. 2010. Genetic studies on drought tolerance and grain shattering PhD Thesis. University of KwaZulu-Natal, South Africa. School of Agricultural Sciences and Agribusiness. pp 205. Farmers preferred rice varieties and their implications for breeding in Uganda 39-55 https://researchspace.ukzn.ac.za/xmliu/bitstream/handle/10413/735/Lamo_J_2010.pdf?sequence=1&isAllowed=y

[14] NARO, *Submission to the variety release committee for the release of 5 irrigated lowland rice varieties (Oryza sativa l.) candidate cultivars in 2014* N.A. R. Organization, Editor. 2014: National Agricultural Research Organization (NARO) Entebbe, Uganda. p. 52 pp.

[15] Haneishi, Y., Maruyama, M., Godfrey, A., Okello, S.E., Tsuboi, T., Takagaki, M., Kikuchi, M. 2013. Exploration of rainfed rice farming in Uganda based on a nationwide survey: Regionality, varieties and yield. Afric. J. of Ag. Res. In print.

[16] Kijima, Yoko, 2014. "Enhancing Rice Production in Uganda: Impact

Evaluation of a Training Program and Guidebook Distribution in Uganda," Working Papers 80, JICA Research Institute.

[17] Odogola WR (2006). Final Survey Report on the Status of Rice Production, Processing and Marketing in Uganda. Kampala, Uganda. 77

[18] Masette M, Candia A, Khahasa E, Okurut S, Tinyiro SE 2013. Preferences of Ugandan consumers for rice varieties and brands on the local market. Uganda Journal of Agricultural Sciences 14(2):1-11

[19] Lamo J. Enhancement of high yielding rice germplasm and breeding capacity of Uganda. pages 327-352 *in* Enhancement of high yielding rice germplasm and development of high-yielding grain quality rice variety. 2019 KAFACI evaluation meeting on the 1st phase of AfricaRice Development Partmeship. Dec. 2-9, 2019, Saint-Louis, Senegal.

[20] Mogga ML, L.J., Asea G, Gibson P, Edema R *Reaction of rice cultivars to a virulent rice yellow mottle virus strain in Uganda.* African Crop Science Society 2012. **20:1**.

[21] Munganyinka E, E., R, Lamo J and Gibson, P. , *The reaction of intraspecific and interspecific rice cultivars for resistance to rice yellow mottle virus disease.*, E.J.o.E. Biology, Editor. 2016. p. 13-18.

[22] Fauquet C, T.J., *Isolation of the Rice yellow mottle virus in Ivory Coas.* Plant Disease Reporter, 1977. **61:6**: p. 443–448. .

[23] Thottappilly, V.J.a.G., *A scoring system for rice yellow mottle virus* I.R.R. Newsletter, Editor. 1987 Los Banos

[24] IRRI, *Standard Evaluation System (SES) for Rice*, n.R.R.I. (IRRI), Editor. 2014: Los Banos. p. 52.

[25] Ndikuryayo C, M. Ochwo-Ssemakula , P. Gibson and J. Lamo 2019. Resistance to Rice yellow mottle virus and performance of selected improved rice genotypes in central Uganda December 2019 Crop Protection 129: 105041. https://doi.org/10.1016/j.cropro.2019.105041

[26] Lamo J, C.G., Ininda J, Ayirebi DPK, Ekebu J, Ekobu M, Alibu S, Okanya S, Oloka B, Otim M, Asea G, and Kang K., *Developing Lowland Rice Germplasm with Resistance to Multiple Biotic Stresses through Anther Culture in Uganda* K.J.o.I. Agriculture, Editor. 2015. p. 415-420.

[27] Lussewa, R., R Edema and J Lamo *Magnitude of genotype x environment interactions for bacterial leaf blight resistance in rice growing areas of Uganda*, A.J.o.C. Science, Editor. 2016.

[28] Zewdu Z, E.R.a.L.J., *Genetic Study of Resistance to Rice Blast in Crosses between Korean and Locally Adapted Rice Genotypes*, A.C.S.a. Technology, Editor. 2017, DOI: 10.4172/2329-8863.1000346

[29] Marco MM, O.-S.M.J.L., Paul G and Edema R,, *Reaction of upland rice genotypes to the brown spot disease pathogen Bipolaris oryzae* A.J.o.R. Development, Editor. 2017.

[30] Marco MM, O.-S.M., Paul G, Edema R, Jimmy, and Lamo J, *Inheritance of resistance to brown spot disease in upland rice in Uganda* J.o.P.B.a.C. Science., Editor. 2016. p. 16.09.16-0613.

[31] Kanaabi M, T.G., Tukamuhabwa P, Andaku J, Ocan D, and Lamo J., *Evaluation of Rice Germplasm Reveals Sources of Bacterial Leaf Streak Disease Resistance in Uganda*. 2018: Journal of Food Security. . p. 163-169.

[32] Belete, D.A., *Inheritance of resistance to African rice gall midge (Orseolia oryzivora) in lowland rice in Uganda*, M. T.M. University, Editor. 2020: Makerere University. p. 120.

[33] Ogah, E.O., Odebiyi, J.A., Omoloye, A.A., Nwilene, F.E., 2012. , *Evaluation of some rice genotypes for incidence of African rice gall midge and its parasitoid (P. diplosisae).* African Crop Science Society, 2012. **20**: p. 137-147.

[34] Charles Ganteh Weelar , J.L., Michael Hilary Otim, Bruno Awio, Mildred Ochwo-Ssemakula *Mode of inheritance of resistance to the stalk-eyed fly (Diopsis longicornis) in rice* International Journal of Agronomic and Agricultural Research, 2017. **10**: p. 9-20.

[35] Mlaki AB, G.P., Edema R, Habineza JC, Mwanje G, Lamo J and Nuwamanya E., *Identification of rice genotypes tolerant to submergence at seedling stage in Uganda*, J.o.P.B.a.C. Science, Editor. 2019. p. 173-184

[36] Marco MM, Ochwo-Ssemakula M J Lamo, Paul G and Edema R, 2017. Reaction of upland rice genotypes to the brown spot disease pathogen Bipolaris oryzae. African Journal of Rural Development, Vol. 2 (1): 2017: pp.127-133

[37] Kanaabi Michael, Tusiime, Geoffrey1., Tukamuhabwa, Phinehas., Zziwa, Simon JL Andaku , Lamo, Jimmy. Characterizing agronomic response of rice genotypes to bacterial leaf streak disease in Uganda. International Journal of Agronomy and Agricultural Research (IJAAR) ISSN: 2223-7054 (Print) 2225-3610 (Online) h ttp://www.innspub.net Vol. 14, No. 1, p. 1-10, 2019.

[38] Kitara IO, L.J., Edema R, Gipson P and Rubaihayo P., *Inheritance of Amylose content and relationship between grain appearance, quality traits and amylose content in rice genotypes in Uganda*, J.o.A. a.V. Science, Editor. 2018. p. 2319-2372.

[39] NARO, *Submission to the variety release committee for the release of rice varieties (Oryza sativa l.) candidate cultivars: DU 363-2 and ARIZE Gold 6444 in 2019*. 2019: National Agricultural Research Organization (NARO) Entebbe Uganda. p. 28 pp.

Chapter 9

# Recent Advances in Crop Establishment Methods in Rice-Wheat Cropping System: A Review

*Sripriya Das, Manoj Kumar Singh, Sneha Kumari and Manimala Mahato*

## Abstract

Traditional practices of growing rice and wheat in Asian countries involve a huge cost in establishment methods adopted by farmers which not only limit the yield and return but also degrade soil and require more water. Adaptation of improved crop establishment methods suitable under adverse climatic conditions is of utmost importance for scientific utilization of natural resources and to maintain the sustainability of rice- wheat cropping system Therefore, an attempt has been made in this chapter to review precision rice establishment methodology viz., direct seeding, non-puddle/unpuddled transplanting, bed transplanting, strip tilled and single pass shallow tilled rice, double transplanting and system of rice intensification (SRI) and wheat establishment methods viz., zero tilled, strip tilled and bed planted wheat. These are recent improved crop establishment techniques that can be used under specific agro-ecological conditions for enhancing yield and resource conservation in Indo-gangetic plains of Eastern India.

**Keywords:** Direct seeded rice, resource use efficiency, single pass shallow tilled rice, SRI, strip tilled rice, unpuddled transplanted rice

## 1. Introduction

Rice–wheat cropping (RWC) system is of immense importance for the food security and livelihood of people residing in South Asian countries [1]. It occupies an area of about 18Mha in Asia, out of which 13.5 Mha lies in the Indo-Gangetic Plains (IGP) and feeds about a billion people (20% of the world population). Rice (*Oryza sativa* L.) is a staple food of more than 50% of the world's population [2] and supplies 20% of total calories required by world and 31% required by the Indian population [3]. Presently, rice is cultivated in 43.79 Mha area with 112.91 Mt. production while wheat is cultivated in 29.58 Mha area with a production of 99.70 Mt. [4]. Introduction of high-yielding varieties along with improved crop management practices, access to irrigation water and chemical inputs during the green revolution period has led to impressive increase in system productivity. But recent evidences indicate a plateau in productivity and decline in total factor productivity because of continuous environmental degradation and socio-economic changes seen in the IGP [5, 6] highly risking the sustainability of the system [1]. Crop establishment methods are important aspects of rice wheat production technology. It refers to the sequence of events starting from sowing of seed of the crop, germination of seed, emergence of the seedling and development of seedling to a stage from where it could be expected to grow to maturity [7]. Precision crop establishment is very vital for realizing optimum plant population and agro-ecological sustainability, lack of which substantially reduces crop yield. Traditional practices not only consume more time and money but also deplete natural resources and may result in unsatisfactory crop stand. Various improved establishment methods for rice and wheat crop are reviewed and discussed in this paper.

## 2. Major challenges in puddle transplanted rice establishment

Increasing futuristic demand of water with increasing population and industries along with decreasing rainfall activity and labor scarcity are the major factors that challenge the sustainability of highwater demanding rice-wheat cropping system especially in South-Asian countries. Although puddling creates proper anaerobic condition for rice growth and reduces weed emergence but puddling and transplanting are highly labour, water, time and energy intensive leading to higher cost of cultivation. Puddling (wet tillage) consumes upto 30% of total irrigation water application in rice in case of light textured soils [8]. Also, it has been reported that on an average wheat yield is reduced by 8% when sown after puddled transplanted rice compared to wheat sown after direct-seeded rice in unpuddled conditions [9] as puddling results in destruction of soil structure and creation of hard pans at shallow depth which affect the performance of succeeding wheat crop [10]. Puddling operation in rice delays wheat planting which results in wheat yield loss of 35–60 kg $day^{-1}$ $ha^{-1}$ in the IGP [11]. Disturbing the flora and fauna of ecosystem regularly in cropping site fails to attain the climax community which provide ample opportunity of invading alien pests. Thus, the adoption of some new crop establishment techniques with higher resources conservation/use efficiency and ecological stability is of vital importance for the sustainability of the agro ecosystem.

## 3. Advances in crop establishment methods of rice

The alternative tillage and crop establishment methods are site-specific and therefore evaluations under wider agro-ecological conditions are needed to have significant adoption. The crop establishment methods which had got renewed interest in case of rice have been discussed below.

### 3.1 Direct seeded rice (DSR)

Direct seeded rice (DSR) involves the establishment of a rice crop from seeds directly sown in the field by any suitable sowing method rather than by transplanting the seedlings from nursery [12]. Three techniques of DSR viz. dry seeding, wet seeding and water seeding are known. Dry seeding involves the sowing of seeds into prepared seed bed under unpuddled and unsaturated soil conditions by broadcasting, drilling or dibbling, which is suitable for rainfed areas with severe water shortages. Dry direct-seeding with 22% increase in grain yield [13] and 35–57% of water saving [14, 15] as compared to flooded system and over 80% NUE [16], is generally adopted for upland rice [17]. The wet seeding method of DSR is suitable for irrigated areas [18] with relatively fair amount of rainfall in which pre-germinated seeds are sown into well puddled field either by broadcasting or by using drum seeder. Drum seeding refers to the process of direct sowing of pre-germinated (sprouted) paddy seeds in puddled and leveled field after draining out the excess water by using an equipment known as drum seeder, which generally consists of 4 hyperboloid shaped drums capable of sowing 8 lines in one pass with 20 cm row to row spacing [19]. However, handling of the equipment is a problem which may lead to uneven seed distribution due to clogging of holes of the drum. In case of water seeding suitable for high rainfall areas, seeds are sown in standing water in fields where ridges and furrows are prepared prior to submergence. The depressions are created to prevent the seeds from getting drifted away and maintain favorable crop geometry.

DSR facilitates saving of resources as well as their efficient utilization and timely sowing of the subsequent crops. Although, the yield obtained in transplanted method of rice is more than that of DSR, but the net return as well as the B-C ratio is higher in case of DSR as reported by Kumar and Batra (**Table 1**) [20]. In the absence of water deficit stress, the faster development of DSR than transplanted rice is consistent over many findings as reported by Alam *et al.* [21]. And drum seeding which is a type of direct seeding is also beneficial in the same way as reported by [22] who found that the B:C ratio was higher in dry seeded rice with drum seeder (1.70) as compared to transplanting after puddling (1.54). However, the main constraint of direct seeding is the preponderance of weeds and proper crop emergence followed by establishment. The risk of yield loss in DSR is much greater (50–91%) as compared to that of conventional transplanted rice [23].

### 3.2 Non puddled/unpuddled transplanted rice

In case of unpuddled transplanting, the field is made ready for transplanting by a single pass strip tillage (or without tillage) followed by inundation of the field for nearly 2 days to make the land sufficiently soft for transplanting [24]. Thus, in this process the travail of puddling is omitted while the advantages of transplanting are obtained. It saves 31–76% of fuel, 25–26% of water [25] and time required for field preparation. Problems of proper establishment of the seedlings at the initial stage of germination and infestation of diseases, pest and weeds are few threats to the rice crop established by this method. Thus, growing rice by this method requires proper care and vigilance. Hossain *et al.* [26] reported greater yield (5.47 t $ha^{-1}$) and lesser fuel consumption (4.38 l $ha^{-1}$) in unpuddled transplanted rice as compared to the puddled transplanted rice. However, similar rice yield under puddled transplanted rice and unpuddled transplanted rice under zero tilled condition was reported in the Eastern gangetic plains [27]. Also, there was a trend of increasing grain yield in zero tilled unpuddled transplanted rice over that of puddled transplanted rice in the

| Particulars | TPR | DSR |
|---|---|---|
| Production (q $ha^{-1}$) | 41.90 | 38.30 |
| Gross Return | 107244.25 | 98142.25 |
| Net Return | 87.28 | 1803.27 |
| Cost of Production (Rs per quintal) | 2517.95 | 2472.94 |
| B-C Ratio | 1.00 | 1.02 |

*Source: Kumar and Batra [20].*

**Table 1.**
*A study showing economics of TPR and DSR in Haryana (Rs $ha^{-1}$).*

second season [28]. The practice of transplanting on unpuddled soil, suitable for low land areas, is a potential technology for those farmers who are skeptical about direct-seeded rice to avoid adverse effect of puddling on succeeding wheat crop.

### 3.3 Bed transplanted rice

Rice is also transplanted in bed with 15 cm height, 35 cm top width, 60 m bottom width and 25 cm furrow length [29] with rice seedlings are transplanted at the edges of beds. This method increases yield by 16% as compared to the conventional method [29]. The yield attributing characters *viz.* plant height, productive tillers/m$^2$, number of grains/panicle and test weight in case of rice grown on beds have been found to be at par with that of rice grown under conventional puddling as reported by Aslam *et al.* [30], however the B:C ratio was higher in case of bed transplanting as compared to conventional transplanting. Two types of nursery bed are possible in this method *viz.* dry bed and wet bed. Bed transplanting has many advantages out of which border effect on majority of the seedlings is most important. Also, irrigation can be applied efficiently in the furrows with comparatively less amount of water. The same beds can be used consequently for 5–6 years which is profitable in monetary terms. Irrigation water productivity (IWP) was significantly higher in beds to the tune of about 13% than flat transplanting during both the years of study by Sandhu *et al.* [31]. However, labour required for bed construction is more in this case. It is generally suitable for medium upland under irrigated condition.

### 3.4 Strip tilled rice

In unpuddled strip tilled rice, 4–6 cm wide and up to 6 cm deep tilled zones are made just after a little rain shower and seedlings are transplanted at a spacing of 25 cm × 20 cm which may vary according to soil conditions [24]. In this method, only 16–25% of the surface soil is disturbed and the rest remain conserved as it is which reduces the mechanical impedance on the soil surface and allows efficient use of resources as fertilizers are applied as band placement. Hossain *et al.* [32] reported that the yield and B-C ratio of rice was increased by 9% and 25%, respectively in the *kharif* season and 13% and 23%, respectively in the *rabi* season for strip tillage as compared to conventionally tilled rice. Adhikari *et al.* [33] reported that the rice grain yield under strip tillage without mulch was significantly higher than rice grown under full tillage with mulch.

### 3.5 Single pass shallow tilled rice

Single pass shallow tillage refers to tilling the entire soil surface upto 4–6 cm depth by using Versatile Multi-crop Planter (VMP) [34] and incorporating the residues into the field in one single go of the equipment. After tillage, irrigation is done to inundate the field for 24 hours before transplanting. Significant differences were not reported for grain yield under single pass shallow tilled (SPST) and conventional tilled rice but the gross margin was significantly higher for SPST as compared to conventional transplanting (**Table 2**) [24]. This method of sowing rice may be followed in both upland and medium land conditions where soil is compacted and impermeable.

### 3.6 Double transplanted rice

Double transplanting is a crop establishment system in which rice seedlings are transplanted twice, first on secondary nursery and then in the main field [35]. In this method, seeds are first sown in the primary nursery and subsequently after

| Parameters | Tillage treatments | |
|---|---|---|
| | Traditional puddled | Single pass shallow tillage |
| Grain yield (t $ha^{-1}$) | 4.91 | 4.98 |
| Gross return (US $ $ha^{-1}$) | 1416 | 1436 |
| Gross margin (US $ $ha^{-1}$) | 1122b | 1209a |

*Source: [24].*

**Table 2.**
*A study showing mean effect of tillage types on grain yield and economic return of rice.*

3 to 4 weeks, rice seedlings from secondary nursery are again uprooted and transplanted in the main field [36]. In situations where the main field is not ready for transplanting at appropriate time due to late onset of monsoon or continuous stagnation of flood water, double transplanting is advantageous producing healthy and taller seedlings that can easily overcome the adverse situation like high water depth at the time of transplanting [37, 38]. Satapathy *et al.*, [35] reported that double transplanting resulted in higher net returns and benefit–cost ratio than normal transplanting which is owing to higher grain yield. Kumar *et al.* [39] also reported higher B-C ratio of 1.99 in double transplanting as compared to single transplanting with a B-C ratio of 1.27. However, this method is quite labor intensive due to the involvement of a second stage nursery and transplanting without proper skill from smaller to larger polybags could give rise to severe transplanting shock. This system is suitable for long duration rice varieties in shallow low land areas.

### 3.7 System of rice intensification

System of Rice Intensification (SRI) is one of most revolutionary method of rice establishment which is being adopted in many countries. It consistently outperforms conventional practices providing new possibilities for food security and poverty reduction [40]. The four main principles of SRI are early, quick and healthy plant establishment, reduced plant density, improved soil conditions through enrichment with organic matter, reduced and controlled water application. Unlike conventional method, in SRI, seedlings are transplanted at 2 leaf and 3$^{rd}$phyllocron stage [41] at 8–12 days age under square planting. Latif and Abdullah [42] reported that the use of irrigation water was reduced by 52.7% in comparison to transplanted rice. In 2002, at the first international conference on SRI, 15 countries reported that the average yield of rice was twice the current average with the use of this system of rice cultivation [43]. Kumar et al. [44] reported higher grain yield and total water productivity of rice grown in SRI method as compared to normal transplanting method. Hossain *et al.* [45] also reported a handsome grain yield of 7.62 t/ha in SRI as compared to 6.59 t/ha in traditional method. SRI is generally suitable for areas where the soil is fertile, fine textured, well drained and maintenance of alternate wetting and drying conditions is possible. However, high labour requirement and problems faced at the time of transplanting of young seedlings are some of the constraints of this method.

## 4. Recent advances in crop establishment methods of wheat

The main constraint faced by wheat crop in the rice-wheat cropping system is delayed planting leading to terminal heat stress due to growing of long duration rice varieties and the time required for land preparation after harvesting of the submerged rice crop. The crop establishment practices mentioned below are devised to manage those problems.

### 4.1 Zero tilled wheat

Zero tillage is an already proven resource conserving technology for wheat crop and it was found that it results in increase in crop yield by 5–7% (140–200 kg $ha^{-1}$) and food production by 0.7% (343000 tonnes $ha^{-1}$) in the Indo Gangetic plains [46]. Seed and fertilizers are placed by opening the furrows with the help of equipments like zero till ferti seed drill or Happy seeder in a single go in standing crop residues by completely avoiding the primary tillage operations. Singh *et al.* [47] reported that the grain and straw yield obtained by sowing of wheat by happy seeder is higher than the farmers practice in one of the two experimental locations (**Table 3**). This method reduces the tillage operations with a single pass and saves fuel, labour, farm machinery cost, water, fertilizers etc. [48], permits earlier wheat planting in rice-wheat system and control the problem of *Phalaris minor* [46]. Pandey *et al.* [49] reported higher grain yield (3440 kg/ha) and B:C ratio (2.38) for zero tilled wheat as compared to conventionally grown wheat with a grain yield of 3224 kg/ha and B:C ratio of 1.81 in Kailali district of Nepal. Since residue retention is a common practice in zero tillage system, so the organic matter content of the soil is also increased and soil compaction is reduced due to enhancement of biological activities in soil. The constrains of adapting zero tillage in wheat under RW system of developing countries are the small size of land holdings of small and marginal farmers and the involvement of lumpy technology (i.e. non-divisible piece of machinery) [5, 6] involving high procurement cost.

### 4.2 Bed planted wheat

In this method of wheat crop establishment, which is synonymous to furrow irrigated raised bed (FIRB), the land is cultivated traditionally and ridges/raised beds and furrows are prepared by using a raised bed planting machine where seeds are planted in rows and irrigation water is applied in furrows. In rice-wheat cropping system, raised beds are newly prepared for wheat and then in the next season rice is grown on the same bed under zero tillage with required repairing of the beds [50]. The most beneficial aspect of this method as mentioned in case of rice is the border effect imparted to maximum number of plants. Bed planted wheat also showed better performance with significantly highest number of tillers per running meter compared to others establishment methods viz. broadcasting and criss cross sowing, in the middle gangetic plain regions during both the experimental years [51]. Mollah *et al.* [50] reported a yield increase of 21% and water saving of 41–46% with a 70 cm wide bed with two rows over conventional method in wheat. However,

| Treatment | Grain yield (q $ha^{-1}$) | | Straw yield (q $ha^{-1}$) | |
|---|---|---|---|---|
| | Jalandhar | Patiala | Jalandhar | Patiala |
| Rotovator | 41.19 | 44.52 | 63.02 | 68.1 |
| Happy Seeder | 43.63 | 49.53 | 66.75 | 75.8 |
| Farmers practice | 42.47 | 46.02 | 64.98 | 70.4 |

*Source: Singh* et al., *2013.*

**Table 3.**
*A study showing influence of sowing methods on grain and straw yield of wheat.*

the requirement of labour for bed preparation and favorable soil texture are some constraints in the adoption of this method.

### 4.3 Strip tilled wheat

In strip till planting, seed and fertilizers are placed simultaneously in a single operation by tilling the planting strips with a width of 4–6 cm [52] and depth of 2–7 cm. Unlike zero tillage, the row zone is completely pulverized with standing crop residues in the field. This method facilitates early establishment of wheat crop, reduces soil erosion from surface and efficient utilization of resources such as labour, fuel, soil etc. The fuel consumption in strip tillage was reduced by 57% and 38%, respectively as compared to conventional tillage and minimum tillage [52]. Usage of strip tillage produces high crop yields with lower production costs and provides better soil erosion control compared to conventional tillage [53]. Hossain *et al*. [32] reported higher yield of wheat in this method as compared to conventional method in all the three experimenting years. This method is recommended for medium land with irrigation facilities. However, higher cost of the strip till machine poses a constraint for the adoption of this method.

## 5. Conclusion

Newly developed techniques are precision establishment techniques, use of need based crop establishment technology conserves the scarce resources and reduces the crop establishment cost of rice-wheat system These techniques are machine based which help in mechanization and optimization of resources. Thus, these techniques should be promoted to obtain higher net return, sustainable intensification, maintenance of soil health and reduction in environmental pollution.

## 6. Future thrust

Weed management is a major challenge in these crop establishment methods as new complexes of weed flora are being observed by the farmers. Therefore, efficient weed management practices should be researched for wider adaptability of these techniques.

## Author details

Sripriya Das[1*], Manoj Kumar Singh[2], Sneha Kumari[2] and Manimala Mahato[1]

1 Department of Agronomy, Bidhan Chandra Krishi Viswavidyalaya, Nadia, West Bengal, India

2 Department of Agronomy, Institute of Agricultural Sciences, Banaras Hindu University, Varanasi, Uttar Pradesh, India

*Address all correspondence to: sripriyadaspaul@gmail.com

## References

[1] Ladha, J. K., J. E. Hill, J. M. Duxbury, R. K. Gupta, and Buresh, R. J. (2003). Improving the productivity and sustainability of rice-wheat systems: Issues and impacts. Madison, Wisconsin, U.S.a., American Society of Agronomy, crop science Society of America, Soil Science Society of America.

[2] Fageria, N.K. (2007). Yield physiology of rice. *Journal of Plant Nutrition* **30**: 843-879.

[3] Anonymous. (2011). Rice Calorie Supply as Percentage of Total Calorie Supply, by Country and Geographical Region, 1961-2005. Available from: (accessed 24.2.11).

[4] Anonymous. (2018). Agricultural Statistics at a Glance,Government of India, Ministry of Agriculture & Farmers Welfare, Department of Agriculture, Cooperation & Farmers Welfare Directorate of Economics & Statistics.

[5] Erenstein, O., J. Hellin, and P. Chandna. (2007a). Livelihoods, Poverty, and Targeting in the Indo-Gangetic Plains: A Spatial Mapping Approach. Research report. New Delhi, International Maize and Wheat Improvement Center and Rice-Wheat Consortium.

[6] Erenstein, O., R. K. Malik, and S. Singh. (2007b). Adoption and impacts of zero tillage in the irrigated rice-wheat systems of Haryana, India.Research Report. New Delhi, International Maize and Wheat Improvement Center and Rice-Wheat Consortium.

[7] Jat, M.L., Gathala, M.K., Ladha, J.K., Saharawat, Y.S., Jat, A.S., Kumar, V., Sharma, S.K., Kumar, V. and Gupta, R. (2009). Evaluation of precision land leveling and double zero-till systems in the rice–wheat rotation: Water use, productivity, profitability and soil physical properties. *Soil & Tillage Research* **105**: 112-121.

[8] Aslam, M., Qureshi, A.S. and Horinkova, V.M. (2002). Water saving strategies for irrigated rice. *Journal of Drainage and Water Management* **6**(1).

[9] Kumar, V., Bellinder, R.R., Gupta, R.K., Malik, R.K. and Brainard, D.C. (2008). Role of herbicide-resistant rice in promoting resource conservation technologies in rice– Wheat cropping systems of India: A review. *Crop Protection* **27**, 290-301.

[10] Hobbs, P. and Morris, M. (1996). Meeting South Asia's future food requirements from rice-wheat cropping system: Priority issues facing researchers in the post-green revolution era. NRG Paper, 96(01).

[11] Pathak, H., J.K. Ladha, P.K. Aggarwal, S. Peng, S. Das, Y. Singh, B. Singh, S.K. Kamra, B. Mishra, A.S.R.A.S. Sastri, H.P. Aggarwal, D.K. Das, and R.K. Gupta. (2003). Trends of climatic potential and on-farm yields of rice and wheat in the indo-Gangetic Plains. Field Crops Research **80**:223-234.

[12] Kaur, J. and Singh, A. (2017). Direct Seeded Rice: Prospects, Problems/ Constraints and Reachable Issues in India. Current Agriculture Research Journal 5(1), 13-32.

[13] Zhu, L. (2008). A report on dry direct seeding cultivation technique of early rice. Journal of Guangxi Agricultural and Biological Science 23:10-11. (In Chinese with English abstract)

[14] Sharma, P.K., Bhushan, L., Ladha, J.K., Naresh, R.K., Gupta, R.K., Balasubramanian, B.V. and Bouman, B.A.M. (2002). Crop-Water Relations in Rice-Wheat Cropping under Different Tillage Systems and Water Management

Practices in a Marginally Sodic, Medium-Textured Soil. Water-wise rice production. International Rice Research Institute, Los Baños, 223-235.

[15] Singh, A.K., Choudhury, B.U. and Bouman, B.A.M. (2002). Effects of Rice Establishment Methods on Crop Performance, Water Use, and Mineral Nitrogen. Water-wise rice production. International Rice Research Institute, Los Baños, 237-246.

[16] Wilson, C., Slaton, N., Norman, R. and Miller, D. (2000). Efficient use of fertiliser. Rice Production Handbook, University of Arkansas, Division of Agriculture, Cooperative Extension Service, 8, 51-74.

[17] Gupta, P.C. and O'Toole, J.C. (1986). Upland Rice-a Global Perspective. IRRI, Manila.

[18] IRRI Rice Knowledge Bank, http://www.knowledgebank.irri.org/

[19] Singh, U.V., Kumar, D. and Moses, S.C. (2016). Perfomance evaluation of manually operated paddy drum seeder in puddle field. *Journal of Agriculture and Veterinary Science***9**(6): 69-83.

[20] Kumar, R. and Batra, S.C. (2017). A comparative analysis of DSR technology Vs. transplanted method in Haryana. *EconomicAffairs***62**(1): 169-174.

[21] Alam, M. J., Humphreys, E., Sarkar, M. A.R. and Yadav, S. (2018). Comparison of dry seeded and puddled transplanted rainy season rice on the high Ganges River floodplain of Bangladesh. *European Journal of Agronomy***96**: 120-130.

[22] Kumar, V., Singh, S., Sagar, V. and Maurya, M.L. (2018). Evaluation of different crop establishment methods of rice on growth, yield and economics of rice cultivation in agro-climatic condition of eastern Uttar Pradesh. *Journal of Pharmacognosy and Phytochemistry***7**(3): 2295-2298.

[23] Rao, A.N., Johnson, D.E., Sivaprasad, B., Ladha, J.K. and Mortimer, A.M. (2007). Advances in Agronomy **93**, 153-255.

[24] Haque, M.E., Bell, R.W., Islam, M.A. and Rahman, M.A. (2016). Minimum tillage unpuddled transplanting: An alternative crop establishment strategy for rice in conservation agriculture cropping systems. *Field Crops Research***185**, 31-39.

[25] Islam, A.K.M.S., Hossain, M.M., Saleque, M.A., Rahman, M.A., Karmakar, B. and Haque, M.E. (2012). Effect of minimum tillage on soil properties, crop growth and yield of aman rice in drought prone northwest Bangladesh. *Bangladesh Agronomy Journal***15**(1): 43-51.

[26] Hossain, M.M., Rabbani, M.A.E., Elahi, H.M.T., Sarkar, S., Saha, C.K., Alam, M.M., Kalita, P.K. and Hansen, A.C. (2017). Options for Rice Transplanting in Puddle and Un-Puddled Soil. 2017 ASABE Annual International Meeting, Spokane, Washington.

[27] Islam, S., Gathala, M.K., Tiwari, T.P., Timsina, J., Laing, A.M., Maharjan, S., Chowdhury, A.K., Bhattacharya, P.M., Dhar, T., Mitra, B., Kumar, S., Srivastwa, P.K., Dutta, S.K., Shrestha, R., Manandhar, S., Sherestha, S.R., Paneru, P., Siddquie, N.-E. A., Hossain, A., Islam, R., Ghosh, A.K., Rahman, M.A., Kumar, U. and Rao, K.K., G'erard, B. (2019). Conservation agriculture based sustainable intensification: increasing yields and water productivity for smallholders of the Eastern Gangetic Plains. *Field Crops Research***238**: 1-17. https://doi.org/10.1016/j.fcr.2019.04.005.

[28] Chaki, A. K., Gaydon, D.S., Dalal, R.C., Bellotti, W.D., Gathala, M.K., Hossain, A., Siddquie, N.E.A. and Menzies, N.W. (2020). Puddled and zero-till unpuddled transplanted rice

are each best suited to different environments- an example from two diverse locations in the eastern Gangetic Plains of Bangladesh. *Field Crops Research***262**: 108031.

[29] Bhuyan, M.H.M, Ferdousi, M.R. and Iqbal, M.T. (2012). Yield and growth response to transplanted aman rice under raised bed over conventional cultivation method. International scholarly research network. Article ID 646859, 8 pages.

[30] Aslam, M., Hussain, S., Ramzan, M. and Akhter, M. (2008). Effect of different stand establishment techniques on rice yields and its attributes. Journal of Animal and Plant Sciences **18** (2-3): 80-82.

[31] Sandhu, S. S., Mahal, S.S., Vashist, K.K., Buttar, G.S., Brar, A.S. and Singh, M. (2012). Crop and water productivity of bed transplanted rice as influenced by various levels of nitrogen and irrigation in Northwest India. *Agricultural Water Management***104**: 32-39.

[32] Hossain, M.I., Sarker, M.J.U. and Haque, M.A. (2015). Status of conservation agriculture based tillage technology for crop production in Bangladesh. Bangladesh Journal of Agricultural Research **40**(2): 235-248.

[33] Adhikari, U., Justice, S., Tripathi, J., Bhatta, M.R. and Khan, S. (2015). Evaluation of Non-puddled and Zero till Rice Transplanting Methods in Monsoon Rice. Paper Presented at International Agricultural Engineering Conference, AIT Bangkok, Thailand.

[34] Haque, M.E., Bell, R.W., Islam, A.K.M.S., Sayre, K. and Hossain, M.M. (2011). Versatilemulti-crop planter for two-wheel tractors: An innovative option forsmallholders. In: Gilkes, R.J., Prakongkep, Nattaporn (Eds.), 5th World Congress of Conservation Agriculture Incorporating 3rd Farming Systems Design Conference, Brisbane, Australia, 102-103.

[35] Satapathy, B.S., Singh, T., Pun, K.B. and Rautarat, S.K. (2015). Evaluation of rice (Oryza sativa) under double transplanting in rainfed lowland rice ecosystem of Asom. Indian Journal of Agronomy **60**(2): 245-248.

[36] Das, S.R. (2006). Indian diversified rice cultivation. Souvenir, 63-70. *Second International Rice Congress*, 9-13 October, 2006, New Delhi. Doi:10.5402/2012/646859.

[37] Ashim, S.S., Thakuria, K. and Kumari, K. (2010). Double transplanting of late transplanted *Sali*rice under lowland situation. Oryza **47**(4): 328-330.

[38] Rautray, S.K. (2007). Strategies for crop production in flood affected areas of Assam. Indian Farming **57**(6): 23-26.

[39] Kumar, V. S., Nayak, A. K., Satapathy, B. S., Munda, S., Tripathi, R, and Khanam, R. (2019). *Indian Farming***69**(05): 36-38.

[40] Thakur, A. K., Mohanty, R. K., Singh, R. and Patil, D. U. (2015). Enhancing water and cropping productivity through integrated system of Rice intensification (ISRI) with aquaculture and horticulture under rainfed conditions. *Agricultural Water Management***161**: 65-76.

[41] Veeramani, P., Singh, R.D. and Subrahmaniyan, K. (2012). Study of phyllochron- system of Rice intensification (SRI) technique. *Agricultural Sciences Research Journal***2**(6): 329-334.

[42] Latif, K.M.A.E. and Abdullah, R. (2012). Rice Yield and Water Saving in the System of Rice Instensification (SRI): Review. Minia International Conference for Agriculture and Irrigation in the Nile Basin Countries, El-Minia, Egypt.

[43] Uphoff, N. (2002). Questions and Answers about the System of Rice Intensification (SRI) for Raising the Productivity of Land, Labour and Water. Cornell International Institute for Food, Agriculture and Development, Cornell University, Ithaca, NY.

[44] Kumar, R. M., Surekha, K., Padmavati, C., Rao, L.V.S., Latha, P.C., Prasad, M.S., Babu, V.R., Ramprasad, A.S., Rupela, O.P., Goud, V., Raman, P.M., Somashekar, N., Ravichandran, S., Singh, S.P. and Viraktamath, B.C. (2009). Research experiences on System of Rice Intensification and future direction. *Journal of Rice Research* **2**(2): 61-71.

[45] Hossain, T., Hossain, E., Nizam, R., Bari, F., A. S. M. and Chakraborty, R. (2018). Response of Physiological Characteristics and Productivity of Hybrid Rice Varieties under System of Rice Intensification (SRI) over the Traditional Cultivation. *International Journal of Plant Biology and Research* **6**(2): 1085.

[46] Erenstein, O., and Laxmi, V. (2008). Zero tillage impacts in india's rice-wheat systems: A review. Soil and Tillage Research **100** (1–2): 1–14.

[47] Singh, A., Kang, J. S. and Kaur, M. (2013). Planting of wheat with happy seeder and rotovator in rice stubbles. Indo-American Journal of Agricultural and Veterinary Sciences 1(2), 32-41.

[48] Yadav, A., Malik, R. K., Chouhan, B. S., Kumar, V., Banga, P. S., Singh, S., Yadav, J. S., Punia, S. S., Rathee, S. S. and Sayre, K. D. (2002). Feasibility of Raising Wheat on Furrow Irrigated Raised Beds in South-Western Haryana. Presented in International Workshop on, Herbicide resistance management and zero tillage in rice-what cropping system, Hisar, Haryana, India.

[49] Pandey, B. P., Khatri, N., Pant, K. R., Yadav, M., Marasini, M, Paudel, G. P. and Bhatta, M. (2020). Zero-till wheat (*Triticum aestivum* L.): A Nepalese perspective. *Fundamental and Applied Agriculture* **5**(4): 484-490.

[50] Mollah, M.I.U., Bhuiya, M.S.U. and Kabir, M.H. (2009). Bed planting- a new crop establishment method for wheat in Rice-wheat cropping system. *Journal of Agriculture and Rural Development* **7**(1&2): 23-31.

[51] Raghuvanshi, N. and Singh, B. N. (2020). Impact of sowing method and nitrogen management on growth of wheat (*Triticum aestivum* L.). International Journal of Chemical Studies **8**(3): 2771-2775.

[52] Hossain, M.I., Gathala, M.K., Tiwari, T.P. and Hossain, M.S. (2014). Strip tillage seeding technique: A better option for utilizing residual soil moisture in rainfed moisture stress environments of north-West Bangladesh. *International Journal of Research Development in Engineering and Technology* **2**(4).

[53] Laufer, D., Koch, H.J. (2017). Growth and yield formation of sugar beet (Beta vulgaris L.) under strip tillage compared to full width tillage on silt loam soil in Central Europe. *European Journal of Agronomy* **82**: 182-189. https://doi.org/10.1016/j.eja.2016.10.017.

Chapter 10

# Rice: Worldwide Production, Utilization, Problems Occurring Due to Climate Changes and Their Mitigating Strategies

*Muhammad Ikram and Haseeb-Ur-Rehman*

## Abstract

The production of rice is least in Pakistan and quite low as compared with other countries. Proper crop management techniques such as intercropping and combining organic manures are useful for better productivity and eco-friendly environment. Whereas studies are needed to evaluate the efficiency of intercrops and incorporation of certain nutrients with these plants. To examine results of intercropping experiment was carried out research by combining nutrient management practices. Five methods were taken including, sole rice, sole Green gram, rice + Green gram (drill), rice + green gram (Ridges), rice + green gram (bed) in the main plot moreover sub-plot included treatments of organic and inorganic supplement. The results show that sole rice followed by intercropping rice along green gram (poultry manure) has better characteristics of growth and yield, higher yield. by changing irrigation methods and farming methods, managing organic additives and fertilizer inputs, and choosing appropriate varieties and planting methods. $CH_4$ decreased by 75% and N2O increased by 58%. The overall rice production of Rice + green gram(ridges) is 2285 kg ha$^{-1}$ followed by rice + green gram(drill) (2060 kg ha$^{-1}$). Rice + green gram(ridges) intercropping and (25 percent Urea+25 percent FYM+ 50 percent PM) were also correlated with better N usage performance and post-harvest soil usable N, phosphorus (P) and potassium (K) Benefit: cost (BC) ratios were also higher in the same treatment. From these results it is obvious that the integration of intercropping and induction of organic manures has a substantial impact on the outcome of rice.

**Keywords:** intercropping, F.M(farmyard manure), P.M(poultry manure), N:P: K, Rice, $CH_4$

## 1. Introduction

As the world population is increasing, the demand of the food supply is also increasing [1]. Estimating the future food demand, it can be considered that the food requirements will increase approximately 100 to 110% in regard to feed the world's growing inhabitants by 2050 [2]. Cereal crops have been utilized as a main component of the human diet for start of life in world. Rice (*Oryza sativa*) belongs to one of the major cereal crops. It is grown for >7000 years in all over the world [3]. The annual production recorded in 2014 > 740Mt. It is consumed as a staple food in many countries of the world [4]. Furthermore, Various components of rice crop such as rice husk, rice brans etc. are being used for the manufacturing of the different products such as rice bran is utilized to obtain the oil and rice husks are being used in manufacturing of different bakery products as a nutrition enhancing element. Rice is inimitable crop in term of its growth, it can be grown in various conditions either these are wet or dry, different kinds of soils, wide range of hydrological circumstances and different climatic conditions. But it is grown mostly in tropics, sub tropic, humid and sub humid areas. Irrigated rice gives average yield of 5 tons/ha globally, but this estimate varies widely in according to seasons, nations and regions. In tropic regions, well expert farmers can get the yield of 5 to 6 tons/ha and 7 to 8 tons/ha in wet and dry seasons accordingly. The decrease in the yield of rice in wet season may be due to the less quantity of solar radiations reaching to the earth [5].

## 2. Classification of rice

| | |
|---|---|
| Kingdom | *Plantae* |
| Sub kingdom | *Tracheobionta* |
| Division | *Magnoliophyta* |
| Class | *Liliopsida* |
| Sub class | *Commelinidae* |
| Order | *Poales* |
| Family | *Poacaeae* |
| Sub family | *Oryzoideae* |
| Tribe | *Oryceae* |
| Genus | *Oryza* |
| Specie | 22 species including *Oryza sativa, Oryza barthii, Oryza glaberrima, Oryza latifolia, Oryza longistaminata, Oryza punctate, Oryza rufipogon* |

### 2.1 Origin

The most grown and utilizing species of rice are glaberrima (known as African rice) and *Oryza sativa* (known as Asian rice). These are considered as a progenitor of the Oryza species. It is estimated that *Oryza sativa* is grown approximately on the area of 1200 km belt including the areas of Himalaya mountains, areas near the Gangs river of India, Bangladesh, Bhutan, northern Burma, crossing the areas of Thailand, passing through Laos and Vietnam, it covers the some area of china as well. The Asian rice is mostly cultivated in south east Asia and south region. Due to the large belt of rice cultivation, the exact origin of domestication and evolution of related species and intermediates of the rice could not find. The domestication of Asian rice took place at many places in various times in south Asian regions [6]. On other hand, O glaberrima was known to be originated from the Niger river areas in west Africa, almost 3000 years before. The wild crop *Oryza barthi* is considered as its progenitor. After that *Oryza glaberrima* was grown in Liberia [7].

### 2.2 Domestication and diversification

There were two ways used for the domestication of the rice cultivation in China and India. In Yangtze japonica, the traits for the domestication were utilized in the beginning and this process was accomplished in China 6,000 to 6,500 years ago. But it was completed approximately after 20,000 years in India. At that time, the Chinese rice was hybridized. When farming trends increase then the growing of rice also moved to the nearby areas among the populations. In this way the rice, the spreading of the rice take place from china to its south regions such as Austronesian and Sino-Tibetan groups. On other hand, In south regions of Asia the distribution of the rice take place after Dravidian and Indo-Aryan have gotten the whole quantity of rice from the northern India [8]. Rice Diversification and domestication processes were exposed to with the help of archaeology, ecology and genetics. Archaeology discovers the evolutionary dates of rice were 4000 BC in india and 5000 BC in china. First detailed study was conducted in Hastinapur city of India on carbonized grains of rice at 1100–750 BC. The samples were taken from Atranjikera a city of Uttar Pradesh almost 1500–1000 BC. After that the cultivation of rice started spreading in different countries of the words such as its cultivation was introduced to the Japan 100-300 BC, west areas of India at 2000 BC, and Philippines and Malaysia by 1400 BC [9].

### 2.3 Distribution of rice

Paddy rice are those rice which do not contain husk on their surface and brown rice are those which contained any outer surface on it. This can be removed by the milling. After that the rice are polished to obtain the white rice. Rice are mostly grown on the flat lands, delta areas and near the rivers, but it is very difficult to describe the specific environmental conditions and areas for its growth. The countries containing the warm temperatures or subtropical climates give the maximum yields of the rice. The rice yield varies in all over the world according to the latitudes such as 50°N, 45°N and 39°S for China, Japan and Australia accordingly. The areas present at the latitude of 40°S and 45°N are considered as extensive rice cultivating areas. Highest yield is observed in between 30°N to 45°N of equators. It is also being grown in those areas which are present at the below levels of sea i.e., Kerala. In Jammu and Kashmir, the rice is also cultivated at 1979 m altitudes. It can also be cultivated in the deep and shallow water while rainy season is on peak [10]. The rice cultivation in widely range is due to its various varieties according to their origin. Rice can be grown in all kinds of environments depending on the nature of the cultivars. Mostly varieties can be grown in the areas where irrigation waters are available. Furthermore, some varieties of rice are also available, which grow in specific season and at specific lands.

### 2.4 Civilization due to rice

Rice play an important role in the human diet from the beginning. It has enhanced the civilization and boosted the national economy in each country growing the rice by exporting it to the rest of countries. It plays important role to fulfill the dietary requirements in increasing population, their culture and civilizations. Rice was swamp grass of the semitropical areas, before its cultivation on large levels in agriculture sectors. Then it became the additional food to the peoples of the tribes, who were dependent on the fishing, hunting and other wild foods. Its yield is less on the local level, but the implementation of the management practices like conservation of nutrients and water, and application of soil fertilizers, weeding and tillage practices, and selection of good varieties of the rice has boosted its yield on greater extend. The maximum rice crop can be yielded by utilization and managements of good practices, which are helpful to mitigate the problems of rice growth. Hence after these adaptations, there was great influence in rice yield per acre, rice growing areas and different cropping systems were developed. In this way, civilization was started flourishing throughout the Asia sub-continent. Now, Asian countries are contributing in their economy and meeting the surplus requirements of the population [11].

## 3. Status of climate change

### 3.1 Crop management

Various methods to reduce greenhouse gas emissions from rice fields can reduce greenhouse gas emissions from rice fields by changing irrigation methods and farming methods, managing organic additives and fertilizer inputs, and choosing appropriate varieties and planting methods. The following section will discuss all these details with suitable options and possibilities in different agro-conditions.

### 3.2 Changing irrigation pattern

Irrigation in rice production process is one of the vital features in regulatory greenhouse gas emissions. According to reports, compared with traditional flooded rice, several water managements schemes (such as different drainage periods in the season, alternating wet and dry soil, recurrent irrigation and controlled irrigation) can minimize GHG emissions, Can be used as an option. Practice under different soil and climate conditions without reducing crop yields. Mid-season drainage Mid-season drainage includes a significant period of interruption of irrigation during crop growth. Usually, a short-term drainage (5–20 days) is performed before the maximum sub-till number stage to prevent grade growth and reduce the number of invalid sub-tills, and the duration is adjusted by the conventional method of regional determination. At the beginning of soil aeration, $CH_4$ emissions may increase in a short time due to the release of $CH_4$ entrained in the soil, and its emissions will continue to decrease even when the field is submerged again. Since additional water can be used to displace the paddy soil, there is also a difference in the efficiency of mid-season drainage in reducing $CH_4$ (15–59%) [12]. Drainage in the spring increases the oxidative conditions of the soil and the uptake of nitrogen [13].

Therefore, this can be achieved by reducing the amount of water applied, because the net reduction in water will ultimately reduce $CH_4$ emissions. Wassmann et al. [14] reported that the aerobic conditions generated by the flux of oxygen discharged into the soil are not conducive to the activities of methanogenic bacteria, therefore, in the medium term, drainage can reduce $CH_4$ emissions by 43%. Timely mid-season drainage management seems to be an important way to obtain net benefits from greenhouse gas emissions [15]. So far, many studies have confirmed its applicability in rice fields based on overall greenhouse gas emissions. Zou et al. [16] recommended mid-season drainage as the best option to reduce greenhouse gas emissions, because they pointed out that the GWP of mid-season drainage was reduced by 27% compared with traditional $CH_4$ and $N_2O$-based floods [17]. Wassmann et al. [18] also reported that the GWP ($CH_4$ and $N_2O$) of the mid-season drainage was reduced by 42% and 72%, respectively, compared with the traditional flood. Since greenhouse gas emissions are greatly affected by the length and time of sewage discharge.

### 3.3 Alternate wetting and drying

Alternating wetting and drying are the periodic drying and re-oil flooding of the rice field. Compared with mid-season drainage, the time interval between wet and dry conditions seems too short to promote the transition from aerobic soil conditions to anaerobic soil conditions [19]. Alternating wetting and drying can significantly reduce $CH_4$ emissions, but the $N_2O$ emissions of this system vary greatly. Drainage and the resulting aerobic soil conditions will oxidize $CH_4$ and avoid $CH_4$ production. Song and Fujiyama [20] reported that compared with traditional flooded rice, alternating wetting and drying may reduce $CH_4$ emissions by 73%. Yagi et al. [21] proposed that optimal irrigation according to the physiological characteristics of crops at different growth stages can limit the frequency of alternating wet and dry conditions, thereby reducing the production and emission of $N_2O$. However, further research is needed in this practice to solve the problem of offsetting $N_2O$ emissions.

### 3.4 Intermittent drainage

Intermittent drainage involves repeated free drainage and irrigation. It has the advantage of improving soil oxidation conditions by enhancing root activity, increasing soil carrying capacity and ultimately reducing water input that leads to anaerobic conditions. It enhances the diffusion of oxygen into the soil, increases the aerobic area and reduces the production of $CH_4$ [22] pointed out that compared with traditional floods, intermittent drainage can reduce $CH_4$ emissions by 44%. Hardy [23] also showed that, compared with permanent floods, intermittent drainage can reduce $CH_4$ emissions by 15%. $N_2O$ emission during intermittent irrigation strongly depends on the flooding conditions of the field. Different water regimes in rice fields lead to sensitive changes in $N_2O$ emissions [17]. Nevertheless [24, 25] reported that compared with traditional floods, the global warming potential ($CH_4$ and $N_2O$) of intermittent irrigation was reduced by 34% and 54%, respectively.

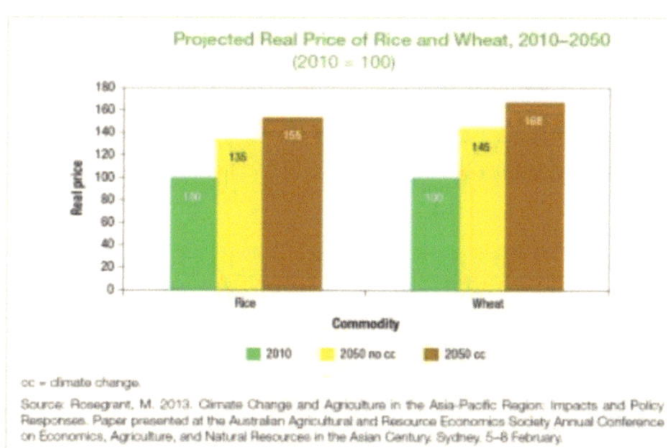


### 3.5 Controlled irrigation

According to reports, compared with irrigated rice, controlled irrigation can minimize the net emissions of greenhouse gases [26, 27]. During the rice growing season, the soil in the controlled irrigated rice field is kept dry (60–80%), but there is no flooding after the rice seedlings are regreened [28], similar to the water in the rice intensive system (Sato) [26] reported that, compared with traditional flooded rice, $CH_4$ emissions from controlled irrigation rice fields were reduced by 79%, and $N_2O$ emissions increased by 10%. The GWP of controlled irrigation was much smaller than that of traditional flooded rice (67%). Peng et al. [27] also reported that the global warming potential ($CH_4$ and $N_2O$) of the irrigation system was reduced by 27% compared to conventional floods. In addition, some authors report that deep-water irrigation (water depth of 10 cm) and permanent humidification, humidification irrigation and water-saving irrigation can also reduce greenhouse gas emissions from rice fields, especially $CH_4$ [29, 30] can be used as a tool to reduce greenhouse gas emissions.

### 3.6 Straw/residues management

Crop production inevitably leads to the production of large amounts of straw/residue that are usually left in the field [31]. With the gradual decrease in the amount of organic fertilizers, the rice soil relies heavily on the recycling of straw to overcome the carbon loss caused by soil cultivation and crop harvesting. Although burning straw can ensure rapid seedbed preparation for farmers and avoid the risk of nitrogen immobilization during the decomposition of residues, carbon and nitrogen are relatively large, but incomplete carbon combustion will produce a large amount of greenhouse gases and affect air quality. Produce adverse effects [31, 32]. In addition, N oxides and other fire-source organic compounds can cause the formation of tropospheric ozone. Rice straw is composed of a variety of organic components, such as cellulose, hemicellulose, lipids, protein, lignin, etc., each component's contribution to the improvement of $CH_4$ emission rate is variable. $CH_4$ The discharge rate is very sensitive to the management method of straw entering the soil. Ali et al. [33] reported that the $CH_4$ emission rate of fresh rice straw is higher compared to the non-crop season in rice fields. In a field survey conducted in Zhejiang Province, China, [34] found that the recorded greenhouse gas emissions of early straw incorporation at the beginning of the fallow period in winter were 11% less than the traditional straw incorporation method in spring. Similarly, [35] showed that in the fallow period (60 days before rice planting), the incorporation of residues is beneficial in terms of greenhouse gas emissions and grain yield, compared to the usual application before transplantation. Abandoning rice straw can also be an effective measure, because compared with the combined use of straw, straw removal reduces the emissions of these three gases [36].

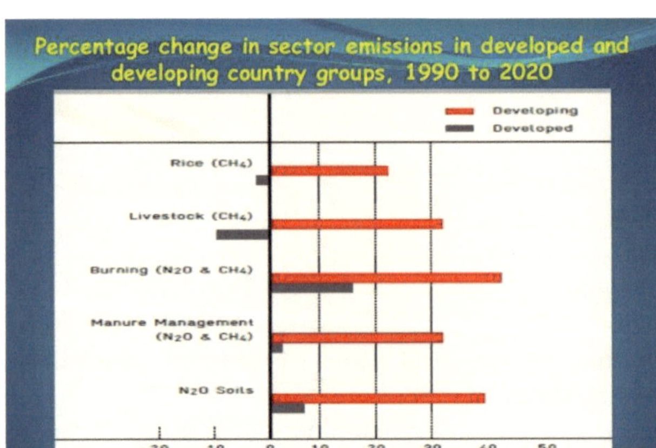

### 3.7 Application of biochar

Biochar is a carbon-rich material produced by pyrolyzing waste biomass under anoxic conditions and high temperatures [37]. The highly porous structure and the increased surface area of carbon-rich fine-grained biochar makes it an ideal soil amendment for carbon sequestration [38, 39]. Galloway et al. [40] pointed out that compared with the plots without biochar, biochar produced by pyrolysis of crop straw can increase carbon sequestration by 22% and reduce $N_2O$ emissions by 35%. Higher levels of biochar are more effective in reducing $N_2O$ emissions from rice fields [41, 42]. Huang et al. [43] observed that the application of biochar at 10 and 40 $ha^{-1}$ reduced $N_2O$ emissions by 58% and 74%, respectively. Jin et al. and Lehmann [44, 45] also recorded that the use of biochar can significantly reduce $N_2O$ emissions. The application of biochar may have a positive impact on soil organic carbon and significantly reduce $N_2O$ emissions, which may be a way to reduce greenhouse gas emissions. However, the long-term effects of biochar on the physical and chemical properties of soil and the rate of soil organic carbon sequestration require further studies to draw reliable conclusions.

### 3.8 Fermentation of manure

The greenhouse gas emission potential of fermented manure in the soil is low because the SOM reservoir will be quickly depleted during the fermentation process. Compared with fresh organic modifiers and a combination of urea and organic modifiers, the application of fermentation residues can reduce $CH_4$ emissions by approximately 60% and 52%, respectively [46]. Several field studies have evaluated various organic corrections for greenhouse gas emissions (especially methane). The difference between fresh materials (straw or fertilizer) is relatively small; however, it has been reported that there is a huge difference between the greenhouse gas emissions triggered by preferred and fresh materials [47, 48]. Using fermented biogas residues can only increase $CH_4$ emissions by 42%, while unfermented manure can increase $CH_4$ emissions by 112–138% [49]. Other carbon benefits obtained by substituting biogas for conventional fossil fuel energy, the use of biogas residues in rice fields can provide soil fertility while reducing $CH_4$ emissions. Nayak et al. [50] concluded that the application of livestock manure in rice fields can greatly reduce $N_2O$ emissions, while increasing $CH_4$ emissions and soil organic carbon sequestration. Patrick and Reddy [51] reported that the application of compost in the rice field reduced $N_2O$ emissions by 50% compared to the application of urea. However, the $CH_4$ emissions during anaerobic composting can offset the output obtained after mixing into the soil, and aerobic composting technology can minimize such emissions. Nayak et al. [50] observed that compared with fresh straw, the emissions of organic amendments produced by aerobic composting of straw were significantly reduced, indicating that it can be used as an environmentally friendly method.

### 3.9 Fertilizer management

Fertilizer management is an important part of reducing the environmental impact of rice fields. Soil fertilizers applied to crops are not always effective [52, 53]. Improving the efficiency of fertilizer use can reduce greenhouse gas emissions, especially $N_2O$ emissions, and indirectly reduce the carbon dioxide emissions of nitrogen fertilizers [54]. Measures to improve fertilizer utilization and reduce greenhouse gas emissions include: accurately adjusting the amount of fertilizer according to crop needs [55–57] and using nitrification inhibitors or slow-release fertilizers [58, 59] adjust the timing of application and select the appropriate source, accurately locate the fertilizer in the soil, avoid excessive application or eliminate the application of nitrogen fertilizer [60].

### 3.10 Adjusting fertilization and matching N supply with demand

Adjusting the nitrogen and phosphorus content to meet crop demand is conducive to crop yields while controlling greenhouse gas emissions. Even in best fertilization practices, large amounts of nitrogen will be released into the atmosphere. In irrigated rice, nearly 48% of applied nitrogen is lost in gaseous form [60]. The responsible mechanism for nitrogen loss is ammonia volatilization, nitrification and denitrification. The specific meaning of all these processes may vary according to natural conditions and crop management practices [45]. The rate of fertilizer controls the emission of greenhouse gases. In general, the emission of greenhouse gases, especially $N_2O$, increases with the increase of nitrogen input [34, 55]. The general strategy for minimizing N loss and reducing $N_2O$ emissions is to avoid excessive use of N in space and time. Reducing the amount of nitrogen fertilizer application to a level that does not reduce crop yields can also reduce the demand for nitrogen fertilizer, and ultimately reduce the indirect emissions of carbon dioxide during the nitrogen fertilizer production process. IPCC (1997) estimated that regardless of the source of N, 1.25% of the applied N would be lost as $N_2O$. Several studies have documented the instantaneous increase in $N_2O$ emissions from rice fields due to the application of nitrogen fertilizers. [25]. It was observed that the application of urea will increase $N_2O$ emissions by 54% compared with no nitrogen fertilizer. Lu et al. [49] reported that $N_2O$ emissions in rice fields increased with the application of nitrogen fertilizer, especially at higher rates. Reducing nitrogen fertilizer has no significant impact on $CH_4$ emissions, while the current average nitrogen fertilizer application for rice can be reduced by 33%, which can reduce $N_2O$ emissions by 27%. Recent field studies report that high nitrogen content can reduce net $CH_4$ emissions from rice systems by roughly 30–50% [59, 60].

Aulakh et al. [34] reported that the increase in nitrogen application reduced $CH_4$ emissions and increased $N_2O$ emissions compared with the control without nitrogen application. Zou et al. [16] also recorded that when the nitrogen application rate increased from 150 kg to 400 kg N $ha^{-1}$, $CH_4$ decreased by 75% and $N_2O$ increased by 58%. A recent meta-analysis showed that the response of $CH_4$ emissions may be related to the N rate, where the addition of N at a low rate tends to stimulate $CH_4$ emissions, but it may alleviate $CH_4$ emissions at high N rates [35, 47]. However, further research is inevitably needed to deal with the compromise between nitrogenous fertilizers and $CH_4$ and $N_2O$. Applying nitrogen fertilizer to the soil near the active root absorption zone can reduce the loss of surface nitrogen and increase plant nitrogen use efficiency, thereby reducing $N_2O$ emissions. Khaliq et al. [30] pointed out that placing chemical fertilizers in a 6–10 cm soil layer can significantly increase nitrogen use efficiency and reduce $N_2O$ emissions. In addition, distributing nitrogen fertilizer at different growth stages of crops can also increase nitrogen use efficiency and reduce nitrogen loss.

## 4. Conclusion

From above all discussion it may concluded that nitrogen have positive impact on rice yield greenhouse gas emissions from rice fields can reduce greenhouse gas emissions from rice fields by changing irrigation methods and farming methods, managing organic additives and fertilizer inputs, and choosing appropriate varieties and planting methods. $CH_4$ decreased by 75% and $N_2O$ increased by 58%. $CH_4$ emissions may be related to the N rate, where the addition of N at a low rate tends to stimulate $CH_4$ emissions.

## Author details

Muhammad Ikram* and Haseeb-Ur-Rehman
Department of Agronomy, Faculty of Agricultural Sciences and Technology,
Bahauddin Zakariya University, Multan, Pakistan

*Address all correspondence to: r4rana88@gmail.com

## References

[1] Petersen LK. Impact of Climate Change on Twenty-First Century Crop Yields in the US. Climate, 2019;7(3):40.

[2] Kontgis C, Schneider A, Ozdogan M, Kucharik C, Duc NH, Schatz J.Climate change impacts on rice productivity in the Mekong River Delta. Appl Geogr.2019; 102:71-83.

[3] Takeshi I, Shimamoto Ko. Perspectives of becoming a model plant: The importance of rice to plant science. Science Direct.1996; 1:9599.

[4] FAOSTAT (2016) http://faostat3.fao.org/download/Q/QC/E. Accessed 13 Feb 2016

[5] Dobermann A, Fairhurst T.Rice: nutritional disorders and nutrient management. Singapore:Potash and Phosphate Institute of Canada. International Rice Research Institute, Los Baños.2000.

[6] Nayar NM. Origins and phylogeny of rices. Elsevier.2014.

[7] Linares OF. African rice (Oryza glaberrima): history and future potential. Proceedings of the National Academy of Sciences.2002; 99(25): 16360-16365.

[8] Fuller DQ. Pathways to Asian civilizations: Tracing the origins and spread of rice and rice cultures. *Rice*.2011;4(3):78-92.

[9] Smith CW. Crop production: evolution, history, and technology. John Wiley and Sons.1995.

[10] Chang TT. The impact of rice on human civilization and population expansion. Interdisciplinary Science Reviews.1987; *12*(1):63-69.

[11] Bloom A, Swisher M. Emissions from rice production. Encyclopedia of Earth.2010; Accessed at 15(6);2011.

[12] Inubushi K. Characteristics and improvements of the poor paddy soils. Paddy fields soils. Introduction to Soil Science.2001

[13] Colton D, Giebermann K, Monk P. The linear sampling method for three-dimensional inverse scattering problems. ANZIAM Journal.2000; 42:434-460.

[14] Wassmann R, Neue HU, Lantin RS, Makarim K, Chareonsilp N, Buendia LV, Rennenberg H. Characterization of methane emissions from rice fields in Asia. II. Differences among irrigated, rainfed, and deepwater rice. In Methane Emissions from Major Rice Ecosystems in Asia.2000;(pp. 13-22). Springer, Dordrecht

[15] Towprayoon S, Smakgahn K, Poonkaew S. Mitigation of methane and nitrous oxide emissions from drained irrigated rice fields. Chemosphere.2005:59(11), 1547-1556.

[16] Zou J, Huang Y, Jiang J, Zheng X, Sass R L.A 3-year field measurement of methane and nitrous oxide emissions from rice paddies in China: Effects of water regime, crop residue, and fertilizer application. Global Biogeochemical Cycles.2005; 19(2).

[17] Osugi A, Itoh H, Ikeda-Kawakatsu K, Takano M, Izawa T. Molecular dissection of the roles of phytochrome in photoperiodic flowering in rice. Plant physiology.2011;157(3): 1128-1137.

[18] Wassmann R, Lantin RS, Neue HU, Buendia LV, Corton TM, Lu Y. Characterization of methane emissions from rice fields in Asia. III. Mitigation options and future research needs. Nutrient Cycling in Agroecosystems. 2000;*58*(1-3):23-36.

[19] Katayanagi N, Furukawa Y, Fumoto T, Hosen Y. Validation of the

DNDC-rice model by using $CH_4$ and N2O flux data from rice cultivated in pots under alternate wetting and drying irrigation management. Soil science and plant nutrition.2012; *58*(3):360-372.

[20] Song JQ, Fujiyama H.Difference in-response of rice and tomato subjected to sodium salinization to the addition of calcium. Soil Science and Plant Nutrition.1996;*42*(3):503-510.

[21] Yagi K, Tsuruta H, Kanda KI, Minami K. Effect of water management on methane emission from a Japanese rice paddy field: Automated methane monitoring. Global Biogeochemical Cycles.1996; 10(2):255-267.

[22] Adhya TK, Bharati K, Mohanty SR, Ramakrishnan B, Rao VR, Sethunathan N, Wassmann R.Methane emission from rice fields at Cuttack, India. Nutrient Cycling in Agroecosystems.2000; 58(1-3):95-105.

[23] HardyBL. Climatic variability and plant food distribution in Pleistocene Europe: Implications for Neanderthal diet and subsistence. Quaternary Science Reviews.2010; *29*(5-6):662-679.

[24] Feng T, Ye R, Zhuang H, Rong Z, Fang Z, Wang Y, Jin Z.Physicochemical properties and sensory evaluation of Mesona Blumes gum/rice starch mixed gels as fat-substitutes in Chinese Cantonese-style sausage. Food Research International.2013;*50*(1):85-93.

[25] Yang N, Zhang H, Chen M, Shao LM, He PJ.Greenhouse gas emissions from MSW incineration in China: Impacts of waste characteristics and energy recovery. Waste management.2012;32(12):2552-2560.

[26] Hou H, Peng S, Xu J, Yang S, Mao Z.Seasonal variations of $CH_4$ and $N_2O$ emissions in response to water management of paddy fields located in Southeast China. Chemosphere.2012; 89(7):884-892.

[27] Peng S, Yang S, Xu J, Gao H. Field experiments on greenhouse gas emissions and nitrogen and phosphorus losses from rice paddy with efficient irrigation and drainage management. Science China Technological Sciences.2011;*54*(6):1581.

[28] Yu S, Ma Y, Menager L, Sun DW. Physicochemical properties of starch and flour from different rice cultivars. Food and bioprocess technology.2012; *5*(2):626-637.

[29] Liu XY, Qu JJ, Li LQ, Zhang AF, Jufeng Z, Zheng JW, Pan GX.Can biochar amendment be an ecological engineering technology to depress N2O emission in rice paddies?—A cross site field experiment from South China. Ecological Engineering.2012; *42*:168-173.

[30] Khaliq A, Shakeel M, Matloob A, Hussain S, Tanveer A, Murtaza G.Influence of tillage and weed control practices on growth and yield of wheat. Philippine Journal of Crop Science (PJCS).2013;*38*(3): 00-00.

[31] Beri V, Sidhu BS, Bahl GS, Bhat AK. Nitrogen and phosphorus transformations as affected by crop residue management practices and their influence on crop yield. Soil Use and Management.1995; 11(2):51-54.

[32] Watanabe A, Katoh K, Kimura M. Effect of rice straw application on $CH_4$ emission from paddy fields: II. Contribution of organic constituents in rice straw. Soil science and plant nutrition.1993;39(4):707-712.

[33] Ali RI, Awan TH, Ahmad M, Saleem MU, Akhtar M. Diversification of rice-based cropping systems to improve soil fertility, sustainable productivity and economics. Journal of Animal and plant sciences.2012; 22(1):108-12.

[34] Aulakh MS, Wassmann R, Bueno C, Kreuzwieser J, Rennenberg H.

Characterization of root exudates at different growth stages of ten rice (*Oryza sativa L.*) cultivars. Plant biology.2001;3(2):139-148.

[35] Banger K, Tian H, Lu Nitrogen fertilizers stimulate or inhibit methane emissions from rice fields. Global Change Biology,2012;18(10):3259-3267.

[36] Canavesi F. ICAR 2006.

[37] Cassman KG, Dobermann A, Walters DT, Yang H. Meeting cereal demand while protecting natural resources and improving environmental quality. Annual Review of Environment and Resources. 2003; 28(1):315-358 Corton TM, Bajita JB, Grospe FS, Pamplona RR, Assis CA, Wassmann R, Lantin RS, Buendia LV. Methane emission from irrigated and intensively managed rice fields in Central Luzon (Philippines). Nutr Cycl Agroecosyst. 2000; 58:37-53.

[38] Dong X, Gong M, Shen J, Wu J.Experimental measurement of vapor–liquid equilibrium for (trans-1, 3, 3, 3-tetrafluoropropene (R1234ze (E))+ propane (R290)). International journal of refrigeration.2011; 34(5):1238-1243.

[39] Freney JR. Emission of nitrous oxide from soils used for agriculture. Nutrient Cycling in Agroecosystems. 1997;49(1-3):1-6.

[40] Galloway JN, Aber JD, Erisman JW, Seitzinger SP, Howarth RW, Cowling EB, Cosby BJ. The nitrogen cascade. Bioscience.2003; 53(4):341-356.

[41] Ghosh S, Majumdar D, Jain MC. Methane and nitrous oxide emissions from an irrigated rice of North India. Chemosphere.2003;51(3):181-195.

[42] Gregorich EG, Rochette P, Vanden B, Ygaart AJ, Angers DA. Greenhouse gas contributions of agricultural soils and potential mitigation practices in Eastern Canada. Soil and Tillage Research.2005; 83(1):53-72.

[43] Huang YC, Simmons C, Kaigler D, Rice KG,Mooney DJ.Bone regeneration in a rat cranial defect with delivery of PEI-condensed plasmid DNA encoding for bone morphogenetic protein-4 (BMP-4). Gene therapy.2005; 12(5):418-426.

[44] Jin H, Li S, Cheng G, Shaoling W, Li X.Permafrost and climatic change in China. Global and Planetary Change.2000; 26(4):387-404.

[45] Lehmann J. A handful of carbon. Nature.2007; 447(7141): 143-144.

[46] Liang W, Yi SHI, Zhang H, Jin YUE, Huang GH. Greenhouse gas emissions from northeast China rice fields in fallow season. Pedosphere.2007; 17(5): 630-638.

[47] Linquist BA, Adviento-Borbe MA, Pittelkow CM,Van Kessel C,Van Groenigen KJ. Fertilizer management practices and greenhouse gas emissions from rice systems: a quantitative review and analysis. Field Crops Research.2012;135:10-21

[48] Liu Y, Guo Y, Gao W, Wang Z, Ma Y, Wang Z.Simultaneous preparation of silica and activated carbon from rice husk ash. Journal of Cleaner Production.2012;32:204-209.

[49] Lu Y, Wassmann R, Neue HU, Huang C. Dynamics of dissolved organic carbon and methane emissions in a flooded rice soil. Soil Science Society of America Journal.2000 64(6); 2011-2017.

[50] Nayak AK,Pal D,Das S.Calcium pectinate-fenugreek seed mucilage mucoadhesive beads for controlled delivery of metformin HCl. Carbohydrate polymers.2013; 96(1):349-357.

[51] Patrick Jr WH, Reddy KR. Nitrification-denitrification reactions in flooded soils and water bottoms: Dependence on oxygen supply and ammonium diffusion. Journal of environmental quality.1976; 5(4):469-472.

[52] Pittelkow CM, Adviento-Borbe MA,Hill J E,Six J,van Kessel C, Linquist BA.Yield-scaled global warming potential of annual nitrous oxide and methane emissions from continuously flooded rice in response to nitrogen input. Agriculture, Ecosystems & Environment.2013;177:10-20.

[53] Schlesinger W H.Carbon sequestration in soils.1999.

[54] Shang Q, Yang X, Gao C, Wu P, Liu J, Xu Y, Guo S. Net annual global warming potential and greenhouse gas intensity in Chinese double rice-cropping systems: a 3-year field measurement in long-term fertilizer experiments. Global Change Biology.2011;17(6): 2196-2210.

[55] Wassmann R, Papen H, Rennenberg H.Methane emission from rice paddies and possible mitigation strategies. Chemosphere. 1993;*26*(1-4):201-217.

[56] Yang H, Du T, Qiu R. Chen J, Wang F, Li Y, Kang S. Improved water use efficiency and fruit quality of greenhouse crops under regulated deficit irrigation in northwest China. Agricultural Water Management. 2017;179:193-20

[57] Yao F, Huang J, Cui K, Nie L, Xiang J, Liu X, Peng S. Agronomic performance of high-yielding rice variety grown under alternate wetting and drying irrigation. Field crops research.2012;126, 16-22

[58] Zhang A, Cui L, Pan G, Li L, Hussain Q, Zhang X, Crowley D. Effect of biochar amendment on yield and methane and nitrous oxide emissions from a rice paddy from Tai Lake plain, China. Agriculture, ecosystems & environment.2010;139(4):469-475.

[59] Zhang W, Hendrix PF, Dame LE, Burke RA, Wu J, Neher DA, Fu S. Earthworms facilitate carbon sequestration through unequal amplification of carbon stabilization compared with mineralization. Nature communications.2013;4: 2576.

[60] Zheng X, Wang M, Wang Y, Shen R, Gou J, Li J, Li L. Impacts of soil moisture on nitrous oxide emission from croplands: a case study on the rice-based agro-ecosystem in Southeast China. Chemosphere-Global Change Science.2000; 2(2):207-224.

Chapter 11

# Genetic Variation and Aflatoxin Accumulation Resistance among 36 Maize Genotypes Evaluated in Ghana

*Abu Mustapha Dadzie, Allen Oppong, Ebenezer Obeng-Bio and Marilyn L. Warburton*

## Abstract

Aflatoxins are carcinogenic secondary metabolites produced predominantly by the fungi *Aspergillus flavus* and *parasiticus.* The toxin contaminate maize grains and threatens human food safety. Survey in Ghana revealed aflatoxin contamination of maize in excess of 941 ppb which is way beyond WHO and USA approved limits of 15 ppb and 20 ppb respectively. Host plant resistance is considered as the best strategy for reducing aflatoxins. This study was designed to (1) identify and select suitable maize lines that combine aflatoxin accumulation resistance and good agronomic traits under tropical conditions and (2) assess the genetic diversity among the exotic and locally adapted maize genotypes using significant morphological traits. Thirty-six maize genotypes, 19 from Mississippi State University, USA and 17 locally adapted genotypes in Ghana were evaluated for aflatoxin accumulation resistance and good agronomic characteristics across six contrasting environments using a 6x6 lattice design with three replicates. Five plants each per genotype were inoculated with a local strain of *Aspergillus flavus* inoculum at a concentration of $9 \times 10^7/3.4$ ml, two weeks after 50% mid silking. Total aflatoxin in the kernels were determined at harvest using HPLC method. Statistical analysis for agronomic traits and aflatoxin levels were performed using PROC GLM procedure implemented in SAS. The result indicated that genotype by environment interaction was significant ($p < 0.05$) for aflatoxin accumulation resistance and many other agronomic traits. Five genotypes (MP715, NC298, MP705, MP719, CML287 and TZEEI- 24) consistently displayed stable resistance across the environments and may serve as suitable candidates for developing aflatoxin resistant hybrids. Cluster analysis showed two distinct groups (locally adapted and exotic genotypes), an indication of re-cycled alleles per region. Broad sense heritability estimates for grain yield and aflatoxin accumulation resistance were moderately high, which could permit transfer of traits during hybrid development.

**Keywords:** Maize, *Aspergillus*, Aflatoxin Accumulation, Genetic Variation

## 1. Introduction

Adaptability and productivity of maize across a wide range of agro- ecologies makes it a suitable food security crop for most parts of the world [1]. However, a major limitation to the contribution of maize towards food supply is the contamination of grains by aflatoxins. Aflatoxins are carcinogenic secondary metabolites produced mainly by *Aspergillus sp.* which contaminates maize grains during pre- and post- harvest seasons and renders the grains unwholesome for consumption by both humans and livestock [2]. In addition to the health risks, aflatoxin contamination is a serious challenge because the pathogen is globally widespread and causes considerable economic losses by down-grading grain quality, nutritional value and taste [3]. Due to the danger it poses to human health, several countries have set for limits to regulate aflatoxin contamination in many agricultural products including maize. Allowable limits set by Japan is 0 ppb, while the European Union and United States of America have limits of 2–4 ppb and 20 ppb respectively [4].

Approaches for the control and reduction of aflatoxin have relied on good agronomic practices, application of biocontrol preparations of atoxigenic strains of *A. flavus* (including aflasafe and aflaguard), and the use of resistant host plant germplasm [5] as well as BT varieties. Host plant resistance is seen as the method of choice since it exploits the accumulation of resistance alleles into single hybrid varieties [6] and is simple for the farmer to use.

Considerable efforts over the years have led to the development and identification of aflatoxin resistant breeding lines. However, some of these lines lack good agronomic characteristics in temperate environments [6] and may additionally lack other disease or insect resistance in tropical environments. They are useful in crosses involving elite or acceptable lines for the incorporation of novel alleles which confers aflatoxin accumulation resistance into hybrid varieties.

Studies on germplasm diversity and characterization have utilized morphological and/or molecular data for grouping of entries and breeding lines into various heterotic groups. These heterotic groupings can be used to rule out many unproductive hybrid crosses and reduce the total number of testcrosses that should be generated to ultimately find the highest yielding hybrids.

In this way, phenotypic evaluation and groupings of inbred lines could be useful in the identification of suitable inbred lines for the development of superior hybrids with high yields and aflatoxin resistance. Assessing the genetic diversity among the exotic and locally adapted maize genotypes would be useful in selecting potential parents with diverse genetic backgrounds that could be utilized in a breeding program for hybrid development.

The objectives of this study were to: (1) identify and select suitable lines that combine aflatoxin accumulation resistance and good agronomic traits under tropical environmental conditions. (2) assess the genetic diversity between exotic and local maize genotypes using significant morphological traits.

## 2. Materials and methods

Thirty-six genetic materials were used for the study and this included 19 exotic inbred lines developed for aflatoxin accumulation resistance by the corn host resistance plant unit (CHPPRU) in Mississippi, USA and seventeen locally adapted genotypes. Pedigree information on germplasm is presented in **Table 1**.

### 2.1 Field experimental sites and layout

Three locations were used for the experiments, namely Fumesua, Akomadan, and Wenchi. Fumesua is in the semi- deciduous forest zone with an altitude of 286 m above sea level and it lies within 6.712 N and 1.523 W. The mean annual rainfall is 1500 mm coupled with mean minimum and maximum temperatures of

| Genotype | Pedigree | Source |
|---|---|---|
| ENTRY-5 | | CIMMYT |
| ENTRY-6 | | CIMMYT |
| ENTRY-70 | | CIMMYT |
| ENTRY-85 | | CIMMYT |
| GH-110 | | CIMMYT |
| ABROHEMAA | | CRI |
| OBATANPA | | CRI |
| HONAMPA | | CRI |
| AHODZIN | | CRI |
| OMANKWA | | CRI |
| TINTIM | 2-B-B: DT-SR-W-C0/1368 × PAC90038–1 × 1368–6-07C04772B 06A11833B x B-B-B-B-B-B:DT-SR-W-C0/1368 × PAC90038–1 × 1368–3-07C04754B 06A11803B | IITA |
| M0826-7F | B-B-B-B-B-B:DT-SR-W-C0/1368 × PAC90038–1 × 1368–3-07C04754B 06A11803B | IITA |
| TZEEI-4 | TEE-W SR BC5 x 1368 STR S7 Inb.85 | IITA |
| TZEEI- 24 | TEE-W SR BC5 x 1368 STR S6 Inb.229B | IITA |
| TZEEI-15 | TZEEI-15 WPopxLDS6(Set A)Inb.44 | IITA |
| TZEEI-6 | TZEEI - 6 WSRBC5x1368STRS7Inb.100 | IITA |
| TZI8 | TZB x TZSR | IITA |
| CML11 | P21-C5-FS219–3–2-2-3-#-7–1-B-4-1-B | CIMMYT |
| CML158Q | EV8762SR-2-1-B-1-B | CIMMYT |
| CML176 | (P63–12–2:1/P67–5–1-1)-1–2:e-e | CIMMYT |
| CML247 | (G24-F119/924-F54)-6–4–1-1-B | CIMMYT |
| CML287 | (P24-F26/P27-F1)-4–1-B-1-1-B | CIMMYT |
| CML322 | LLMBR-17-B-5-3-1-4-B | CIMMYT |
| CML343 | LAPOSTA SEQ-C3-FS17–1–2-3-2-1-B | CIMMYT |
| CML5 | PobZ1C5HC133-1-B_B | CIMMYT |
| CML108 | Pop. 44 | CIMMYT |
| Hi27 | [CM104(India)BC6 (is and MV source) | THAILAND |
| Ki3 | Ki 3 (86329) | THAILAND |
| MP705 | from Mp SWCB-4 | MISSISSIPI |
| MP715 | Line derived from Tuxpan | MISSISSIPI |
| MP719 | (Mp715 x Va35) -1-3-4-2-3-1-1-1-B | MISSISSIPI |
| NC334 | (SC76*B52); sister line of NC332 | NORTH CAROLINA |
| NC298 | PX105A.H5 x Agroc.155 | NORTH CAROLINA |
| NC340 | 340 (31105) | NORTH CAROLINA |
| NC356 | TROPHY SYN | NORTH CAROLINA |

**Table 1.**
*List of germplasm and pedigree.*

21°C and 31°C, respectively. Soils around Fumesua are classified as Asuansi series, a ferric acrisol. Akomadan is situated within the forest savanna transition zone and it lies within 7.396 N and 1.973 W. It has a bimodal rainfall distribution pattern same as Fumesua. Wenchi on the other hand lies within 7.733 N and 2.100 W, a transitional savanna zone with bimodal rainfall pattern similar to the other two locations.

For the three locations, major season begins in March and usually ends in July whilst the minor season begins from September and ends in November. The experiments were conducted during the major season of 2017 and minor season of 2017/2018 in all three locations for genotype evaluation. Plantings were staggered at weekly interval between sites. Experimental design used was a 6 x 6, square - lattice with three replications. Single row plots, each 5 m long, spaced 0.70 m apart with 0.4 m spacing between plants in each row were used in all the environments. Three seeds of the lines were planted in each hole and thinned to two plants per hill at two weeks after emergence to give a population density of 66,667 plants per hectare. Weeds were controlled through the use of Atrazine and Gramozone as pre- and post- emergence herbicides at 5 liters/ha each of Primextra and praraquat and subsequently supported by manual weeding.

### 2.2 Data collected

Data was collected on the following specific parameters:

- DS = Number of days from planting to the time 50% silking was observed.

- DP = Number of days taken for 50% of the plants to begin to shed off pollen.

- ASI = Anthesis-silking interval (ASI) was calculated as the difference between days to 50% silking and 50% anthesis.

- PH = Plant height was determined by measuring the distance from the base of the plant to the height of the first tassel branch.

- EH = Ear height was measured as the distance from the soil surface to the node bearing the upper ear.

- RL = Root lodging (RL) was estimated as the percentage of plants leaning more than 30 degrees from the vertical.

- SL = Stalk lodging (SL) was determined as the proportion or percentage of plants with broken stalk below the ear or the stalk bending more than 45 degrees from the vertical position.

- EA = Ear aspect was estimated based on a scale of 1 to 5, where 1 = clean, uniform, large, and well-filled ears. 2 = moderately uniform and well filled, 3 = ears with mild disease/insect damage and fully-filled grains with one or two irregularities in cob size, 4 = ears with severe disease/insect damage, scanty grain filling, few ears, non-uniformity of cobs, while 5 = ears with totally undesirable features, very few or no grains.

- PA = Plant aspect was determined based on the general assessment of the plant architecture as they appear in the plot and was rated on a scale of 1–5 where, 1 = excellent overall phenotypic appeal, 2 = very good overall phenotypic appeal, 3 = good overall phenotypic appeal, 4 = poor overall phenotypic appeal and 5 = very poor overall phenotypic appeal.

- EPP = Ear number per plant was obtained by dividing the total number of ears per plot by the number of plants harvested.
- SG = Stay- green (chlorophyll concentration) was measured by randomly selecting any five plants per plot and determining chlorophyll concentration from ear leaf at approximately 4 weeks after anthesis and 2 weeks after *A. flavus* inoculation. WAA gadget with a portable SPAD meter (CCM-200 plus-opti sciences) was used to measure the chlorophyll content.
- Blight (BD) and Maize streak diseases (MSVD) were also scored on a scale of 1 to 5, where 1 = absence of disease and 5 = severe infection.
- ID = Insect damage was scored on a scale of 1–5 depending on the extent of damage caused by insects to the ear on plot by plot basis. Scale of 1 = highly resistant, 2 = resistant, 3 = moderately resistant, 4 = susceptible, 5 = highly susceptible.
- ER = Ear rot was also rated on a scale of 1 to 5. 1 = highly resistant, 2 = resistant, 3 = moderately resistant, 4 = susceptible, 5 = highly susceptible.
- HC = Husk cover or open-tip was rated on a scale of 1–5 where, 1 = very tight husk extending beyond the tip and 5 = exposed ear tip.
- Grain yield estimation = Harvested ears from each plot were shelled to determine the percentage grain moisture using moisture meter and then subsequently determine the grain yield in kg ha$^{-1}$ from the shelled grain weight based on 80% shelling percentage and adjustment of moisture content to 15%.

Grain yield was calculated as follows:

$$GY = fwt \times \frac{(100-m)}{85} \times \frac{10000}{(8\times\phi)} \times 0.8 \tag{1}$$

where,
GY = grain yield (kg ha$^{-1}$),
fwt = field weight of harvested ears per plot (kg),
m = grain moisture content at harvest.
10,000 = land area per hectare (m$^2$),
8 = land area per plot (0.70 m x 0.4 m),
$\phi$ = number of hills/plot (11) and 0.80 = 80% shelling percentage.
Broad-sense heritability (H$^2$) was estimated as:

$$H^2 = \sigma_G^2 / \left(\sigma_E^2 / re + \sigma_{GE}^2 / e + \sigma_G^2\right), \tag{2}$$

Where;
$\sigma_G^2$ = variation due to genotype, $\sigma_E^2$ = variation due to environment, $\sigma_{GE}^2$ = variation due to genotype by environment interactions, r = number of replications and e = number of environments.

### 2.3 Source of inoculum and isolation of *A. flavus*

Aflatoxin contaminated maize samples from Ejura main farms were cut into 3 mm pieces with a sterile scalpel blade, after being surface-sterilized in 1% hypochlorite for 2 minutes, then placed on Potato Dextrose Agar (PDA) and incubated at room temperature for 5 days.

After incubation, colonies of different morphology, shape, and color were observed. A pure culture of each colony was obtained through serial dilution where 1 agar plug containing mycelia was serially diluted into 9 mls of distilled water till a concentration of 1 x $10^5$ was achieved.

One ml of the final dilution was transferred onto water agar (2% agar) and incubated at 31°C in unilluminated growth chamber. Identification slides were prepared by picking spores with isolation needle onto a slide containing a drop of distilled water. *A. flavus* was subsequently identified by observing colony characteristics, conidial morphology as described previously [7, 8].

Isolates that produced large smooth conidial surface and either an average sclerotial diameter > 400 μm or without sclerotia were identified as L-type *A. flavus* using Leica Microscope X 40. Identified isolates were subsequently maintained on potato dextrose agar (PDA) as described by Jha [9]. Maintenance of colonies were done by sub-culturing of the colony onto PDA plates and incubated at room temperature for 5 days.

### 2.4 Inoculum preparation

Identified toxigenic isolate was used to prepare the inoculum as described by Windham [10]. The procedure involved multiplication of the isolate on sterile corn cob grit in 500-ml flasks each containing 50 g of grits and 100 ml of sterile distilled water and incubated at 28°C for 3 weeks. Conidia in each flask was washed from the grits using 500 ml of sterile distilled water containing 20 drops of Tween 20 per liter and then filtered through four layers of sterile cheesecloth. The concentrations of conidia was determined with a hemacytometer and adjusted with sterile distilled water to 9 x $10^7$ conidia per ml. Excess inoculum not used immediately was refrigerated at 4°C.

### 2.5 Inoculation method (wounding)

The side needle technique described by Scott and Zummo [11] which utilizes an Idico tree-marking gun fitted with a 14-gauge needle was used for inoculations 14 days after mid silk. Ears were inoculated by inserting the needle under the husks on the upper 1/3 of the ear and 3.4 ml of a spore suspension of 9 x $10^7$ conidia/ml was injected over the kernels. A total of 5 ears per genotype were used for the inoculation study.

### 2.6 Aflatoxin analysis

Ears that did not touch the ground were harvested from plots at maturity, approximately 60 days after mid-silk. The cobs were shelled and samples ground using a Romer mill (Romer Industries, Inc., Union, MO) according to manufacturer's instructions. Aflatoxin was extracted using the method described by Sirhan [12] with modifications. Maize samples were homogenized into suspension using a Preethi Mixer Grinder.

A weight of 2 g of slurry was weighed into a 15 ml centrifuge tube and topped-up with a 4 ml of 60:40 (v/v) methanol:acetronitrile solution, and vortexed for

3mins. 1.32 g of anhydrous $MgSO_4$ and 0.2 g of NaCl were added to the mixture, and vortexed for additional 1 min. The tube was centrifuged for 5 min at 4000 rpm and the upper organic layer filtered through a 0.45 µm nylon syringe prior to injection. A volume of 100 µl of the filtered extract was injected into the HPLC.

A Cecil-Adept Binary Pump HPLC coupled with Shimadzu 10 AxL fluorescence detector (Ex: 360 nm, Em: 440 nm) with Phenomenex Hyper Clone BDS C18 Column (150 x 4.60 mm, 5 µm) was used for analysis. The mobile phase used was methanol: water (40:60, v/v) at a flow rate of 1 ml/min with column temperature maintained at 40°C. To 1 liter of mobile phase were added 119 mg of potassium bromide and 350 µl of 4 M nitric acid (required for postcolumn electrochemical derivatisation with Kobra Cell, R-Biopharm Rhone). Aflatoxin Mix (G1, G2, B1, B2) standards (ng/g) were prepared from Supelco® aflatoxin standard of 2.6 ng/µL in methanol. Concentration of B1 and G1 were 0.5, 1, 2, 8, 16 ng /g per 100 µl injection of each standard.

Concentration of B2 and G2 were 0.15, 0.3, 0.6, 2.4, 4.8 ng/g per 100 µl injection of each standard. Limit of Detection and Limit of quantification of total aflatoxin were established at 0.5 ng/g and 1 ng/g respectively. The unit (ng/g is equivalent to ppb). Aflatoxin concentration was estimated as:

$$ng/g = A \times (T / I) \times (1 / W) \tag{3}$$

where A = ng of aflatoxin as eluate injected, T = final test solution eluate volume (µl), I = volume eluate injected into LC (µl), W = mass (g) of commodity represented by final extract.

#### *2.6.1 Validation of HPLC method*

Recovery studies were conducted to check for precision and accuracy. Blank samples were spiked at 5 (five) replicated maize samples at 13 ng/g, 26 ng/g and 104 µg/g with recoveries 91 ± 1.75%, 98 ± 1.33% and 102 ± 1.87% respectively. Blanks that were run periodically contained no detectable amount of target analyte. Trueness was further validated using a certified reference material (TR-A1000) from Triology laboratory, USA. The value obtained, 20.17 ± 1.14 µg/kg from ten replicates was within the recommended range of the certified value of 21.0 ± 2.9ug/kg. Coefficient of variation was less than 15% for replicates.

### **2.7 Statistical analysis**

Analysis of Variance (ANOVA) was performed on plot means for grain yield and all other agronomic traits for each environment and across environments using PROC GLM procedure of SAS software, version 9.4 [13]. Data on aflatoxin contamination was transformed as Ln (y + 1) where y is the aflatoxin level whilst Ln is Log base e.

This transformation was done to reduce the heterogeneity of variance of contamination levels. Genotype or entry means were adjusted for block effects and analyzed according to lattice design [14]. Each environment was defined as season x location x *A. flavus* inoculation treatment. Effects of environment were considered as random while genotypes were classified as fixed effects. Additionally, genetic correlations between aflatoxin accumulation and selected agronomic traits were performed using the meta menus program implemented in SAS to examine the relationships among the traits and also predict strategies to enhance their improvement.

#### *2.7.1 Clustering analysis using agro-morphological traits*

Classification of genotypes was based on significant agro- morphological traits. The significant traits were standardized and used to generate Euclidean genetic distance co-efficient whiles Ward's minimum variance method implemented in SAS software version 9.4 [13] was used for the clustering.

## 3. Results

Environmental effect was significant ($p < 0.01$) for all agronomic and aflatoxin accumulation resistance traits except open-tip while genotypic mean squares were significant for all measured traits (**Tables 2** and **3**). Genotype by environment interactions were significant ($p < 0.05$) for all traits except days to 50% pollen and silking and streak incidence.

Broadsense heritability showed relatively low to high estimates for agronomic traits, ranging from 18.90% for open-tip to 62.70% for grain yield. For the disease traits, estimated broad sense heritability ranged from a relatively low values of 9.70% for rust incidence to 24.40% for maize streak virus disease incidence. Other traits with relatively moderate to high heritability estimates were ear (67.30%) and plant heights (78.00%).

### 3.1 Aflatoxin accumulation resistance and agronomic performance of germplasm

Generally, performance of the thirty-six genotypes showed significant ($p < 0.05$) differences in aflatoxin accumulation (**Table 4**). Aflatoxin accumulation ranged from a minimum of 14.85 ppb for MP705 to a maximum of 140.60 ppb for HONAMPA (local check). Grain yield varied from 565.63 kg $ha^{-1}$ for MP715 (inbred) to 4721.03 kg $ha^{-1}$ for AHODZEN (OPV) with a mean of 1853.22 kg $ha^{-1}$ (**Table 5**). Days to 50% pollen ranged from 48 days to 62 days whilst days to 50% silking ranged between 50 and 65 days. Anthesis silking interval varied from 2 to 4 days with an average of 3 days.

The number of ears per plant ranged from approximately 1 to 2 whilst means for cob aspect, plant aspect and open-tip were 2.01, 1.97 and 1.49, respectively. Generally, an observed mean of 1.24 for insect damage (**Table 6**) was an indication of partial tolerance of the germplasm utilized, nonetheless, OMANKWA and NC340 appeared moderately susceptible to insect damage.

Analysis of stay-green characteristics revealed NC298 as the genotype with prolonged green pigmentation whilst CML11 had less and reduced pigmentation (**Table 6**). Means observed for rust, blight and streak resistance indicated a fairly tolerant germplasm. Mean scores obtained for ear rot showed appreciable tolerance of the germplasm whilst plant height ranged between 100.73 cm and 176.25 cm. Ear height also varied from 52.24 cm to 92.50 cm.

### 3.2 Location effect on aflatoxin accumulation resistance

A combined analysis of aflatoxin accumulation resistance among genotypes evaluated across the three locations in two seasons was significant ($p < 0.05$) and variable (**Table 7**). The general observation showed a relatively high aflatoxin accumulation among genotypes evaluated in Wenchi (transitional savanna zone) whilst those evaluated across Akomadan (forest transitional zone) and Fumesua (rain forest zone) recorded relatively low amount of the toxin. Aflatoxin levels

| Sources of variation | DF | Grain Yield (kg/Ha) | Days to 50% pollen | Days to 50% Silking | Anthesis Silking interval (Days) | Ear Per Plant | Cob Aspect (1–5) | Plant Aspect (1–5) | Open-Tip (1–5) | Stay-green |
|---|---|---|---|---|---|---|---|---|---|---|
| GENOTYPE | 35 | 27259961*** | 172.76*** | 203.57*** | 3.81*** | 0.93*** | 1.32*** | 1.38*** | 0.95*** | 348.37*** |
| ENV*GENOTYPE | 175 | 4510728*** | 40.23 ns | 45.85 ns | 1.28*** | 0.38* | 0.78** | 0.98** | 0.52*** | 99.54** |
| REP(ENV) | 12 | 4284472*** | 82.21* | 90.67* | 0.66 ns | 0.49 ns | 0.47 ns | 0.75 ns | 0.95*** | 346.31*** |
| BLOCK(ENV*REP) | 90 | 1072299 ns | 44.77 ns | 49.85 ns | 1.04* | 0.26 ns | 0.64 ns | 0.84* | 0.3*** | 103.92** |
| POOLED ERROR | 326 | 1117600 | 43.35 | 46.95 | 0.75 | 0.29 | 0.51 | 0.58 | 0.25 | 69.39 |
| **$H^2$** | | **62.70** | **28.40** | **30.40** | **28.40** | **20.20** | **25.20** | **23.40** | **18.90** | **30.60** |

*, **, *** *Significant at 0.05, 0.01 and 0.001 probability levels, respectively, and ns: not significant.*
*$H^2$ = Broad-sense heritability*

**Table 2.**
*A combined mean squares of grain yield and agronomic traits among 36 maize genotypes evaluated across six environments.*

| Source of variation | DF | Aflatoxin | Insect | Rust | Blight | Streak | Ear rot | Plant height | Ear height | Root | Stalk |
|---|---|---|---|---|---|---|---|---|---|---|---|
| | | ln (y + 1) | Damage (1–5) | (1–5) | (1–5) | (1–5) | (1–5) | (cm) | (cm) | Lodging (1–5) | Lodging (1–5) |
| ENV | 5 | $17.73^{***}$ | $8.03^{***}$ | $1.82^{***}$ | $9.62^{***}$ | $43.76^{***}$ | $259.46^{***}$ | $80382.63^{***}$ | $21788.88^{***}$ | $27.77^{***}$ | $35.01^{***}$ |
| GENOTYPE | 35 | $4.43^{***}$ | $0.89^{***}$ | $0.32^{**}$ | $0.28^{**}$ | $1.95^{***}$ | $28.37^{***}$ | $6003.46^{***}$ | $2090.11^{***}$ | $0.34^{***}$ | $1.98^{***}$ |
| ENV*GENOTYPE | 175 | $1.06^{***}$ | $0.36^{**}$ | $0.31^{***}$ | $0.24^{***}$ | 0.63 ns | $81.89^{**}$ | $932.37^{***}$ | $369.87^{***}$ | $0.20^{***}$ | $1.64^{**}$ |
| REP(ENV) | 12 | 0.34 ns | 0.18 ns | 0.25 ns | $0.65^{***}$ | $0.98^{*}$ | 6.00 ns | $3405.97^{***}$ | $1366.06^{***}$ | $0.24^{*}$ | $1.99^{**}$ |
| BLOCK(ENV*REP) | 90 | 0.69 ns | 0.20 ns | 0.34 ns | $0.25^{*}$ | $0.81^{**}$ | 31.27 ns | $889.42^{***}$ | $350.16^{***}$ | $0.18^{*}$ | $1.36^{**}$ |
| Error | 326 | 0.52 | 0.24 | 0.18 | 0.16 | 0.52 | 0.33 | 320.92 | 135.07 | 0.13 | 0.92 |
| **$H^2$** | | **61.70** | **21.80** | **9.70** | **10.00** | **24.40** | **14.80** | **78.00** | **67.30** | **15.20** | **11.80** |

**, **, *** Significant at 0.05, 0.01 and 0.001 probability levels, respectively, and ns: not significant.*
*$H^2$ = Broad-sense heritability*

**Table 3.**
*A combined mean squares for aflatoxin accumulation, disease and agronomic traits among 36 maize genotypes evaluated across six environments.*

| **Genotype** | **Aflatoxin levels** | |
|---|---|---|
| | **ln (y + 1)** | **Geometric means (ppb)** |
| ABROHENEMAA | 2.73 | 24.60 |
| AHODZEN | 3.03 | 20.95 |
| CML 108 | 3.70 | 41.62 |
| CML11 | 3.05 | 25.83 |
| CML158 | 2.79 | 20.16 |
| CML176 | 3.29 | 29.65 |
| CML247 | 3.28 | 47.33 |
| CML287 | 2.87 | 20.17 |
| CML322 | 3.05 | 25.59 |
| CML343 | 2.99 | 24.19 |
| CML5 | 3.35 | 28.92 |
| ENTRY-5 | 3.20 | 34.86 |
| ENTRY6 | 3.32 | 31.80 |
| ENTRY-70 | 2.82 | 21.02 |
| ENTRY-85 | 2.93 | 23.37 |
| GH-110 | 4.04 | 56.83 |
| Hi27 | 3.37 | 34.21 |
| HONAMPA (Check) | 4.93 | 140.60 |
| Ki3 | 3.19 | 31.60 |
| M0826-12F | 3.42 | 46.12 |
| M0826-7F | 3.04 | 23.58 |
| MP705 | 2.13 | 14.85 |
| MP715 | 2.49 | 15.89 |
| MP719 | 2.56 | 16.86 |
| NC298 | 3.49 | 34.59 |
| NC334 | 3.26 | 29.94 |
| NC340 | 3.36 | 28.69 |
| NC356 | 2.78 | 19.49 |
| OBAATANPA | 3.82 | 45.45 |
| OMANKWA | 4.01 | 55.30 |
| TINTIM | 3.28 | 26.95 |
| TZEEI- 24 | 2.85 | 21.93 |
| TZEEI- 4 | 3.32 | 28.43 |
| TZEEI- 6 | 3.13 | 24.92 |
| TZEEI-15 | 3.49 | 32.64 |
| TZI8 | 3.25 | 25.87 |
| **MIN** | 2.13 | 14.85 |
| **MAX** | 4.92 | 140.60 |
| **SED** | 1.05 | 2.86 |

**Table 4.**
*Mean aflatoxin accumulation levels among 36 genotypes across six environments.*

| Genotype | Grain yield (kg/ha) | Days to 50% Pollen | Days to 50% silking | Anthesis silking interval | Ear Per Plant | Cob Aspect (1–5) | Plant Aspect (1–5) | Open-Tip (1–5) | Insect Damage (1–5) |
|---|---|---|---|---|---|---|---|---|---|
| AHODZEN | 4162.52 | 54 | 56 | 2 | 0.94 | 1.75 | 1.71 | 1.34 | 1.22 |
| CML11 | 951.41 | 54 | 57 | 3 | 0.67 | 2.27 | 2.30 | 0.93 | 1.22 |
| CML158 | 1022.67 | 57 | 60 | 3 | 1.26 | 2.11 | 1.81 | 1.22 | 1.17 |
| CML176 | 944.15 | 58 | 62 | 4 | 0.68 | 2.67 | 2.33 | 1.06 | 1.00 |
| CML247 | 1452.77 | 58 | 61 | 3 | 0.84 | 2.09 | 2.38 | 1.67 | 1.28 |
| CML287 | 1024.73 | 58 | 61 | 3 | 0.60 | 2.51 | 2.07 | 1.68 | 1.11 |
| CML322 | 1064.86 | 55 | 59 | 4 | 1.02 | 2.56 | 2.01 | 1.55 | 1.06 |
| CML343 | 1445.98 | 59 | 62 | 3 | 1.28 | 1.95 | 1.82 | 1.28 | 1.00 |
| CML5 | 1082.19 | 56 | 59 | 3 | 1.07 | 2.08 | 2.10 | 1.27 | 1.44 |
| CML108 | 1930.27 | 54 | 56 | 2 | 1.05 | 1.68 | 1.61 | 1.63 | 1.33 |
| ENTRY-5 | 1475.85 | 53 | 55 | 2 | 0.78 | 2.17 | 2.09 | 1.45 | 1.17 |
| ENTRY-70 | 1313.41 | 54 | 57 | 3 | 0.90 | 1.90 | 2.14 | 1.58 | 1.28 |
| ENTRY-85 | 1794.45 | 49 | 52 | 3 | 1.03 | 1.79 | 2.08 | 1.72 | 1.39 |
| ENTRY- 6 | 1409.59 | 54 | 57 | 3 | 0.94 | 2.06 | 2.05 | 1.26 | 0.90 |
| GH-110 | 4025.51 | 54 | 56 | 2 | 1.00 | 1.67 | 1.65 | 1.28 | 1.11 |
| Hi27 | 1150.69 | 54 | 57 | 3 | 0.87 | 2.33 | 2.09 | 1.72 | 1.39 |
| HONAMPA (Check) | 3577.67 | 52 | 55 | 3 | 0.99 | 1.51 | 1.58 | 1.67 | 1.28 |
| Ki3 | 1075.05 | 53 | 56 | 3 | 1.02 | 2.25 | 1.78 | 1.68 | 1.50 |
| M0826-12F | 4772.76 | 53 | 56 | 3 | 0.94 | 1.76 | 1.60 | 1.61 | 1.50 |
| M0826-7F | 4448.18 | 52 | 54 | 2 | 0.91 | 2.10 | 1.94 | 1.73 | 1.56 |
| MP705 | 1130.63 | 51 | 55 | 4 | 0.55 | 2.37 | 2.93 | 1.05 | 1.11 |
| MP715 | 565.63 | 62 | 65 | 3 | 0.76 | 2.60 | 2.56 | 1.39 | 1.33 |

| Genotype | Grain yield (kg/ha) | Days to 50% Pollen | Days to 50% silking | Anthesis silking interval | Ear Per Plant | Cob Aspect (1–5) | Plant Aspect (1–5) | Open-Tip (1–5) | Insect Damage (1–5) |
|---|---|---|---|---|---|---|---|---|---|
| MP719 | 1223.68 | 57 | 61 | 4 | 0.68 | 2.31 | 2.16 | 1.55 | 1.22 |
| NC298 | 690.59 | 52 | 54 | 2 | 1.62 | 2.02 | 2.42 | 1.09 | 0.89 |
| NC340 | 1323.46 | 52 | 54 | 2 | 0.96 | 1.95 | 1.64 | 1.60 | 1.67 |
| NC356 | 1341.91 | 52 | 55 | 3 | 0.82 | 2.12 | 2.16 | 0.96 | 0.94 |
| NC334 | 718.99 | 52 | 55 | 3 | 1.95 | 1.77 | 2.03 | 0.95 | 0.89 |
| OBAATANPA | 2751.56 | 56 | 58 | 2 | 0.96 | 1.75 | 1.91 | 1.33 | 1.11 |
| OMANKWA | 4688.35 | 48 | 50 | 2 | 1.00 | 1.89 | 1.93 | 1.68 | 1.67 |
| TINTIM | 2624.39 | 52 | 54 | 2 | 1.41 | 1.61 | 1.73 | 1.50 | 1.21 |
| TZEEI-15 | 1735.97 | 49 | 52 | 3 | 0.92 | 1.99 | 1.90 | 1.49 | 1.22 |
| TZEEI- 24 | 1062.38 | 50 | 53 | 3 | 1.78 | 0.89 | 1.35 | 2.32 | 0.89 |
| TZEEI-4 | 1382.63 | 48 | 51 | 3 | 1.15 | 0.93 | 1.56 | 2.60 | 1.28 |
| TZEEI- 6 | 906.995 | 51 | 53 | 2 | 0.97 | 2.16 | 1.94 | 1.57 | 1.22 |
| TZI8 | 855.43 | 55 | 58 | 3 | 0.84 | 2.47 | 1.84 | 1.62 | 1.56 |
| MEAN | 1853.22 | 54 | 57 | 3 | 1.00 | 2.01 | 1.97 | 1.49 | 1.24 |
| MIN | 565.63 | 48 | 50 | 2 | 0.55 | 0.89 | 1.34 | 0.94 | 0.88 |
| MAX | 4772.76 | 62 | 65 | 4 | 1.95 | 2.67 | 2.92 | 2.60 | 1.67 |
| SED | 909.22 | 2.03 | 2.01 | 0.30 | 0.18 | 0.39 | 0.35 | 0.22 | 0.18 |

**Table 5.**
*Grain yield and agronomic performance of 36 genotypes across six environments.*

| Genotype | Stay-Green | RUST Incidence | BLIGHT Incidence | MSVD | Ear rot | Plant Height | Ear Height |
|---|---|---|---|---|---|---|---|
| | | (1–5) | (1–5) | (1–5) | (1–5) | (cm) | (cm) |
| ABROHENEMAA | 33.02 | 1.26 | 1.21 | 2.07 | 1.22 | 162.37 | 82.88 |
| AHODZEN | 35.32 | 1.18 | 1.09 | 1.43 | 1.06 | 169.90 | 92.22 |
| CML11 | 20.77 | 1.04 | 1.03 | 2.34 | 1.06 | 111.76 | 57.27 |
| CML158 | 25.85 | 1.29 | 1.42 | 1.90 | 1.11 | 124.12 | 65.27 |
| CML176 | 31.90 | 1.53 | 1.43 | 2.61 | 1.50 | 136.93 | 60.45 |
| CML247 | 25.02 | 1.16 | 1.16 | 2.43 | 1.33 | 127.34 | 63.52 |
| CML287 | 28.31 | 1.10 | 1.13 | 2.19 | 1.22 | 126.15 | 65.64 |
| CML322 | 31.28 | 1.08 | 1.42 | 2.28 | 1.50 | 113.03 | 56.65 |
| CML343 | 28.19 | 1.14 | 1.44 | 1.92 | 0.94 | 133.30 | 60.77 |
| CML5 | 26.05 | 0.99 | 1.10 | 1.72 | 1.33 | 118.77 | 67.20 |
| CML108 | 35.06 | 1.12 | 0.99 | 1.58 | 1.17 | 142.35 | 75.60 |
| ENTRY-5 | 25.50 | 1.12 | 1.16 | 1.82 | 1.06 | 127.47 | 67.17 |
| ENTRY-70 | 36.31 | 1.51 | 1.36 | 2.32 | 1.06 | 139.52 | 66.30 |
| ENTRY-85 | 31.37 | 1.22 | 1.16 | 2.09 | 1.22 | 145.64 | 78.10 |
| ENTRY-6 | 31.46 | 1.38 | 1.31 | 2.79 | 1.21 | 133.66 | 68.08 |
| GH-110 | 34.78 | 1.39 | 1.18 | 2.78 | 0.94 | 152.22 | 78.67 |
| HI27 | 30.51 | 1.38 | 1.37 | 2.02 | 1.89 | 137.04 | 72.24 |
| HONAMPA (Check) | 42.58 | 1.34 | 1.23 | 1.87 | 1.11 | 157.23 | 88.38 |
| Ki3 | 32.90 | 1.31 | 1.26 | 2.61 | 1.44 | 140.08 | 73.96 |
| M0826-12F | 28.77 | 1.29 | 1.40 | 1.94 | 1.06 | 165.72 | 86.19 |
| M0826-7F | 32.67 | 1.19 | 1.19 | 1.76 | 1.06 | 153.81 | 81.38 |

| Genotype | Stay-Green | RUST Incidence | BLIGHT Incidence | MSVD | Ear rot | Plant Height | Ear Height |
|---|---|---|---|---|---|---|---|
| MP705 | 27.56 | 1.22 | 1.50 | 1.66 | 1.78 | 104.46 | 52.24 |
| MP715 | 23.42 | 1.23 | 1.32 | 2.68 | 1.06 | 117.52 | 67.66 |
| MP719 | 28.36 | 1.36 | 1.59 | 2.36 | 1.33 | 143.20 | 77.84 |
| NC298 | 42.04 | 0.99 | 1.38 | 1.77 | 1.17 | 113.90 | 56.48 |
| NC340 | 35.60 | 1.56 | 1.31 | 1.73 | 1.39 | 167.62 | 84.26 |
| NC356 | 22.44 | 1.03 | 1.21 | 2.13 | 1.22 | 108.73 | 53.81 |
| NC334 | 23.93 | 1.08 | 1.27 | 1.95 | 0.72 | 100.73 | 53.09 |
| OBAATANPA | 29.11 | 1.20 | 1.02 | 1.96 | 0.89 | 176.25 | 92.52 |
| OMANKWA | 30.82 | 1.27 | 1.15 | 1.88 | 1.28 | 149.15 | 79.97 |
| TINTIM | 34.70 | 1.49 | 1.13 | 1.78 | 1.13 | 164.12 | 85.77 |
| TZEEI-15 | 31.02 | 1.23 | 1.26 | 1.96 | 1.39 | 140.47 | 70.29 |
| TZEEI- 24 | 33.73 | 3.37 | 1.10 | 1.11 | 1.69 | 137.70 | 64.63 |
| TZEEI- 4 | 30.73 | 3.07 | 1.38 | 1.23 | 1.77 | 160.21 | 85.91 |
| TZEEI- 6 | 31.43 | 1.18 | 1.33 | 1.59 | 1.22 | 155.36 | 81.96 |
| TZI8 | 30.88 | 1.16 | 1.13 | 1.61 | 1.33 | 129.41 | 64.03 |
| MEAN | 30.65 | 1.35 | 1.25 | 2.00 | 1.25 | 138.53 | 71.62 |
| MIN | 20.77 | 0.98 | 0.99 | 1.11 | 0.72 | 100.73 | 52.24 |
| MAX | 42.6 | 1.56 | 1.59 | 2.80 | 1.88 | 176.25 | 92.5 |
| SED | 5.06 | 0.14 | 0.17 | 0.32 | 0.32 | 13.52 | 7.62 |

**Table 6.**
*Means of 36 genotypes for disease and agronomic traits across six environments.*

| | Wenchi | | | Fumesua | | | Akomadan | |
|---|---|---|---|---|---|---|---|---|
| Genotype | Aflatoxin Ln (y + 1) | Geometric Means(ppb) | Genotype | Aflatoxin Ln (y + 1) | Geometric Means(ppb) | Genotype | Aflatoxin Ln (y + 1) | Geometric means(ppb) |
| Resistant | | | Resistant | | | Resistant | | |
| MP715 | (3.08) | 21.79 | MP715 | (2.68) | 14.70 | NC298 | (2.64) | 14.02 |
| MP719 | (3.09) | 22.13 | MP705 | (2.72) | 15.20 | MP715 | (2.67) | 14.51 |
| MP705 | (3.12) | 22.71 | MP719 | (2.77) | 15.90 | MP705 | (2.68) | 14.73 |
| CML287 | (3.17) | 23.94 | CML287 | (2.86) | 17.40 | MP719 | (2.68) | 14.69 |
| CML158 | (3.27) | 26.39 | CML158 | (2.96) | 19.30 | CML287 | (2.77) | 15.95 |
| TZEEI-24 | (3.35) | 28.54 | TZEEI-24 | (3.07) | 21.70 | NC356 | (3.13) | 22.94 |
| TZEEI-4 | (3.42) | 30.78 | TZEEI-4 | (3.13) | 22.90 | TZEEI-24 | (3.04) | 21.09 |
| NC298 | (3.42) | 30.82 | ENTRY-85 | (3.16) | 23.70 | ENTRY-85 | (3.20) | 24.67 |
| ENTRY-85 | (3.43) | 31.05 | NC356 | (3.20) | 24.70 | CML158 | (2.97) | 19.47 |
| NC356 | (3.46) | 32.11 | CML322 | (3.25) | 25.90 | TZEEI-4 | (3.12) | 22.86 |
| CML343 | (3.53) | 34.25 | ENTRY-70 | (3.30) | 27.20 | CML343 | (3.24) | 25.63 |
| ENTRY-70 | (3.54) | 34.34 | CML5 | (3.31) | 27.50 | CML322 | (3.21) | 24.78 |
| CML322 | (3.54) | 34.43 | NC298 | (3.32) | 27.60 | CML5 | (3.30) | 27.23 |
| CML5 | (3.61) | 37.24 | CML343 | (3.36) | 28.90 | ENTRY-70 | (3.30) | 27.14 |
| CML11 | (3.67) | 39.41 | TZEEI-6 | (3.49) | 32.80 | CML11 | (3.44) | 31.28 |
| Worst | | | Worst | | | Worst | | |
| CML108 | (4.58) | 97.66 | M0826-12F | (4.48) | 89.00 | GH-110 | (4.52) | 92.03 |
| GH-110 | (4.67) | 107.29 | GH-110 | (4.52) | 92.00 | CML247 | (4.60) | 100.46 |
| CML247 | (4.67) | 107.41 | TZEEI-15 | (4.68) | 108.40 | CML108 | (4.65) | 105.18 |
| TZEEI-15 | (4.82) | 124.15 | CML247 | (4.89) | 133.80 | TZEEI-15 | (4.66) | 106.09 |

| | Wenchi | | | Fumesua | | | Akomadan | |
|---|---|---|---|---|---|---|---|---|
| **Genotype** | **Aflatoxin Ln (y + 1)** | **Geometric Means(ppb)** | **Genotype** | **Aflatoxin Ln (y + 1)** | **Geometric Means(ppb)** | **Genotype** | **Aflatoxin Ln (y + 1)** | **Geometric means(ppb)** |
| HONAMPA | (5.25) | 190.03 | HONAMPA | (5.19) | 179.70 | HONAMPA | (5.26) | 193.13 |
| Min | 3.08 | | | 2.68 | | | 2.64 | |
| Max | 5.25 | | | 5.19 | | | 5.26 | |
| CV% | 21.7 | | | 22.6 | | | 24.3 | |
| LSD(0.05) | 1.42 | | | 1.40 | | | 1.57 | |

*CV = Co-efficient of variation, Digits in parenthesis are transformed means.*

**Table 7.**
*Top 15 aflatoxin resistant and worst genotypes across six environments in three locations.*

vin Wenchi ranged between 21.80 ppb for MP715 to 190 ppb for HONAMPA. In Fumesua, the accumulation level ranged between 14.70 ppb for MP715 to 179.7 ppb for HONAMPA whilst in Akomadan, it ranged between 14.51 ppb to 193.13 ppb for same genotypes.

MP715 appeared to be the most stable and resistant line across the six environments within the three locations whilst HONAMPA consistently performed poorly as the worse or most susceptible genotype (**Table 7**). The ranking order of resistance in terms of aflatoxin accumulation varied among genotypes from one location to the other. However, some particular genotypes consistently appeared in the top ten resistant genotypes irrespective of location.

### 3.3 Genetic correlation among selected traits

Significant and positive correlations were also observed between aflatoxin accumulation and traits such as cob and plant aspects and insect damage while stay-green, open-tip and ears per plant were not significantly correlated (**Table 8**).

| | Open Tip | Cob Aspect | Stay -Green | Plant_ Aspect | Insect Damage | Ear Per Plant |
|---|---|---|---|---|---|---|
| Aflatoxin ln (y + 1) | | | | | | |
| (r) | 0.02 | 0.24 | 0.07 | 0.47 | 0.30 | - 0.04 |
| p (0.05) | ns | 0.01 | ns | 0.003 | 0.05 | ns |

*r = Correlation co-efficient, P = Probability.*

**Table 8.**
*Genetic correlation between aflatoxin accumulation and selected agronomic traits.*

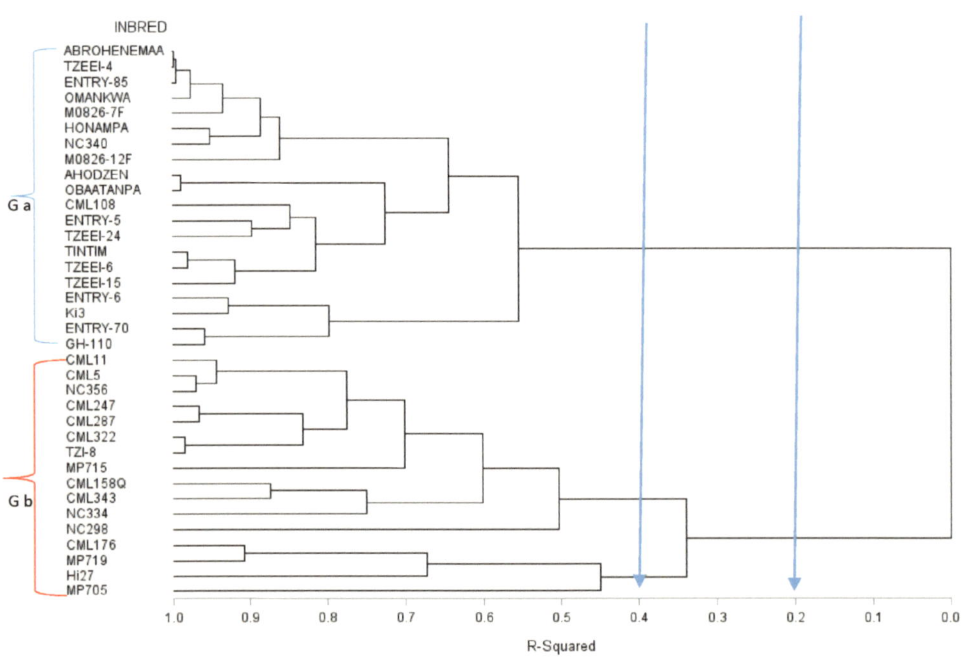


**Figure 1.**
*Dendrogram based on agro-morphological traits showing relationships among 36 maize genotypes using the neighbor joining procedure.*

### 3.4 Heterotic grouping based on multiple agro-morphological traits

Significant agro-morphological traits assigned genotypes into respective heterotic groups. Three groups were observed at 40% co-efficient of determination but at 20%, two major clusters (Ga and Gb) were revealed. Group Ga was made up of 20 members which were mostly local genotypes except Ki3, CML108 and NC340. All the exotic lines (16) clustered in group Gb. (**Figure 1**).

## 4. Discussion

High genetic variability observed for aflatoxin resistance accumulation was an indication of the presence of novel or favorable alleles for population improvement. Furtherance to this, genotypes identified with reduced aflatoxin accumulation could be exploited in the development of superior hybrids that combine resistance to aflatoxin accumulation and high yields as described previously by Warburton and Williams [6]. The observed significant phenotypic variation for the major, minor and across seasons and locations among the genotypes for aflatoxin accumulation reduction and other agronomic traits suggested that progress could be made in developing well adopted lines with good aflatoxin accumulation resistance.

The significant environmental and genotypic effects detected for aflatoxin accumulation resistance and other agronomic traits indicated variability among the genotypes under different environments. Also the significant genotype x environment interactions observed across the seasons and locations indicates the need for evaluation of genotypes across several environments in order to determine most stable genotypes for aflatoxin accumulation resistance and other agronomic traits. According to Comstock and Moll [15], genotype x environment interactions determined in multi-location trials implied reduced correlation between genotypic and phenotypic values. Zuber [16] identified significant environmental effects on aflatoxin accumulation among commercial hybrids and OPVs in the United States of America. It was not uncommon to observe G x E effects on the genotypes evaluated across contrasting environments in Ghana.

Broad sense heritability among traits ranged from moderate to very high estimates during the major season where adequate rains and less disease pressure was observed. However, significantly lower range of estimates were detected during the minor season (data not shown) confirming earlier reports by [17] who demonstrated significant environmental influence on heritability estimates in cashew. Moderate to high heritability estimates realized in this study, suggested that possible gains are achievable in hybrid maize development for high yields and aflatoxin accumulation resistance.

Further evaluation of the agronomic traits revealed a range of genotypic influence on several parameters studied. Significant among them were days to 50% pollen and silking which clearly categorized the genotypes into the three well defined and established classes of extra early, intermediate and late types as reported by Badu- Apraku [18] who categorized maize genotypes into the different maturity groups. Such information is critical and necessary to guide planting periods in breeding nurseries designed to cross among the maturity groups for trait introgression and further improvement for hybrid development.

The combined analysis showed that most genotypes belonged to the intermediate group while a few were extra early or late maturing. For instance, genotype MP715 has been reported in previous studies [19] as highly resistant to aflatoxin accumulation with delayed silking ranging between 70 and 80 days when grown in a temperate environment, however in this study it ranged between 62 and 67 days under tropical conditions in Ghana, where it is evidently adapted. On the other hand, genotypes identified with delayed silking dates included MP719, TZI8, NC334, OBAATANPA, CML176, CML247, CML287, CML343 and CML5.

Aflatoxin accumulation levels during the major season were comparatively lower across environments although levels in Wenchi were slightly higher. Unlike the major season, the accumulation levels in the minor season were considerably higher across environments with Wenchi still ranking highest. This observation agrees with the findings of several authors [2, 6, 20, 21] who reported the existence of positive correlation of drought and heat on aflatoxin accumulation level. The Guinea savanna transition environment, which appears to be relatively drier, was conducive for aflatoxin production.

The ranking of top ten resistant genotypes across environments revealed a consistent set of genotypes (although in different ranking order of resistance per environment) during both major and minor seasons. Genotypes which consistently displayed stable resistance across the environments included MP715, NC298, MP705, MP719, CML287 and TZEEI- 24 while the rest appeared less stable. Two local extra early lines (TZEEI-24 and TZEEI-4) were identified as sources of potential resistance to aflatoxin but their level of resistance was not as good as the Mississippi lines specifically bred for resistance and that further evaluations of these two locally adapted lines may be required to confirm their levels of resistance to aflatoxin accumulation.

It was also obvious from this study that, majority of the inbred lines outperformed the OPVs and the populations' in-terms of reduced aflatoxin accumulation levels which was in agreement with the previous findings of Zuber [16] who reported superiority of hybrids (inbred combinations) over OPVs in-terms of measured aflatoxin accumulation resistance across locations and years in the United States.

The levels of total aflatoxin accumulated by the resistant genotypes in this study are comparable to the levels previously reported by William and Windham [5] and Brown [2] where a set of hybrids were evaluated for aflatoxin accumulation resistance. Information obtained from the total aflatoxin accumulation levels among the genotypes could therefore guide the selection of appropriate parental candidates for future aflatoxin resistance breeding in Ghana. Breeding for resistance involves several approaches of which trait correlations is paramount. Indirect selection of one trait simultaneously improves other traits that are significantly correlated.

Zuber [16] discovered strong significant correlation between insect damage and aflatoxin accumulation during evaluation of OPVs and released commercial hybrids in the United States of America. The correlation observed between insect damage and aflatoxin accumulation in this study was similar to that of Zuber [16] as well as the observations made by Williams [22] and Ni [23]. Significant positive correlations were also observed between aflatoxin accumulation levels and plant aspect and cob aspect whilst ears per plant showed a weak negative correlation. Stay-green and open- tip did not correlate with aflatoxin accumulation levels as recounted in other studies [24, 25].

The study of genetic relationship among genotypes which was based on significant agronomic traits assigned all genotypes into three main groups when 40% of the variation among the genotypes was explained. On the contrary, only two main groups were realized when 20% of the genotypic variation was explained by the significant traits used for the grouping. In the case of the two groups, all exotic genotypes were assigned into one group except Ki3, CML108 and NC340 which clustered together with the local genotypes. This observation was not surprising since most of the local genotypes were sourced from CIMMYT and may have similar pedigree or ancestry records.

It appears that the top resistant genotypes which also clustered in one common group perhaps originated from a common ancestry of Tuxpe˜no germplasm native to Mexico which exhibits tropical characteristics coupled with aflatoxin accumulation resistance [6].

Although the analysis of the multiple phenotypic traits assigned genotypes into distinct groups, it showed a low corroboration when compared to other molecular methods (data not shown). This is probably because the expression of most agronomic traits are influenced by the environment.

## 5. Conclusions

Six most stable aflatoxin accumulation resistant genotypes across six environments have been identified. They included MP715, NC298, MP705, MP719, CML287 and TZEEI- 24. Furthermore, it was evident from the study that traits such as cob and plant aspect correlated significantly and positively with aflatoxin accumulation levels whilst grain yield had significant negative correlation.

Genotypic effects on several traits were consistently significant across environments and that the environments used in this study were discriminatory enough to aid the identification and selection of consistent genotypes for aflatoxin accumulation resistance. Significant genotype by environment interaction aided in the identification of relatively stable genotypes for specific important agronomic and aflatoxin accumulation resistant traits. The Wards clustering method assigned genotypes into two main groups (exotic and local) based on the significant agro-morphological traits including grain yield. Broad sense heritability estimates for grain yield and aflatoxin accumulation resistance were moderately high to enable permissible transfer of traits to progeny.

## Acknowledgements

This research was funded by USAID through the National Academies of Sciences' (NAS) programme for Partnership for enhanced engagement in Research (PEER). Grant No. Project 4-040, 2015 for "the development of high yielding aflatoxin resistant maize for improved nutrition and health in Ghana", the West Africa Centre for Crop Improvement for the PhD training in Plant Breeding. We would also like to thank all the staff of Plant health and Maize Breeding Division of CSIR –Crops Research Institute Kumasi, Ghana for their support. The help of Messers Akyeampong, and Josuah Obeng and William Appaw is deeply appreciated. Special thanks to Mr. Richard Yeboah for assisting with the data collection.

## Conflict of interest

Authors wish to state that, there is no conflict of interest in relation to the writing of this manuscript.

## Author details

Abu Mustapha Dadzie[1*], Allen Oppong[2], Ebenezer Obeng-Bio[1] and Marilyn L. Warburton[3]

1 Cocoa Research Institute of Ghana, New Tafo Akim, Ghana

2 CSIR-Crops Research Institute' Kumasi, Ghana

3 USDA ARS CHPRRU, Mississippi State, MS, USA

*Address all correspondence to: xmusto@yahoo.co.uk

## References

[1] Badu-Apraku B, Adewale S, Paterne A, Gedil M, Asiedu E. Identification of QTLs controlling resistance/tolerance to striga hermonthicain an extra-early maturing yellow maize population. Agronomy. 2020; doi:10.3390/agronomy1008116.

[2] Brown RL, Williams WP, Windham GL, Menkir A, Chen ZY. Evaluation of African-Bred Maize Germplasm Lines for Resistance to Aflatoxin Accumulation. Agronomy. 2016; 6, 24; doi:10.3390 / 6020024.

[3] Dadzie MA, Oppong A, Ofori K, Blay E, Eleblu J, Beatrice E, Obeng –Bio E, Appiah-Kubi Z, Warburton M.L Distribution of Aspergillus flavus and aflatoxin accumulation in stored maize grains across three agro-ecologies in Ghana. *Food Control*. 2019; https://doi.org/10.1016/j.foodcont.2019.04.035

[4] Šošo VM, Marija MŠ, Nevena TB, Slavica MVM. Identification of aflatoxigenic fungi using polymerase chain reaction-based assay. BIBLID: 2014; 1450-7188, 45, 259-269

[5] Williams WP, Windham GL. Aflatoxin accumulation in a maize diallel cross. *Agriculture*. 2015; 5:344-52.

[6] Warburton ML, Williams WP. Aflatoxin Resistance in Maize: What Have We Learned Lately? *Advances in Botany:* 2014; http://dx.doi.org/10.1155/2014/352831.

[7] Dadzie MA, Oppong A, Ofori K, Eleblu JSY, Beatrice IE, Blay ET, Obeng –Bio E, Appiah-Kubi Z, Warburton ML Distribution and Genetic diversity among Aspergillus flavus isolates across three agro-ecologies essential for maize cultivation in Ghana. Plant Pathology. 2019; 68: 1565-1576.

[8] Bediako KA, Dzidzienyo D, Ofori K, Offei S K, Asibuo JY, Adu Amoah R, Obeng J. Prevalence of Fungi and Aflatoxin Contamination in Stored Groundnut in Ghana. Food control. 2019; doi: 10.1016/j.foodcont.2019.04.034.

[9] Jha DK. Laboratory Manual on Seed Pathology. Vikas Publishing House (PVT) Ltd; 1995. P.13-30.

[10] Windham, G.L., W.P. Williams, P.M. Buckley, & Abbas, H.K. Inoculation techniques used to quantify aflatoxin resistance in corn. Journal of Toxicology. 2003; 22: 313-325.

[11] Scott GE, Zummo N. Kernel infection and aflatoxin production in maize by Aspergillus flavus relative to inoculation and harvest dates. Plant Disease. 1994; 78 (2), 123-125.

[12] Sirhan AY, Tan GH, Wong RCS. Determination of aflatoxins in food using liquid chromatography coupled with electrospray ionization quadrupole time of flight mass spectrometry (LC-ESI-QTOF-MS/MS). Food Control. 2013; 31, 35-4.

[13] SAS Institute, Inc. Statistical analysis software (SAS) user's guide. 2012; SAS Institute, Inc., Cary, NC.

[14] Cochran WG, Cox GM, Experimental designs. John Wiley and Sons; 1960. New York, USA.

[15] Comstock RE, Moll RH, Genotype-environment interactions. 1963; pp. 164-194. In Statistical Genetics in Plant Breeding, Edited by Hanson WD, and Robinson HF, NAS-NRC Publ. 982, Washington, D.C.

[16] Zuber MS, Darrah L, Lillehojetal EB. "Comparison of open pollinated maize varieties and hybrids for preharvest aflatoxin contamination in the southern United States. *Plant Disease*. 1983; 67, 185-187.

[17] Dadzie AM, Adu-Gyamfi PKK, Akpertey A, Ofori A, Opoku SY, Yeboah J, Akoto EG, Padi FK, Obeng-Bio E Assessment of juvenile growth and yield relationship among dwarf cashew types in Ghana. Journal of Agricultural Science. 2020; https://doi.org/10.5539/

[18] Badu-Apraku B, Lum AF. The pattern of grain yield response of normal and quality protein maize cultivars in stress and nonstress environments. *Agronomy Journal.* 2010; 102, 2, 381-394.

[19] Williams WP, Windham GL. "Registration of Mp718 and Mp719 germplasm lines of maize,"*Journal of Plant Registrations.* 2012; 6 (2), 200-202.

[20] Windham GL, & Williams WP, Evaluation of corn inbreds advanced breeding lines for resistance to aflatoxin contamination in the field. *Plant Disease*. 2002; 86: 232-234. 10.1094.

[21] Henry WB, Windham GL, Rowe DE, Blanco MH, Murray SC, & Williams WP. Diallel analysis of diverse maize germplasm lines for resistance to aflatoxin accumulation. Crop Science. 2013; 53:394-402.

[22] Williams WP, Windham GL, Buckley PM, Perkins JM. "Southwestern corn borer damage and aflatoxin accumulation in conventional and transgenic corn hybrids," *Field Crops Research*. 2005; vol. 91, no. 2-3, pp. 329-336.

[23] Ni X, Wilson JP, Buntin GD, Guo B, Krakowsky MD, Lee RD. Spatial patterns of aflatoxin levels in relation to ear-feeding insect damage in pre-harvest corn. Toxins (Basel). 2011; 3:920-931. doi:10.3390/toxins3070920.

[24] Tubajika KM, Damann KE. Sources of Resistance to Aflatoxin Production in maize. *Food Chemistry*. 2000; 49: 2652-2656.

[25] Brown RL, Bhatnagar D, Cleveland TE, Chen ZY, Menkir A. Development of Maize Host Resistance to Aflatoxigenic Fungi. INTECH. 2013; http://dx.doi.org/10.5772/54654.

Chapter 12

# Microbiological Control: A New Age of Maize Production

*Damilola Omobowale Seyi-Amole and Abiodun A. Onilude*


## Abstract

Maize is one of the world's most widely grown and consumed cereal. It is known for its multipurpose use; it provides food and fuel to humans, feeds to animals and used as raw material in manufacturing industries. Globally, maize production is a large and significant market which produced 1,116.41 million tons in year 2020 and it's expected to increase by 1.57% in year 2021. Pests and disease of maize cause significant damage to maize thereby reducing its's yield and quality. There are many methods of controlling maize disease and pests; they include cultural, biological and chemical methods etc. Recent research studies have discovered an alternative agricultural practices that are sustainable and safe as compared to chemical control of pests and disease. However, biological control has gained large acceptance and its believed to yield positive outcome as compared to chemical control. Various microorganisms are used to control pathogens of maize and thus, there is a need to understand better their interactions with plants. Furthermore, microorganism known as entomopathogens are used to control arthropods. They are biopesticides that play integral role in Pest Management. This section focuses on microbiological control of pathogens and arthropods, their mechanisms of action, applications and the future of entomopathogenic microorganisms and microbiological control of pathogens.


**Keywords:** maize, pathogens, pests, microbiological control, entomopathogens

## 1. Introduction

Corn, also referred to as Maize, *Zea mays*, is an annual grass in the family Poaceae and is the third most widely grown cereal after wheat and rice throughout the world [1]. It is a staple food crop which has a total production of 1.09 billion metric tons achieved in 2018/2019, [2] and still a vital source of energy and protein in humans' diet and animals, hence ensuring food security globally [3]. The United States was recorded to be the largest corn producer in the world with an estimated volume of 345 million metric tons in 2019/20 which is approximately one third of corn produced globally. In that year, China and Brazil were the next top corn producing countries after the United States [4].

The origin of corn is quite unknown but history revealed that corn was first domesticated in Mexico's Tehuacan Valley. There are several types of corn which include sweet corn, popcorn, pod corn, flint corn, flour corn, waxy corn and dent corn. In the United States corn is known to be an important crop and in the past few years, the country's corn farmers experienced constant increases in annual revenues [4].

However, during preharvest and postharvest operations, insect pests and microorganisms attack maize, thereby reducing both the qualitative and quantitative value of maize [5]. In addition to the reduction of production yield, some pathogens produce toxins that are detrimental to both man and animals' health, they also reduce the nutritive value of maize and thus negatively impacting world food security [6]. A vast number of pathogenic microorganisms (fungi, bacteria, virus) and insects damage maize grains and plant; leading to worldwide annual losses of 9.4%. Insects are known to the the most important cause of deterioration and low yield of maize followed by fungi [7, 8]. Maize pests happens to be one of the major challenges of growing maize and some of the major threat to maize mainly include insect pests (stalk borers and armyworms) and soil pests (wireworms and rootworms). The damaged caused by the western corn rootworm (*Diabrotica virgifera virgifera*) in Europe and in USA is estimated to be more than $1 billion annually [9]. Roberts et al. [10] also reported the annual losses attributed to plant diseases to be about 40 billion dollars worldwide either directly or indirectly.

There are three significant and most noxious soil-borne pathogens that infest maize in the field namely; *Fusarium* species, *Rhizoctonia* spp. and *Verticillium* spp. [11, 12]. Furthermore, three fungal pathogens that are mostly found in stored grains are *Aspergillus* spp., *Penicillium* spp., *Fusarium* spp. [13, 14] and some xerophytic species, a number of them are known to produce toxins that causes adverse health problems including death [14–16]. The control of these microorganisms are difficult to quell due to their ability to utilize various infection modes to overcome maize immune system, possession of important structures for pathogenesis that are resistance to adverse conditions and the development of some resistance genes that ae understudied [17]. Over the years, pests and diseases management have depended majorly on the use of pesticides and agricultural practices such as crop rotation and irrigation for control of pests and diseases [18, 19]. However, the potency and environmental concerns such as its possibility of destroying beneficial microorganism and insects that promote plant growth and health, bioaccumulation of the chemicals on crops and their harvest, as well as pathogen resistance to some pesticides, have encouraged the pursuit for an alternative that is ecofriendly, less expensive, more sustainable in the management of pests and diseases [20, 21]. Amidst these alternatives, biological control method seems to be the preferable and acceptable option. Biological control using microorganism is an important tool for controlling and managing plant pests and diseases in sustainable agriculture [22].

Microbial biological control agents (MBCAs) are applied to crops for biological control of plant pathogens, they use various modes of action. Their mode of action may include nutrient competition, antagonist relationship (hyperparasitism and antibiosis) against the pathogen or by inducing resistance or priming plants without any direct interaction with the targeted pathogen [23]. In addition to using microorganism as biocontrol of pathogens, microorganisms known as entomopathogens are used in the control arthropods such as insects, mites, and ticks that infest and deteriorate maize. Diverse species of bacteria, fungi, nematodes, and viruses are used in pest management. The use of entomopathogens as biopesticides in pest management is referred to as microbial control, which can be an integral part of integrated pest management (IPM) [24].

In rhizosphere of plants, microorganisms do interact and display different associations, some may be mutualistic, commensal or even pathogenic [25–27]. Interestingly, maize' rhizosphere contains some specific microorganisms that are beneficial to its growth [28, 29]. Positive interactions in rhizospheres are known to be of importance all through the plant's life-cycle [30]. In recent years, there have been an increased interest on the issue of inoculating rhizobacteria into the agricultural soil because they are known to increase productivity and quality of agriculturally important crops and help to the stabilize agroecosystems [31]. Inoculation of maize with various plant growth-promoting rhizobacteria (PGPR) strains, however could result in significant increases in plant biomass, root and shoot length and uptake of essential plant nutrients. The use of plant growth-promoting rhizobacteria (PGPR) is a promising alternative method to external chemical inputs to improve crop yield in sustainable agricultural systems [32]. PGPR's modes of action include nutrient uptake, stress protection, induced resistance and plant growth promotion by production of phytohormones [33–35].

With respect to the severe maize' annual losses, and threat to food security caused by pathogens and insect pests, thus the need for Microbiological control methods to minimize losses caused by pathogens and insect pests. This scope of this chapter concentrates on the use of microbiological agents; an alternative, safe, less toxic, and less disruptive method of controlling the growth and development of pathogens and insect pests of maize, and optimizing maize production.

## 2. Maize

### 2.1 Maize production

Maize is known to be one of the world's most important cereal crops. It has a wide genetic diversity and diverse uses which accounts for its cultivation in a vast range of agro-ecological environments. Apart from the consumption of maize by man and animals, maize is also used to produce corn ethanol and other maize products, such as corn starch and corn syrup [36].

Andean countries of South America, Mexico, Central America and the Caribbean, Africa and South and Southeast Asia are known to consume maize as human food much higher than half of its maize production. Interestingly, maize accounts for at least 15 percent of the total calories daily intake in almost all the countries in Africa and Latin America. The economy of the developed and developing countries is significantly impacted by maize production [37]. The world market has recorded an enormous growth in maize production in the most especially in countries with temperate environment where hybrids and high yielding agronomic practices are used. The main maize exporters are: United States, Argentina, France, China P.R., Hungary, Canada, South Africa. China is a relatively new exporter being the main suppliers of Asian neighbor countries. There was a prediction for developing countries by Ortiz et al. [13] that there will be a growing demand for maize alone as food to increase by around 1.3% per annum until 2020. Furthermore, another prediction by Rosegrant et al. [38] stated a double demand for maize by 2050 in the developing world, and maize is predicted to become the crop with the greatest production globally, and in the developing world by 2025.

### 2.2 Maize losses

Abiotic and Biotic factors (pests, pathogens and weeds) significantly contribute to grain loses and thus affects food supply. About one-third of potential crop yield is lost to pre-harvest pests, pathogens and weeds [39]. Coupled with pre-harvest losses, the losses occurring during transport, pre-processing, storage, processing, packaging, marketing and plate waste are also important. An average of 35% of potential crop yield is lost to pre-harvest pests worldwide [40].

There are different number of ways pests reduce crops productivity; their effects include, stand reducers (damping-off pathogens), photosynthetic rate reducers (fungi, bacteria, viruses), leaf senescence accelerators (pathogens), light stealers (weeds, some pathogens), assimilate sappers (nematodes, pathogens, sucking arthropods), and tissue consumers (chewing animals, necrotrophic pathogens) [41].

Post-harvest loss occurs between harvest and consumptions. The major physiological, physical and environmental causes of post-harvest losses are high crop perishability; mechanical damage; excessive exposure to high ambient temperature, relative humidity and rain; contamination by spoilage fungal and bacteria; invasion by birds, rodents, insects and other pests; and inappropriate handling, storage and processing techniques [42]. Post-harvest losses lead to high food prices thus reducing food in the market. Reducing post-harvest losses in maize is an important element in any strategic planning to make more food available without increasing the burden on the natural environment.

## 2.3 Major pathogens of maize

*Fusarium* species are among the most common fungal pathogens causing diseases in maize. This genus is ubiquitous in nature and contains various toxigenic species, with *F. graminearum* and *F. verticillioides* being the most commonly found pathogens in maize. They infect several parts of maize at any stage of development, processing and storage thereby reducing maize quality and production yields. They do produce mycotoxins (fumonisin, deoxynivalenol, and zearalenone.) that are poisonous to man when consumed [43, 44].

### *2.3.1* F. verticillioides

*F. verticillioides* causes Fusarium ear (characterized by discolored and a reduced number of grains), and stalk rot which leads to global significant losses of maize [45]. It is one of the most prevalent disease causing agent in maize (*Zea mays* L.). Fusarium ear/stalk rot is common during hot and dry weather, both pre and post-harvest conditions. Fumonisins are carcinogenic [46, 47], and are produced in large amounts in maize and contaminates maize based food and feed, therefore they are of high importance to farmers [43, 44].

### *2.3.2* F. graminearum

Maize kernels contaminated with *F. graminearum* results in a moldy kernels called Gibberella ear rot. This organism also produces mycotoxin (deoxynivalenol and zearalenone), toxic to humans and farm animals when consumed. This fungus often starts infecting the tip of an ear when it starts silking during the cool and wet weather, [48].

### *2.3.3* Aspergillus flavus

*A. flavus* is a phytopathogenic fungus that causes diseases in several agricultural crops and at the same time producing aflatoxins which is a toxic metabolite produced during its secondary metabolism [49, 50]. *A. flavus* is the disease causing agent of Aspergillus ear rot; a global disease of maize. Aflatoxins are hazardous to both humans and animal' health if ingested via contaminated food and feed. In humans, aflatoxins have been directly linked to hepatocellular carcinoma, since they are metabolized in the liver [51].

#### *2.3.4* Curvularia lunata

*C. lunata* is a foliar fungal pathogen that causes Curvularia leaf spot of maize, especially during the hot and humid seasons [52]. *C. lunata* produces a furanoid type toxin, both *in vitro* and *in planta*, which can possibly lead to leaf lesions which invariably lead to a reduction in maize yields [53].

#### *2.3.5 Other pathogens of maize*

Some other economically important pathogens that infest maize and their corresponding diseases are listed as follows: *Pythium* spp., *Rhizoctonia* spp., and *Acremonium* spp. (Root and Stalk rot), *Puccinia sorghi* and *P. polysora* (Leaf rusts), *Helminthosporium turcicum* or *Setosphaeria turcica* (Leaf blights), *H. maydis* (Maydis leaf blight), *Cercospora zaeamaydis* (Gray leaf spot (GLS)) [54]. *Sclerophthora macrospora* (Downy mildew/Yellow tuft), *Sphacelotheca* reiliana (Head smut of maize), *S. macrospora* (Downy mildew/ Yellow tuft), *Trichometasphaeria turcica* Luttr. (Northern Leaf Blight), *Ustilago maydis* (Corn smut), *P. coronata* (Crown crust), Maize streak virus (Maize streak disease), Sugarcane Mosaic Virus (SCMV) is another viral pathogen and causal agent of mosaic disease in maize and other graminaceous plants [55].

### 2.4 Major insect pests of maize

Globally, insect pests are categorized into two classes; (1) field pests such as stalk borer (*Busseola fusca*), maize leafhoppers (*Cicadulina mbila*) and mole crickets (*Gryllotalpidae*), African bollworm (*Helicoverpa armigera*), African armyworm (*Spodoptera exempta*) and black cutworms (*Agrotis ipsilon*) and (2) storage pests like the maize weevil (*Sitophilus zeamais*), larger grain borer (*Prostephanus truncatus*) (Hon), red flour beetle (*Tribolium castaneum*) and dried bean beetles (*Callosobruchus maculatus*) and Indianmeal moth (*Plodia interpunctella*) [54].

The most important arthropod pests of maize in Europe is known as European corn borer, *Ostrinia nubilalis* (Hbn., Lepidoptera: Crambidae). The lepidopteran larvae (i.e., caterpillars) known as stalk borers, ear or leaf feeders, and coleopteran larvae (i.e., beetle grubs) that feed on roots. The European corn borer is a nick-named the "billion dollar bug" because it cost growers over a billion dollars annually in insecticides and lowers crop yields [56, 57]. It is known globally to cause enormous economic damage. While in America the borer mostly found include the genera Zeadiatraea, Diatraea and Elasmopalpus. The western corn rootworm (*Diabrotica virgifera virgifera* LeConte), a chrysomelid beetle is known to be the most destructive for maize production in the USA, Hungary and other central and eastern European countries [58]. While in Africa the following pests are associated to this region; *Chilo*, *Sesamia*, and *Busseola*, and in Southeast Asia *Chilo*, *Sesamia* and *Ostrinia furnicalis* are present in their maize fields. While damage is mainly caused by the larvae feeding on roots, adults feeding on silk and ears may cause additional losses, particularly in maize production for grain, seed or food (sweet maize). Sap sucking pests, like aphids (Aphididae) and leafhoppers (Cicadellidae), as well as the frit fly (*Oscinella frit* L.) cause limited economic damage as compared to the european corn borer. Other pests of regional importance include armyworms such as *Pseudaletia unipuncta* (Haworth, Lepidoptera: Noctuidae), Diptera species such as *Delia platura* (Meig.), *Geomyza* spp. and *Tipula* spp., Coleoptera species such as *Oulema melanopus* L., *Glischrochilus quadrisignatus* (Say), *Tanymecus dilaticollis* Gyll. and *Melolontha melolontha* L., spider mites (*Tetranychus* spp.) and thrips (Thysanoptera) [59].

## 2.5 Maize disease and Pest management

### 2.5.1 Planting resistant varieties

One of the most reliable method of controlling plant disease is planting of resistant varieties [60]. It is one of the most attractive approaches and can be considered as an ideal method if good quality plants are adapted to the growing regions with sufficient levels of tolerance and durable resistance This method is considered ideal and mostly used in many crops because its less expensive as compared to pesticides cost and residual effects on man, animals and the environment. Although its economical as compared to pesticides, these resistant varieties often take decades to develop and GM-plants suffer from extremely high regulatory approval cost and consumer acceptance. Its ultimately used by farmers provided quality plants are selected and adapted to exhibit adequate levels of tolerance and substantial resistance to pathogens [61]. Inspite of its advantages, it is faced with some backlash as regards the time in developing Genetically Modified (GM) plants, cost of approval and acceptance rate by customers. There have been also cases where resistance breakdown was recorded in several crops coupled with pathogens mutating their virulence gene, inconsistent uniformity in the genetics of the plants. Such cases were observed in cotton leaf curl disease [62].

### 2.5.2 Chemical control

Agrochemicals have been adapted over the years to secure food production and improve crop yield thus protecting crops from pests and pathogens. Since the 1960s, there have been an increase in pesticides use. They help in preventing losses and damages of crops; it has now become an integral component in Integrated Pest Management (IPM) [63]. It cannot be overemphasized the advancement that pesticides have brought to the agricultural sector as regards improving crop quality and annual agricultural output [64]. Nevertheless, the development of resistance genes by pathogens and pests coupled with the growing concern of accumulation off these chemicals in feeds and the ecosystem has been a great concern to farmers [65, 66].

### 2.5.3 Biological control of pathogens

Heimpel and Mills [67] defined biological control of plant diseases to be the suppression of the populations of plant pathogens by the use of living organisms. In plant pathology, beneficial organisms (crops, insects and microorganisms) are selected to diminish the effects of pathogenic organisms and improve the crop yield microorganisms. Other examples of biological control include the application of natural products and chemical compounds extracted from different sources, such as plant extracts, natural or modified organisms or gene products control [68]. This method was developed to minimize the dependence on agrochemical use and the risks for human health and the environment [69].

There are various interactions between plants, biological control agent and pathogens, they include mutualism, commensalism, neutralism, competition, amensalism, parasitism, protocooperation and predation [70–72]. The interactions between the microbes and plants occurs naturally at both macroscopic and microscopic level [68].

### 2.5.4 Cultural/traditional insect Pest control

Timely harvesting, proper harvesting and processing methods are the best strategy for controlling insect pest in maize. Proper sanitation, removal of old stock, avoid storing infected crops inside the storage facility. Other methods used by farmers to reduce infestation of maize by insect pest include the use of material such as ashes (it is known to abrasive and lethal effect on the insects' cuticle), sand, crushed limestone, mineral and oil in which physical barrier effects are responsible for the control of insects, storing dried maize that are properly dried or re-drying when infestation is detected, the use of sheaths in storing maize for protection by the husk, the use of repulsive local herbs and plants to scare off the pests (Nim ground seed, leaves of acanthaceae, acardiaceas, annonaceae, myrtaceae, other plants extract [73].

### 2.6 Microbiological control of pathogens

In modern agriculture, biological control of pathogens using microorganism is playing a major role in disease control of crops. Beneficial microorganisms are used as biopesticides and is known to be the most effective methods for safe crop-management practices [74].

The rhizosphere was discovered by Hiltner [75] to be the layer of soil dominated by the root, and is much richer in bacteria than the surrounding bulk soil. The plant rhizosphere is regulated by the synergistic relationship between the soil, plant root, and the microbes present and is controlled by the soil pH, texture, complexity and plant roots exudates mainly composed of sugars, amino acids and various nutrients [27]. The rhizosphere is a zone of soil that surrounds the plant root, is a niche colonized by numerous organisms and is considered as one of the most complex ecosystem on Earth [76].

There are some heterogeneous group of bacteria known as Plant growth-promoting rhizobacteria (PGPR), they are free-living soil bacteria mostly found in the rhizosphere, at the rhizoplane or in association with roots. They are used as biocontrol agent for the control of plant pests and disease by suppressing the activity and growth of phytopathogenic organisms, and also help to improve the extent or quality of plant growth directly or indirectly [77] by providing nutrients, synthesizing phytohormones, solubilizing phosphate, reducing stress, alleviating soil contamination with heavy metals [78–83] or improving the microbial community structure of the rhizosphere [84, 85]. The following genera of bacteria have been reported as PGPR: *Agrobacterium*, *Arthrobacter*, *Azoarcus*, *Azotobacter*, *Azospirillum*, *Bacillus*, *Burkholderia*, *Caulobacter*, *Chromobacterium*, *Enterobacter*, *Erwinia*, *Flavobacterium*, *Klebsiella*, *Micrococcous*, *Rhizobium*, *Pantoea*, *Pseudomonas* and *Serratia* [86, 87] which have shown prospect as biocontrol agents against various fungal pathogens [87].

### 2.7 Relationships that promotes biocontrol

#### *2.7.1 Microbial antagonisms*

Microorganisms that have the ability to grow in plant rhizophere are considered to be ideal for use as biological control agents. The rhizophere provides a leading edge defense for plants roots against disease causing microorganisms by suppressing pathogens growth and infestation. Pathogen-antagonizing metabolites produced by beneficial microbes that colonize the plant root, help to suppress phytopathogens' growth and thus preventing them from penetrating the root system [87]. Furthermore, this antagonistic relationship displayed between the beneficial microbes and pathogens often results to significant disease control, in which the established metabolites produced by active beneficial microbes protects plants either by directly antagonizing pathogen activity directly, by outcompeting pathogens or by stimulation of host plant defenses (priming) [88], also displays its antagonism against pathogens by antibiosis which is the secretion of diffusible antibiotics, volatile organic compounds, and toxins, as well as the development of extracellular cell wall degrading enzymes such as chitinase, β-1,3-glucanase, beta-xylosidase, pectin methylesterase and many more [87, 89].

### *2.7.2 Plant-microbe mutualistic interaction*

Microbes that inhabit plant rhizophere are nourished with nutrients obtained from plant roots in the form of root exudate and lysates. The plant-microbe interaction is not only beneficial to the microbe but it also improves plant nutrition, growth and proliferation and do enhances plant's ability to prevail over biotic and abiotic stress. This associoation gives the plant a good competitive advantage due to the presence of rhizophere [90]. Various endophytic bacteria and free-living rhizobacteria that inhabit the root surface and rhizosphere secrete metabolite substances that suppress deleterious pathogen growth and activity which invariably leads to the control plant diseases caused by fungi or bacteria [91–94].

Furthermore, microorganisms can be directly involved in plant growth promotion, by acting as agents for stimulation of plant growth and management of soil fitness, for example through the production of auxin [95].

### *2.7.3 Production of allelochemicals/antimicrobial compounds*

Allelochemicals/antimicrobial compounds produced biological control bacteria helps improve the plant-microbe rhizophere niche. Example of such compounds include iron-chelating siderophores, antibiotics, biocidal volatiles, lytic enzymes (chitinases and glucanases), and detoxification enzymes. These chemical may have detrimental effect on target pathogens, some help the plant to induce resistance against pathogen infestation and attack while some assist in nutrient absorption which promotes plant growth [96–98]. For example, rhizobacteria include antibiotic-producing strains such as *Bacillus* sp. producing iturin A and surfactin, *Agrobacterium* spp. producing agrocin 84, *Pseudomonas* spp. producing phenazine derivatives, pyoleutorin and pyrrolnitrin, and *Erwinia* sp. producing herbicolin A [99, 100], that are tenacious in the rhizosphere [101, 102]. The mycoparasitism of phytopathogenic fungi of the Trichoderma and Streptomyces genera have important roles in secretion of chitinases and glucanases [103]. A common feature of successful biocontrol strains and a crucial factor for plant root pathogen suppression is the production of antibiotic compounds and fluorescent siderophores that enable effective competition for iron [104].

*Trichoderma* spp., are universally known as BCAs and used to prevent plant pathogens and increase plant immunity in field and greenhouse conditions [105]. This is due to its ability to interact with plants (maize, cotton, cucumber) through production of auxin like compounds and secondary metabolites [106–108]. BCAs of *Trichoderma* spp. have ultimate functions in promoting the plant beneficial microbial community and decreasing the pathogen attack through the specific interactions with host-pathogen. In maize, growth-promoting and antifungal compounds-producing bacteria have been shown to have inhibitory effects on southern leaf blight disease caused by the fungus *Cochliobolus heterostrophus* [109, 110].

### *2.7.4 Induced systemic resistance (ISR)*

Van Peer et al. [111] first discovered rhizobacteria-induced systemic resistance or ISR, also referred to in its early stage as priming. It is as an enhanced defensive capacity of the whole plant to multiple pathogens induced by beneficial microbes in the rhizosphere [112] or elicited by specific environmental stimuli which lead to potentiation of the plant's innate defense against biotic challenges [113]. Non-pathogenic rhizobacteria are capable of activating defense mechanisms in plants in a similar way to pathogenic microorganisms, including reinforcement of plant cell walls, production of phytoalexins, synthesis of PR proteins and priming/ISR [112]. Plants that possess ISR displays stronger and/or faster activation of defense mechanisms after a subsequent pathogen or insect attack or as a response to abiotic stress, when inoculated with rhizobacteria [114].

### 2.8 Entomopathogens

Entomopathogens are microorganisms that are pathogenic to arthropods such as insects, mites, and ticks. Various species of naturally occurring bacteria, fungi, nematodes, and viruses infect a several arthropod pests and play an important role pest management. Some entomopathogens are produced in large scale as in vitro (bacteria, fungi, and nematodes) or in vivo (nematodes and viruses) and sold commercially. In some scenario, they are also produced on small scale for non-commercial local use. The use of entomopathogens as biopesticides is an alternative method to chemical control and a novel approach pest management, which can be a profound part of integrated pest management (IPM) against several pests [24].

#### *2.8.1 Entomopathogenic fungi*

They typically cause infection when spores come in contact with the arthropod host. Fungal spores germinate and breach the insect cuticle through enzymatic degradation and mechanical pressure to gain entry into the insect body provided the environmental conditions such as moderate temperatures and high relative humidity are in place. Once inside the body of the insect, the fungi multiply, invade the insect tissues, emerge from the dead insect, and produce more spores [24]. Fungal pathogens have an eclectic host range and are especially suitable for controlling pests that have piercing and sucking mouthparts reason being that spores do not have to be ingested. However, entomopathogenic fungi are also effective against a variety of pests such as wireworms and borers that have chewing mouthparts [24].

The potential use entomopathogenic fungus has been reported by some researchers. For example, *Beauveria bassiana* (Bal.) Vuillemin (Deuteromycotina: Hyphomycetes) can be used against the following stored-grain insects: rice weevil (*Sitophilus oryzae*), corn weevil (*S. zeamais*), granary weevil (*S. granarius*), lesser grain borer (*Rhyzopertha dominica*), red and confused flour beetles (*Tribolium castaneum* and *T. confusum*), *Oryzaephilus surinamensis*, and *Prostephanus truncatus* [115–121], and for another entomopathogenic fungus *Metarhizium anisopliae* (Metch.) Sorokin (Deuteromycotina: Hyphomycetes) against the following stored-grain insects: rice weevil (*S. oryzae*), lesser grain borer (*Rhyzopertha dominica*), and red flour beetle (*Tribolium castaneum*) [120–125].

#### *2.8.2 Entomopathogenic bacteria*

Entomopathogenic bacteria are well known for their ability to produce a plethora of protein insecticidal toxins [126]. Bacterial toxins acting as virulence factors have been shown to range from very specific to broad insecticidal spectrum ever since it was first discovered in the 19th century. When compared with chemical insecticides, bacterial toxins displayed high diversity of simultaneous action, contributing to the sustainability of bacteria-based bio-pesticides by limiting insect resistances. *Bacillus thuringiensis* (Bt) has been profoundly used in biocontrol of insects and it represents approximately 95% of microorganisms used in biocontrol [127].

*B. thuringiensis* produces protein-based δ-endotoxins known as "Cry", which are lethal for several species of various insect orders [128]. Presently, about 170 different "Cry" toxins have been identified, which are effective against several coleoptera, lepidoptera, and diptera species [129]. These proteins are produced upon sporulation, and are contained in crystal inclusions. Once ingested, crystals inclusions are solubilized by the insect proteases in the midgut, inadvertently activating the "Cry" proteins [130]. A vast number of research work has produced various of Bt-based insecticides, ranging from wettable powder or liquid formulation to transgenic crops, thereby facilitating their use in organic farming and integrated pest management (IPM) programs.

#### *2.8.3 Entomopathogenic viruses*

As compared to entomopathogenic bacteria, entomopathogenic viruses are also required to be ingested by the insect host and are therefore ultimate in controlling pests that have chewing mouthparts. Diverse lepidopteran pests are important hosts of baculoviruses including nucleopolyhedroviruses (NPV) and granuloviruses (GV). These related viruses have various types of inclusion bodies in which the virus particles (virions) are implanted. Virus particles attack the nucleus of the midgut, fat body or other tissue cells, compromising the integrity of the tissues and liquefying the cadavers. Before the insect pathogen dies, infected larvae climb higher in the plant canopy, which helps in dispersing virus particles from the cadavers to the lower parts of the canopy. This conduct assists in the proliferation of the virus to cause infection in healthy larvae. Viruses are host specific and can cause remarkable reduction of host populations. Examples of some commercially available viruses include *Helicoverpa zea* single-enveloped nucleopolyhedrovirus (HzSNVP), *Spodoptera exigua* -enveloped nucleopolyhedrovirus (SeMNPV), and *Cydia pomonella* granulovirus (CpGV) [24, 131].

#### *2.8.4 Entomopathogenic nematodes*

They are microscopic, soil-inhabiting worms that are detrimental to insects. Diverse species of *Heterorhabditis* and *Steinernema* are obtainable in multiple commercial formulations, majorly for managing soil insect pests. Infective juveniles of entomopathogenic nematodes actively explore their hosts and penetrate through natural openings such as the mouth, spiracles, and anus or the intersegmental membrane. Immeddiately the get into the host body, the nematodes extricate symbiotic bacteria that kill the host through bacterial septicemia. *Heterorhabditis* spp. carry *Photorhabdus* spp. bacteria and *Steinernema* spp. carry *Xenorhabdus* spp. bacteria. *Phasmarhabditis hermaphrodita* is also available for controlling slugs in Europe, but not in the USA [24].

### 2.9 Application of biocontrol

#### *2.9.1 Seed dressing*

A suitable method for suppressing plant pathogens in the spermosphere and rhizosphere is dressing seeds with biocontrol agents [132]. Recently, bacterial inoculants have been used to antagonize soil-borne plant pathogens such as *Fusarium verticillioides (Fv)* and to promote plant growth. *Bacillus subtilis* and *Pseudomonas cepacia* have been used to control root rot caused by *Fv* in Argentina [133]. *Bacillus amyloliquefaciens* or *Microbacterium oleovorans* can reduce the fumonisin content in harvest grains during three evaluated seasons [134]. *Burkholderia* spp. stimulate plant growth and suppress disease caused by *Fv* in maize [45], and species like *Bacillus amyloliquefaciens* and *Enterobacter hormaechei* reduce the *Fv* infection and fumonisin accumulation in maize kernels [135]. Another example, is the application of *Gliocladium virens* and *Trichoderma viride* isolates on corn seeds for the reduction of Pythium and Fusarium-induced damping-off [136].

#### *2.9.2 Rhizophere inoculation*

Inoculation of rhizophere with biocontrol agents by alters the rhizosphere microbiota, thereby antagonizing soil-borne plant pathogens and promote plant growth. *Bacillus subtilis* and *Pseudomonas cepacia* have been used to control root rot caused by Fv in Argentina [133].

#### *2.9.3 Conventional spraying*

Entomopathogens viz., fungi, bacteria, virus and nematodes have an important place in the biological control because they have a wide host range, are harmless to the environment and human, and could be applied with conventional sprayers. They can be used more against stored product pests with the development of new biotechnical methods such as collecting pests in some stations to meet them with entomopathogens [137].

### **2.10 Advantages of microbiological control**

#### *2.10.1 Reduced use of Insecticides*

Many farmers have adopted the use of microbiological control agents (MCAs). Bt maize is an example of MCA, it has provided maize farmers testimonies coupled with both economic and environmental advantages. Many farmers quote unique opportunities to protect yield and reduce handling (and use) of insecticides to explain their rapid adoption of Bt maize [138].

#### *2.10.2 Protected yields*

Over the years, maize farmers had challenges in controlling corn borers because insecticides are not successful after larvae have tunneled into the stalk. In 1990, entomologists experimented the use of Bt maize and found out the "bullet proof" effect it gave to corn borer. Until then, plant breeders were able to increase host plant resistance, but none of these plants were "bullet proof". That has been the reason why farmers chose to use Bt maize which resulted in higher yields due to this reduced insect injury [139].

#### *2.10.3 Improved grain quality*

The use of Bt maize also helps to reduce the occurrence of ear mold on the field. This is as a result of the reduction of insect attacks that provides a site for infection by molds, Bt-protected maize can have lower levels of toxins produced by molds (i.e., mycotoxins), especially fumonisin and deoxynivalenol [140, 141]. Consequences of contamination with mold may be serious, as fumonisins can cause fatal leukoencephalomalacia in horses, pulmonary edema in swine, and cancer in laboratory rats. Economic analysis suggests that USA farmers save $23 million annually through reduced mycotoxins [142] and mycotoxin reduction also could be a significant health benefit in other parts of the world where maize is a diet staple [143].

## 3. Conclusion

The presented chapter outlines the use microbiological control, an ecofriendly, non-toxic, effective and biodegradable alternative to chemical pesticides. It is also an effective strategy for pest and disease management but it requires developing beneficial microorganisms that are native to the soils where maize is grown [144]. However, for biological methods to reach their full potential, an increased research effort is required. Future functional studies are still needed to fully unravel this intricate alternative approach to pest and disease management of maize and thus help boost maize yield and improve food security.

## Conflicts of interest

All authors declare no conflict of interest.

## Author details

Damilola Omobowale Seyi-Amole[1]* and Abiodun A. Onilude[2]

1 Lutheran Hospital, Fort Wayne, Indiana, United States of America

2 Department of Microbiology, Microbial Physiology and Biochemistry Laboratory, University of Ibadan, Ibadan, Nigeria

*Address all correspondence to: doseyiamole@gmail.com

## References

[1] Golob P, Kutukwa N, Devereau A, Bartosik RE, Rodriguez JC. Maize. In: Hodges R, Farrell G, editors. Crop post-harvest. Science and Technology: Ames, Iowa. Blackwell Publishing Ltd; 2004 Volume 2. Chapter two.

[2] Shahbandeh M. Corn - Statistics & Facts. Statista. 2020. Available from: https://www.statista.com/topics/986/corn/#dossierSummary__chapter4/.

[3] Rehman ZV. Storage Effects on Nutritional Quality of Commonly Consumed Cereals. Food Chemistry. 2006; 95:53-57.

[4] Shahbandeh M. Corn production worldwide 2019/2020, by country. Statista. 2021. Available from: https://www.statista.com/statistics/254292/global-corn-production-by-country/

[5] IITA. International Institute for Tropical Agriculture. Growing in Nigeria. Commercial Crop Production Guide Series. Information and Communication Support for Agricultural Growth in Nigeria. USAID. 2012. p 1-8.

[6] Marin S, Homedes V, Sanchus V, Ramos AJ. Magan N. Impact of *Fusarium moniliforme* and *F. proliferatum* Colonization of Maize on Calorific Losses and Fuminisin Production Under Different Environmental Conditions. Journal of Stored Product Research.1999;35:15-26.

[7] Uzma S, Shahida A. Efficacy of fungicides, sodium hypochlorite and neem seed powder to control seed borne pathogens of maize. Pak. J. Bot. 2007;39(1):285-292.

[8] Ali RH, Mahdi M, Ali RB, Hojjatollah S. Mycoflora of Maize Harvested from Iran and Imported Maize. Pakistan Journal of Biological Sciences. 2007;10(24):4432-4437.

[9] Gray ME, Sappington TW, Miller NJ, Moeser J, Bohn MO. Adaptation and invasiveness of western corn rootworm: Intensifying research on a worsening pest. Annu. Rev. Entomol. 2009; 54:303-321. doi: 10.1146/annurev.ento.54.110807.090434.

[10] Roberts MJ. Schimmelpfennig, DE, Ashley E, Livingston MJ, Ash MS, Vasavada U. The Value of Plant Disease Early-warning Systems: A Case Study of USDA's Soybean Rust Coordinated Framework, United States Department of Agriculture, Economic Research Service; 2006.

[11] Katan J. Diseases caused by soilborne pathogens: Biology, Management and Challenges. J. Plant. Pathol. 2017; 99:305-315.

[12] Raaijmakers JM, Paulitz TC, Steinberg C, Alabouvette C, Moënne-Loccoz Y. The rhizosphere: a playground and battlefield for soilborne pathogens and beneficial microorganisms. Plant and soil. 2009;321(1-2):341-361.

[13] Ortiz R, Taba S, Chavez Tovar VH, Mezzalama M, Xu Y, Yan J, Crouch JH. Conserving and Enhancing Maize Genetic Resources as Global Public Goods– A Perspective from CIMMYT. Crop Science. 2010;50: 13-28.

[14] Castllellarie C, Marcos FV, Mutti J, Cardoso L, Bartosik R. Toxigenic fungi in Corn (maize) stored in hermetic plastic bags. National institute of agricultural Technologies Mardel Plata University Argentina. 2010; 115-297.

[15] Lerda P, Blaggi BM, Peralta N, Ychari S, Vazquez M, Bosio G. Fumonisins in foods from Cordoba (Argentina), presence and genotoxicity. Food and Chemical Toxicology. 2005; 43:691-698.

[16] Vess KA, Smith GW, Haschelh WM. Fumonisins: Toxicokinetics mechanism of action and toxicity. Animal feed science and Technology. 2007;137: 299-325.

[17] Mendes R, Kruijt M, De Bruijn I, Dekkers E, Van Der Voort M, Schneider, JHM, Piceno YM, DeSantis TZ, Andersen GL, Bakker PAHM, et al. Deciphering the Rhizosphere Microbiome for Disease-Suppressive Bacteria. Science. 2011; 332:1097-1100.

[18] Rao B, Schmitthenner AF, Ellet CW. Prevalence and virulence of *Pythium* spp. associated with root rot of corn in poorly drained soil. Phytopathology. 1978; 68:1557-1563.

[19] Sumner DR. Root disease, populations of soil fungi, and yield decline in con-tinuous double-crop corn. Plant Dis.1990; 74:94-98.

[20] Berg G, Köberl M, Rybakova D, Müller H, Grosch R, Smalla K. Plant microbial diversity is suggested as the key to future biocontrol and health trends. FEMS Microbiol. Ecol. 2017;93(5).

[21] Budi S, Van Tuinen D, Arnould C, Dumas-Gaudot E, Gianinazzi-Pearson V, Gianinazzi S. Hydrolytic enzyme activity of *Paenibacillus* sp. strain B2 and effects of the antagonistic bacterium on cell integrity of two soil-borne pathogenic fungi, Appl. Soil Ecol. 2000;15;191-199.

[22] Azcón-Aguilar C. Barea J. Arbuscular mycorrhizas and biological control of soilborne plant pathogens–an overview of the mechanisms involved. Mycorrhiza.1997;6:457-464.

[23] Köhl J, Kolnaa R, Ravensberg WJ. Mode of Action of Microbial Biological Control Agents Against Plant Diseases: Relevance Beyond Efficacy. Front. Plant Sci. 2019. Available from: https://doi.org/10.3389/fpls.2019.00845

[24] Dara, SK. Entomopathogenic microorganisms: modes of action and role in IPM. E-Journal of Entomology and Biologicals; 2017.

[25] Cordovez V, Dini-Andreote F, Carrión VJ, Raaijmakers JM. Ecology and Evolution of Plant Microbiomes. Annu. Rev. Microbiol. 2019; 73:69-88.

[26] Lambers H, Mougel C, Jaillard B, Hinsinger P. Plant-microbe-soil interactions in the rhizosphere: An evolutionary perspective. Plant Soil. 2009; 32:83-115.

[27] Mendes R, Garbeva P, Raaijmakers JM. The rhizosphere microbiome: Significance of plant beneficial, plant pathogenic, and human pathogenic microorganisms. FEMS Microbiology Reviews. 2013;37(5):634-663.

[28] Peiffer JA, Spor A, Koren O, Jin Z, Tringe SG, Dangl JL, Buckler ES, Ley RE. Diversity and heritability of the maize rhizosphere microbiome under field conditions. Proc. Natl. Acad. Sci. 2013; 110:6548-6553.

[29] Walters WA, Jin Z, Youngblut N, Wallace JG, Sutter J, Zhang W, González-Peña A, Peiffer J, Koren O, Shi Q. Large-scale replicated field study of maize rhizosphere identifies heritable microbes. Proc. Natl. Acad. Sci. 2018;115;7368-7373.

[30] Philippot L, Raaijmakers JM, Lemanceau P, Van Der Putten WH. Going back to the roots: The microbial ecology of the rhizosphere. Nat. Rev. Genet. 2013; 11:789-799.

[31] Javoreková S, Cinkocki R, Maková J, Hricáková N. Isolation and identification of rhizobacteria from maize *(Zea mays* L.) In: Luvisols and documentation their plant growth promoting traits. J Microbiol Biotech Food Sci. 2020;10(3):505-510. doi: 10.15414/jmbfs.2020.10.3.505-510.

[32] Yadav AN, Verma P, Singh B, Chauhan VS, Suman A, Saxena AK. Plant growth promoting bacteria: *biodiversity* and multifunctional attributes for sustainable agriculture. Advances in Biotechnology and Microbiology. 2017;5(5):1-16. Available from: http://dx.doi.org/10.19080/aibm.2017.05.5556671.

[33] Bakker PAHM, Doornbos RF, Zamioudis C, Berendsen RL, Pieterse CMJ. Induced Systemic Resistance and the Rhizosphere Microbiome. Plant Pathol. J. 2013; 29:136-143.

[34] Berg G, Rybakova D, Grube M, Köberl M. The plant microbiome explored: Implications for experimental botany. J. Exp. Bot. 2015; 67:995-1002.

[35] Lugtenberg B, Kamilova F. Plant-Growth-Promoting Rhizobacteria. Annu. Rev. Microbiol. 2009;63;541-556.

[36] Foley J. "It's Time to Rethink America's Corn System". Scientific American. Retrieved February 18, 2019.

[37] Mejía D. Maize: Post-Harvest Operation Organisation: Food and Agriculture Organization of the United Nations (FAO), AGST. Edited by AGST/FAO: Danilo Mejía, PhD, FAO (Technical) Last reviewed: 15/05/2003.

[38] Rosegrant MW, Msangi S, Ringler C, Sulser TB, Zhu T, Cline SA. International Model for Policy Analysis of Agricultural Commodities and Trade (IMPACT): Model Description. International Food Policy Research Institute: Washington, DC; 2008.

[39] Popp J, Pető K, Nagy P. Pesticide productivity and food security. A review. Agron. Sustain. Dev. 2013; 33:243-255.

[40] Oerke E. Crop losses to pests. J. Agric. Sci. 2006; 144:31-43.

[41] Boote KJ, Jones JW, Mishoe JW, Berger RD. Coupling pests to crop growth simulators to predict yield reductions. Phytopathology. 1983; 73:1581-1587.

[42] Zorya S, Morgan N, Diaz Rios, L, Hodges R, Bennett B, Stathers T, Mwebaze P, Lamb J. Missing food: the case of postharvest grain losses in sub-Saharan Africa. Technical Report. The International Bank for Reconstruction and Development / The World Bank, Washington DC, USA. Economic Sector Work Report No. 60371-AFR; 2011.

[43] Fandohan P, Hell K, Marasas WFO, Wingfield MJ. Infection of maize by *Fusarium* species and contamination with fumonisin in Africa. Afr. J. Biotechnol. 2003;2:570-579.

[44] Munkvold GP. Epidemiology of Fusarium diseases and their mycotoxins in maize ears. Eur. J. Plant Pathol. 2003; 109:705-713. doi: 10.1023/A:1026078324268.

[45] Hernández-Rodríguez A, Heydrich-Pérez M, Acebo-Guerrero Y, Velazquez-del Valle MG, Hernández-Lauzardo AN. Antagonistic activity of Cuban native rhizobacteria against *Fusarium verticillioides* (Sacc.) Nirenb. in maize (*Zea mays* L.). Appl Soil Ecol. 2008;39(2):180-186. doi:10.1016/j.apsoil.2007.12.008

[46] Marasas WF. Fumonisins: Their implications for human and animal health. Nat. Toxins. 1995; 3:193-198. doi: 10.1002/nt.2620030405.

[47] Stockmann-Juvala H, Savolainen K. A review of the toxic effects and mechanisms of action of fumonisin B1. Hum. Exp. Toxicol. 2008; 27:799-809. doi: 10.1177/0960327108099525.

[48] Woloshuk C, Wise K. Diseases of Corn: Gibberella Ear Rot. [accessed on 26 February 2021]. Available from:

http//www.extension.purdue.edu/extmedia/BP/BP-77-W.pdf.

[49] Diener UL, Davis ND. Aflatoxin formation by *Aspergillus flavus*. In: Goldblatt LA. editor. Aflatoxin: Scientific Background, Control and Implications. New York, USA: Academic Press; 1969. p. 360-391.

[50] Payne GA. Aflatoxins in maize. Crit. Rev. Plant. Sci. 1992; 10:423-440. doi: 10.1080/07352689209382320.

[51] Neal GE. Participation of animal biotransformation in mycotoxin toxicity. Rev. Med. Vet. 1998; 149:555-560.

[52] Ito MF, Paradela F, Soave J, Sugimori MH. Leaf spot caused in maize (*Zea mays* L.) by *Culvularia lunata* (wakker) boedijn. Summa Phytopathol. 1979; 5:181-184.

[53] Liu T, Liu LX, Huang XL, Jiang X, Zhou PH. A new furanoid toxin produced by *Culvularia lunata*, the causal agent of maize Curvularia leaf spot. Can. J. Plant Pathol. 2009; 31:22-27. doi: 10.1080/07060660909507568.

[54] ASSP. Agriculture service support programme. Agricultural sector development program (ASDP). Integrated pest management plan (IPMP). The united republic of Tanzania. Final report; 2004.

[55] Pechanova O, Pechan T. Maize-Pathogen Interactions: An Ongoing Combat from a Proteomics Perspective. Int J Mol Sci. 2015;16(12): 28429-28448. doi: 10.3390/ijms161226106.

[56] Brookes G. The existing and potential impact of using GM insect resistant (GM IR) maize in the European Union. Dorchester, UK: PG Economics; 2009. Available from: http://www.pgeconomics.co.uk/pdf/btmaize.

[57] Hellmich RL, Hellmich KA. Use and Impact of Bt Maize. Nature Education Knowledge. 2012;3(10):4

[58] Boriani M, Agosti M, Kiss J, Edwards CR. Sustainable management of the western corn rootworm, Diabrotica virgifera virgifera LeConte (Coleoptera: Chrysomelidae), in infested areas: experiences in Italy, Hungary and the USA. Bulletin OEPP/EPPO Bulletin. 2006; 36:531-537. doi: 10.1111/j.1365-2338.2006.01055.x

[59] Meissle M, Mouron P, Musa T, Bigler F, Pons X, Vasileiadis VP, Otto S, Antichi D, Kiss J, Pálinkás Z, Dorner Z, Van Der Weide R, Groten J, Czembor E, Adamczyk J, Thibord JB, Melander B, Cordsen Nielsen G, Poulsen RT, Zimmermann O, Verschwele A, Oldenburg E. Pests, pesticide use and alternative options in European maize production: current status and future prospects. Journal of Applied Entomology. 2010. Available from: https://doi.org/10.1111/j.1439-0418.2009.01491.x.

[60] Maloy OC. Plant disease management, Plant Health Instr. 2005;10.

[61] Kiyosawa S. Genetics and epidemiological modeling of breakdown of disease resistance, Annu. Rev. Phytopathol. 1982; 20:93-117.

[62] Peressotti E, Wiedemann-Merdinoglu S, Delmotte F, Bellin D, Di Gaspero G, Testolin R, Merdinoglu D, Mestre P. Breakdown of resistance to grapevine downy mildew upon limited deployment of a resistant variety, BMC Plant Biol. 2010;10:147.

[63] Hirooka T, Ishii H. Chemical control of plant diseases, J. Gen. Plant Pathol. 2013; 79:390-401.

[64] Ragsdale NN. Hylin JW, Sisler HD, Witt JM, Alford H. Health and Environmental Factors Associated with

Agricultural Use of fungicides. USDA/ States National Pesticide Impact Assessment Program Fungicide Assessment Project 117; 1991. Available from: http://cipm.ncsu.edu/piappud/.

[65] Ma Z, Michailides TJ. Advances in understanding molecular mechanisms of fungicide resistance and molecular detection of resistant genotypes in phytopathogenic fungi. Crop Prot. 2005; 24:853-863.

[66] Carvalho FP. Agriculture pesticides, food security and food safety. Environ. Sci. Policy. 2006;9;685-692.

[67] Heimpel GE, Mills N. Biological Control - Ecology and Applications. Cambridge: Cambridge University Press; 2017.

[68] Pal KK, Gardener BM. Biological control of plant pathogens. Plant Health Instr. 2006; 2:1117-1142.

[69] Cook RJ, Baker KF. The Nature and Practice of Biological Control of Plant Pathogens. American Phytopathological Society;1983.

[70] Bronstein JL. Our current understanding of mutualism. Q. Rev. Biol. 1994; 69:31-51.

[71] Price PW. 1977. General concepts on the evolutionary biology of parasites. Evolution. 31: 405-420.

[72] Price PW, Bouton CE, Gross P, McPheron BA, Thompson JN, Weis AE. Interactions among three trophic levels: influence of plants on interactions between insect herbivores and natural enemies. Annu. Rev. Ecol. Syst. 1980;11:41.

[73] Boa E, Chernoh E, Jackson G. Pest and disease manual. In: African Soil Health Consortium, Nairobi. Sones K, editor. 2015. ISBN: 9781780648187.

[74] Schafer KS, Kegley S. Persistent toxic chemicals in the US food supply, J. Epidemiol. Community Health 2002;56:.813-817.

[75] Hiltner L. Uber neuere Erfahrungen und Probleme auf dem Gebiete der Bodenbakteriologieunter besonderden berucksichtigung und Brache.Arb. Dtsch. Landwirtsch. Gesellschaft. 1904; 98:59-78.

[76] Raaijmakers JM, Paulitz TC, Steinberg C, Alabouvette C, Moënne-Loccoz Y. The rhizosphere: a playground and battlefield for soilborne pathogens and beneficial microorganisms.Plant and soil. 2009;321 (1-2):341-361. http://dx.doi.org/10.1007/s11104-008-9568-6

[77] Ahmad F, Ahmad I, Khan MS. Screening of free-living rhizospheric bacteria for their multiple plant growth promoting activities. Microbiol Res. 2008;163(2):173-181. doi:10.1016/j.micres.2006.04.001.

[78] Grobelak A, Napora A, Kacprzak M. Using plant growth-promoting rhizobacteria (PGPR) to improve plant growth. Ecological Engineering, 2015;84:22-28. Available from: http://dx.doi.org/10.1016/j.ecoleng.2015.07.019.

[79] Kloepper JW, Schroth MN. Plant growth-promoting rhizobacteria on radishes. Proceedings of the 4th international conference on plant pathogenic bacteria. Angers. 1978;2: 879-882.

[80] Bardi L, Malusá, E. Drought and nutritional stresses in plant: Alleviating role of rhizospheric microorganisms. In: Haryana, N, Punj S, editors. Abiotic Stress: New Research. Hauppauge, NY: Nova Science Publishers; 2012. p.1-57.

[81] Cawoy H, Debois D, Franzil L, Pauw ED, Thonart P, Ongena M. Lipopeptides as main ingredients for inhibition of fungal phytopathogens by

*Bacillus subtilis/amyloliquefaciens*. Microb Biotechnol. 2015;8: 281-295.

[82] Wang X, Mavrodi DV, Ke L, Mavrodi OV, Yang M, Thomashow LS, et al. Biocontrol and plant growth-promoting activity of rhizobacteria from Chinese fields with contaminated soils. Microb Biotechnol. (2015;8: 404-418.

[83] Vessey JK. Plant growth promoting rhizobacteria as biofertilizers. Plant Soil. 2003; 255:571-586.

[84] Malusa E, Vassilev N. A contribution to set a legal framework for biofertilisers. Appl Microbiol Biotechnol. 2014;98: 6599-6607.

[85] Bruto M, Prigent-Combaret C, Muller D, Moënne-Loccoz Y. Analysis of genes contributing to plant-beneficial functions in plant growth-promoting rhizobacteria and related Proteobacteria. Sci Rep. 2014; 4:6261. doi:10.1038/srep06261.

[86] Ahemad M, Kibret M. Mechanisms and applications of plant growth promoting rhizobacteria: current perspective. J King Saud Univ Sci. 2014; 26(1):1-20. doi:10.1016/j.jksus.2013.05.001.

[87] Shoda M. Bacterial control of plant diseases, J. Biosci. Bioeng. 2000; 89:515-521.

[88] Nihorimbere V, Ongena M, Smargiassi M, Thonart P, Beneficial effect of the rhizosphere microbial community for plant growth and health. Biotecnol. Agron. Soc. 2011; 15:327.

[89] Compant S, Duffy B, Nowak J, Clément C, Barka EA. Use of plant growth promoting bacteria for biocontrol of plant diseases: principles, mechanisms of action, and future prospects. Appl. Environ. Microbiol. 2005; 71:4951-4959.

[90] Haney CH, Samuel BS, Bush J, Ausubel FM. Associations with rhizosphere bacteria can confer an adaptive advantage to plants, Nat. Plants. 2015; 1.

[91] Gray E, Smith D. Intracellular and extracellular PGPR: commonalities and distinctions in the plant-bacterium signaling processes. Soil Biol. Biochem. 2005; 37: 395-341

[92] Kiely P, Haynes J, Higgins C, Franks A, Mark G, Morrissey J, O'gara F. Exploiting new systems-based strategies to elucidate plant-bacterial interactions in the rhizosphere. Microbiol. Ecol. 2006; 51:257-266.

[93] Chet I, Chernin L. Biocontrol, microbial agents in soil. In: Bitton G, editor. Encyclopedia of Environmental Microbiology. New York, USA: Willey; 2002. p. 450-465.

[94] Bais HP, Weir TL, Perry LG, Gilroy S, Vivanco, JM. The role of root exudates in rhizosphere interactions with plants and other organisms. Annu. Rev. Plant Biol. 2006; 57:233-266.

[95] Welbaum GE, Sturz AV, Dong Z, Nowak J. Managing soil microorganisms to improve productivity of agro-ecosystems. Crit. Rev. Plant Sci. 2004; 23:175-193.

[96] Glick BR. The enhancement of plant growth by free-living bacteria. Can. J. Microbiol. 1995; 41:109-117.

[97] Sturz, A, Christie B. Beneficial microbial allelopathies in the root zone: the management of soil quality and plant disease with rhizobacteria. Soil Tillage Res. 2003; 72:107-123.

[98] Bais HP, Park SW, Weir TL, Callaway RM, Vivanco JM. How plants communicate using the underground information superhighway, Trends Plant Sci. 2004;9; 26-32.

[99] Compant S, Reiter B, Sessitsch A, Nowak J, Clément C, Barka EA, Endophytic colonization of Vitis vinifera L. by plant growth-promoting bacterium *Burkholderia* sp. strain PsJN. Appl. Environ. Microbiol. 2005;71: 1685-1693.

[100] Viswanathan R, Samiyappan R. Induction of systemic resistance by plant growth promoting rhizobacteria against red rot disease in sugarcane. Sugar Tech. 1999;1: 67-76.

[101] Brooks DS, Gonzalez CF, Appel DN, Filer T. Evaluation of endophytic bacteria as potential biological-control agents for Oak Wilt. Biol. Control. 1994; 4:373-381.

[102] Carvalhais LC, Dennis PG, Fedoseyenko D, Hajirezaei MR, Borriss R, von Wirén N. Root exudation of sugars amino acids, and organic acids by maize as affected by nitrogen, phosphorus, potassium, and iron deficiency. J. Plant Nutr. Soil Sci. 2011;174; 3-11.

[103] Oku S, Komatsu A, Nakashimada Y, Tajima T, Kato J. Identification of Pseudomonas fluorescens chemotaxis sensory proteins for malate, succinate, and fumarate and their involvement in root colonization. Microbes Environ. 2014;29: 413-419.

[104] O'Sullivan DJ, O'Gara F. Traits of fluorescent Pseudomonas spp. involved in suppression of plant root pathogens. Microbiol. Rev. 1992; 56:662-676.

[105] Harman GE, Howell CR, Viterbo A, Chet I, Lorito M. *Trichoderma* species - opportunistic, avirulent plant symbionts. Nature Rev. Microbiol. 2004; 2:43-56. doi:10.1038/nrmicro797.

[106] Harman GE. Multifunctional fungal plant symbionts: New tools to enhance plant growth and productivity. New Phytol. 2011; 189:647-649. doi:10.1111/j.1469-8137.2010. 03614.x

[107] Salas-Marina MA, et al. Colonization of Arabidopsis roots by *Trichoderma atroviride* promotes growth and enhances systemic disease resistance through jasmonic acid/ ethylene and salicylic acid pathways. Eur. J. Plant Pathol. 2011; 131:15-26. doi:10.1007/s10658- 011-9782-6.

[108] Contreras-Cornejo HA, Macías-Rodríguez L, Del-Val E, Larsen J. Ecological functions of *Trichoderma* spp. and their secondary metabolites in the rhizosphere: interactions with plants. FEMS Microbiol. Ecol. 2016;92.

[109] Huang C, Yang K, Liu Y, Lin Y, Chen C. Suppression of Southern Corn Leaf Blight by a Plant Growth-Promoting Rhizobacterium *Bacillus cereus* C1L. Annals Applied Biology. 2010; 157:45-53. https://doi.org/10.1111/j.1744-7348.2010.00408.x.

[110] Ye Y, Li Q, Fu G, Yuan G, Miao J, Lin W. Identification of Antifungal Substance (Iturin A2) Produced by Bacillus subtilis B47 and Its Effect on Southern Corn Leaf Blight. Journal of Integrative Agriculture. 2012; 11:90-99. Available from: https://doi.org/10.1016/S1671-2927(12)60786-X.

[111] Van Peer R, Niemann G, Schippers B. Induced resistance and phytoalexin accumulation in biological control of Fusarium wilt of carnation by Pseudomonas sp. strain WCS 417 r. Phytopathology. 1991; 81:728-734.

[112] Conrath U, Beckers GJ, Langenbach CJ, Jaskiewicz MR. Priming for enhanced defense. Annu. Rev. Phytopathol. 2015;53; 97-119.

[113] Van Loon L. Induced resistance in plants and the role of pathogenesis-related proteins. Eur. J. Plant Pathol. 1997; 103:753-765.

[114] Pieterse CM, C. Zamioudis, R.L. Berendsen, D.M. Weller, S.C. Van Wees, P.A. Bakker, Induced systemic resistance

by beneficial microbes, Annu. Rev. Phytopathol. 20; 1452:347-375.

[115] Adane K, Moore D, Archer SA. Preliminary studies on the use of *Beauveria bassiana* to control Sitophilus zeamais (Coleoptera: Curculionidae) in the laboratory. J. Stored Products Res. 1996.;32:105-113.

[116] Padin SB, Dal-Bello G, Fabrizio M. Grain loss caused by *Tribolium castaneum, Sitophilus oryzae,* and *Acanthoscelides obtecus* in stored durum wheat and beans treated with *Beauveria bassiana*. J. Stored Products Res. 2002;38: 69-74.

[117] Hidalgo E, Moore D, Le Patourel G. The effect of different formulations of *Beauveria bassiana* on *Sitophilus zeamais* in stored maize. J. Stored Products Res. 1998;34: 171-179.

[118] Moino Jr A. Alves SB, Percira RM. Efficacy of *Beauveria bassiana* (Bals.) Vuilemin isolates for control of stored-grain pests. J. Appl. Entomol. 1998;122: 301-305.

[119] Rice, WC, Cogburn RR. Activity of the entomopathogenic fungus *Beauveria bassiana* (Deuteromycota: Hyphomycetes) against three coleopteran pests of stored grains. J. Econ. Entomol. 1999;92: 691-694.

[120] Dal-Bello G, Padin S, Lopez-Lastra C, Fabrizio M. Laboratory evaluation of chemical-biological control of the rice weevil (*Sitophilus oryzae* L.) in stored grains. J. Stored Products Res. 2001;37: 77-84.

[121] Sheeba G, Seshardi S, Raja N, Janarthanan S, Ignacinutha S. Efficacy of *Beauveria bassiana* for control of the rice weevil Sitophilus oryzae (L.) (Coleoptera: Curculionidae). Appl. Entomol. Zool. 2001;36: 117-120.

[122] Batta YA, Safieh DIA. A study of treatment effect with *Metarhizium anisopliae* and four types of dusts on wheat grain infestation with red flour beetles (*Tribolium casteneum* Herbs Coleoptera: Tenebrionidae). J Islamic Univ Gaza. 2005;13(1): 11-22.

[123] Batta YA. Control of the lesser grain borer (*Rhyzopertha dominica* Fab., Coleoptera: Bostrichidae) by treatments with residual formulations of *Metarhizium anisopliae* (Metchnikoff) Sorokin (Deuteromycotina: Hyphomycetes). J. Stored Products Res. 2005;41: 221-229.

[124] Batta YA, Abu Safieh DI. A study of treatment effect with *Metarhizium anisopliae* and four types of dusts on wheat grain infestation with red flour beetles (*Tribolium castaneum* Herbs., Coleoptera: Tenebrionidae). J. Islamic Univer. Gaza. 2005;13: 11-22.

[125] Michalaki M, Athanassiou C, Kavallieratos N, Batta Y, Balotis G. Effectiveness of *Metarhizium anisopliae* (Metchinkoff) Sorokin applied alone or in combination with diatomaceous earth against *Tribolium confusum* (Du Val) larvae: influence of temperature, relative humidity and type of commodity. Crop Prot. 2006;25: 418-425.

[126] Henkel JS, Baldwin MR, Barbieri JT. Toxins from bacteria. EXS. 2010; 100:1-29.

[127] Schünemann R, Knaak N, Fiuza LM. Mode of action and specificity of *Bacillus thuringiensis* toxins in the control of caterpillars and stink bugs in soybean culture. ISRN Microbiol. 2014; 2014:135675. doi: 10.1155/2014/135675.

[128] Höfte H, Whiteley HR. Insecticidal crystal proteins of *Bacillus thuringiensis*. Microbiol. Rev. 1989; 53:242-255.

[129] Frankenhuyzen KV. Insecticidal activity of *Bacillus thuringiensis* crystal proteins. J. Invertebr. Pathol. 2009; 101:1-16. doi: 10.1016/j.jip.2009.02.009.

[130] Palma L, Muñoz D, Berry C, Murillo J, Caballero P. *Bacillus thuringiensis* toxins: An overview of their biocidal activity. Toxins. 2014; 6:3296. doi: 10.3390/toxins6123296.

[131] Gelernter WD, Toscano NC, Kido K, Federici BA. Comparison of a nuclear polyhedrosis virus and chemical insecticides for control of the beet armyworm (Lepidopter: Noctuidae) on head lettuce. J. Econ. Entomol. 1986;79: 714-717.

[132] Pereira P, Nesci A, Etcheverry M. Effects of biocontrol agents on *Fusarium verticillioides* count and fumonisin content in the maize agroecosystem: impact on rhizospheric bacterial and fungal groups. Biol Control. 2007;42(3):281-287. doi:10.1016/j.biocontrol.2007.05.015.

[133] Cavaglieri L, Orlando J, Rodríguez MI, Chulze S, Etcheverry M. Biocontrol of *Bacillus subtilis* against *Fusarium verticillioides* in vitro and at the maize root level. Res Microbiol. 2005;156(5-6):748-754. doi:10.1016/j.resmic.2005.03.001.

[134] Pereira P, Nesci A, Castillo C, Etcheverry M. Field studies on the relationship between *Fusarium verticillioides* and maize (*Zea mays* L.): effect of biocontrol agents on fungal infection and toxin content of grains at harvest. Int. J Agron. 2011;2011:7. doi:10.1155/2011/486914 .

[135] Pereira P, Nesci A, Castillo C, Etcheverry M. Impact of bacterial biological control agents on fumonisin B1 content and *Fusarium verticillioides* infection of field-grown maize. Biol Control. 2010;53(3):258-266. doi:10.1016/j.biocontrol.2010.02.001.

[136] Mao W, Lewis JA, Hebbar PK, Lumsden RD. Seed treatment with a fungal or a bacterial antagonist for reducing corn damping-off caused by species of Pythium and Fusarium. Plant Dis. 1996; 81:450-454.

[137] Pilcher R. The use of entomopathogens in the controlling of insect pests of stored product. Scientific Papers. Series A. Agronomy. 2016; 59: 235-240.

[138] Pilcher CD, et al. Biotechnology and the European corn borer: Measuring historical farmer perceptions and adoption of transgenic Bt corn as a pest management strategy. Journal of Economic Entomology. 2002; 95:878-892.

[139] Gómez-Barbero M, Berbel J, Rodríguez-Cerezo E. Bt corn in Spain - the performance of the EU's first GM crop. Nature Biotechnology. 2008; 26:384-386.

[140] Dowd PF. Indirect reduction of ear molds and associated mycotoxins in Bacillus thuringiensis corn under controlled and open field conditions. J Econ Entomol. 2000; 93:1669-1679.

[141] Munkvold GP. Epidemiology of Fusarium diseases and their mycotoxins in maize ears. Eur. J. Plant Pathol. 2003; 109:705-713. doi:10.1023/A:1026078324268.

[142] Wu F, Miller JD, Casman EA. The economic impact of Bt corn resulting from mycotoxin reduction. Journal of Toxicology-Toxin Reviews. 2004; 23:397-424. doi:10.1081/txr-200027872.

[143] Wu, F. Mycotoxin reduction in Bt corn: Potential economic, health, and regulatory impacts. Transgenic Research. 2006; 15:277-289. doi:10.1007/s11248-005-5237-1.

[144] Etcheverry M, Scandolara A, Nesci A, Vilas Boas Ribeiro MS, Pereira P, Battilani P. Biological interactions to select biocontrol agents against toxi-genic strains of *Aspergillus flavus* and *Fusarium verticillioides* from maize. Mycopathologia. 2009;167(5): 287-295. doi:10.1007/s11046-008-9177-1.

## Section 3

# Bio-Waste

Chapter 13

# The Creation of Furniture Products from Rice Stubble

*Somchai Seviset and Songwut Egwutvongsa*

## Abstract

The objective of this research was to develop the transformation procedure of rice stubble in the dry season of Thailand. This would be significantly used for the creative building processes of furniture products for earning increased economic value for the people and the communities. Fiber was applied in the transformation procedure for the rice stubble, including the cementing of the formation procedure in the boiling and soaking methods. From the results from using both procedures in this research, it was found that knock-down furniture sheets had the ability to be formed as standard furniture products with JIS A 5908–1994 for the customer groups who had the most level of satisfaction to rice stubble furniture. Therefore, this resulted in a positive result affecting the reduction of the ratio of burning of rice stubble in terms of preparing the area for planting in the next season on a large scale, including decreasing the occurrence of the PM. 2.5 problem in Thailand in every winter season. As a result, this research could be considered as another choice as a proposal to present the solution guidelines for solving the PM. 2.5 problem sustainably.

**Keywords:** Products from rice stubble, the creation of furniture, furniture design, rice stubble, material transformation

## 1. Introduction

From predicting the result of the in-season rice production for Thailand for the period of 2016–2020, it was found that there was an increasing ratio of 0.14% for each year. Additionally, this demonstrated that the amount of rice production was approximately 25.522 million tons, or an increasing production ratio of 6.06%. As such, it could be seen that currently Thailand is a world-class rice producing country with the amount of exported rice being second to India, or with approximately 7.58 million tons, which is the increasing trend for the exporting ratio in every year [1].

This further presented the promotion by the Thai government for boosting agriculturists to cultivate rice in the North-eastern region and the Central region of Thailand. Thus, nowadays, this cultivation has contributed to the country gaining large areas of rice fields, especially for the remaining rice stubble, which is approximately 72.321 million tons from the agriculturists' harvesting in each season. Moreover, this stubble is utilized for the mixing of straw mushroom cultivation or soil adjustment before the next season. However, this still leaves large amounts of rice stubble, so the agriculturists would remove it from their areas; such as, bringing

**Figure 1.**
*Problem of burning rice stubble that results in PM 2.5 in Thailand.*

the soil for cultivating other plants; namely, beans and other short harvesting cycle vegetables. As a consequence, this would enable the agriculturists to increase their income after doing the rice cultivation in the dry season without burning the rice stubble in the North-eastern or Central regions in the same way. In addition, this material always contributes to dust called polycyclic aromatic hydrocarbons (PAHs), or PM 2.5 that is a size less than 2.5 microns, which is the dust from burning monoculture plants [2]. This would eventually be fed into the supply chain of the agricultural food production amounting to more than 209,937 tons per year, as well as being released into the world's atmosphere to become an annual environmental effect at a high level. Furthermore, this has had an impact on the Mekong River's region. Unfortunately, according to the checking of the quality index from the Pollution Control Department for the period of 2016–2018, it was found that the value of the Air Quality Index (AQI) was way below the set standard of the World Health Organization (WHO), which in turn would have a negative impact on people's health in the area of the rice fields of the Northern and North-eastern regions of Thailand [3].

These problems have also continued to occur in every dry season to become severe environmental and health problems for the people in these areas (**Figure 1**).

Therefore, the government sector has a policy to promote resolving these problems sustainably by encouraging the people to bring the remaining rice stubble to increase the income from agriculture, including encouraging the agriculturists to bring the rice stubble to be transformed into community products to be sold and allow them to earn more agricultural income. Therefore, this could reduce the chances of burning the rice fields by the agriculturists prior to cultivating in the next season. This was conducted according to the National Strategy 2018–2037. Additionally, it could be considered as having a leading role in the determination of the direction toward the United Nations' Sustainable Development Goals (SDGs). This could also be initiated to prepare for the structural adjustment for Thailand 4.0 in the future [4]. Thus, the government has aimed at the alteration based on the potential of the communities and local areas to utilize rice stubble in the rice fields to be transformed into a positive aspect of the sustainability development of the lifestyle and economy of the agriculturist communities [5]. As the result, this has depended on having the appropriate technologies with the potential of the community's people and expression through a participatory procedure until gaining a modern method with the tools and the specificness for the suitable uniqueness for increasing their own lifestyles' sustainability together with rice cultivation [6].

## 2. Objectives

To develop the transformation procedure from the remaining rice stubble to be wood substitute material.

To create furniture products for children from the remaining rice stubble and predict the solutions with the customer groups' requirements affecting the new products.

## 3. Scope

1. Scope of the Procedural Development from the Remaining Rice Stubble as a Wood Substitute Material

   This would have the objective to develop the transformation procedure of remaining rice stubble in the cultivation areas of the North-eastern region.

   The population of agriculturists cultivating the fields in the North-eastern region is 3,741,346 people [7].

   The group sampling consisted of 100 agriculturists of the North-eastern region with the level of discrepancy being 100% by using stratified random sampling that divided the cultivation areas of each province in Thailand [8].

   A structured questionnaire with a checklist of a five-point Likert scale to rate the levels of satisfaction in the transformation procedure for the newly developed rice stubble was used as the research tool [9]. Additionally, this used the statistical value from the confidence value of 10 group samplings with the value of Cronbach's alpha at a level of 0.873 by using the data analysis of the statistical descriptive value; such as, the mean, standard deviation (S.D.), and one way or single factor ANOVA [10].

2. Scope of the Production of Furniture Products for Children from the Remaining Rice Stubble and Predicting the Solution with the Customer Groups' Requirements That Affected the New Products

   This aimed to present the designing procedure for the furniture products by using wood substitute material from rice stubble to be an invention of the communities for producing small industries inside the households.

   The population comprised 9,183 customers interested in furniture who visited the product booths in the Furniture and Electronic Fair 2018 in Khon Kaen province, Thailand. Stratified random sampling was used to find 100 people as the discrepancy level of 100% [8].

   A structured questionnaire with a checklist of a five-point Likert scale to rate the level of satisfaction of newly developed rice stubble furniture that utilized wood substitute material was used as the research tool [9]. This used the statistical value from the confidence value of 10 group samplings with the value of Cronbach's alpha at a level of 0.786 by using data analysis of the statistical descriptive value; such as, the mean, standard deviation (S.D), and dependent t-test [10].

## 4. Framework

The framework properties with the causes and results from studying the phenomenon [11], and the development of the transformation procedure from remaining rice stubble in Thailand were applied. Moreover, wood substitute material for the furniture industry was used by dividing into the influencing factors of designing and creating (**Table 1**) [12].

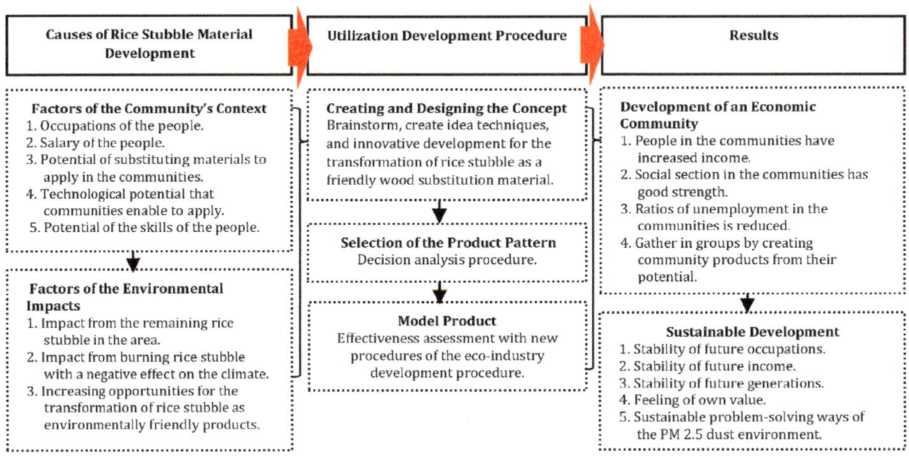


**Table 1.**
*Research framework to present the development.*

## 5. Results

### 5.1 Development of the result of the transformation procedure from the remaining rice stubble as wood substitute material

Development of the requirements of rice stubble were studied in the communities of the North-eastern region of Thailand. This could promote the lifestyle of the people in the communities to gain more opportunities from earning income with the increasing valuable furniture products, including increasing the opportunity to burn the rice stubble that affected the overall environment in the country. However, this was based on the problem-solving method and the management from the high quantity of rice stubble in each season. Normally, the agriculturists would always manage to do this through three methods: 1. To let the rice stubble decompose naturally, 2. burn the rice stubble, and 3. till the rice stubble into the ground [13]. Similarly, these three methods may have different effects on the agriculturists and society according to the selection of the guidelines for removing the rice stubble in the next season prior to cultivating the rice (**Figure 2**).

This research process aimed at applying the benefits of the rice stubble in terms of developing manufactured products easily without the complexity of the production procedure. Similarly, this also promoted the people in the communities in a positive way they reduced rice stubble burning in the rice fields in the dry season by creating the good well-being of the people in the future. Consequently, this could

**Figure 2.**
*Agriculturists cultivate rice in the dry season of north-eastern areas of Thailand.*

increase quality environmental conditions and boost the effective economy and other fields.

Hence, overall, this would generate integrated development in the system and sustainability for the communities based on the potential of the people; namely, lifestyle, environment, and economy. Additionally, this was considered as sustainable development based on the input factors that caused the alteration and output factor to represent the early successful result, intermediate successful result, and final successful result.

Therefore, this would take place from people in the model communities who were participating in the study by applying the problem-solving methods to stimulate the alteration sustainably and continuously (**Table 2**).

**Table 3** presented the requirements to bring the rice stubble in the agriculturists own cultivation areas to be applied to the transformation process. The group sampling had the requirement to increase the value in the transformation for handicraft products that had the most level of requirement (mean = 4.530; S.D. = 0.559), followed by the transformation to be agricultural products that had an excellent level of requirement (mean = 4.100; S.D. = 0.541), the tilling or implantation had an excellent level of requirement (mean = 3.560; S.D. = 0.729), and burning had a moderate level of requirement (mean = 2.990; S.D. = 1.159). Therefore, overall, this showed that the community groups had the requirement of the feeling to bring the rice stubble to be applied as a new method more than the original one; namely, tilling or burning. In fact, these two methods caused direct effects to the agriculturists, except for having the requirement to increase the rice stubble more than in

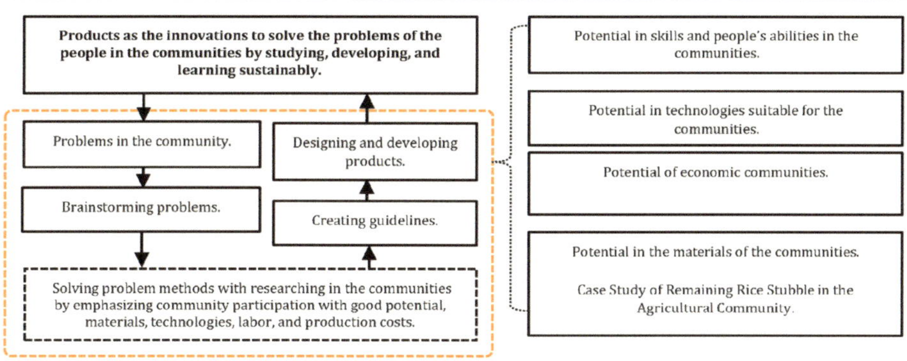


**Table 2.**
*Creating the direction of the method for rice stubble furniture products in rice cultivation in the north-eastern region.*

| Method to apply | Mean | S.D. | S.E. | Level |
|---|---|---|---|---|
| 1. Transforming to be industrial hand products. | 4.530 | 0.559 | 0.056 | Most |
| 2. Transforming to be agricultural products. | 4.100 | 0.541 | 0.054 | Excellent |
| 3. Tilling or implantation. | 3.560 | 0.729 | 0.073 | Excellent |
| 4. Burning. | 2.990 | 1.159 | 0.116 | Moderate |
| **Overall** | **3.795** | **0.975** | **0.049** | **Most** |

**Table 3.**
*Opinions of agriculturists affected by bringing the rice stubble to apply to the transformation process, (n = 100).*

the past. Thus, the researcher brought the requirement result from the community groups prior to applying with the transformation procedure of the rice stubble fiber.

From **Table 4**, the two-tailed test for the significance value found that the value F = 0.000 showed the mean of at least one pair with a level of significance of .05 by using multiple comparison. As such, there were four method patterns that used the statistical test of Fisher's least significant difference (LSD) to make a comparison between the differences of the mean with the pairs. Similarly, according to the test result, it was found that every compared pair had the value Sig. in the test with the least significance of .05, and these four methods could be applied with the opinions of the agriculturist community groups with a level of significance of .05.

Most of the agriculturist groups had similar opinions to the stimulation trend for bringing the rice stubble to apply with the transformation. Likewise, this could generate higher economic value with a positive image to the income after applying it in beneficial ways.

The rice farmer group had the requirement to bring the rice stubble in their cultivation area to do an activity in an agricultural way; such as, planting mushrooms or making fertilizer. Then, these farmers had few requirements due to the less economic incentives for applying with the transformation than doing in other ways. As such, this was affected by the price of agricultural crops in the areas which existed to have low crop prices, so the farmers group did not have the incentive for being interested in the use of rice stubble in agricultural crops. Additionally, in the tilling process, it showed that the farmers in the research area would not bring the materials to apply with the transformation because this would result in the increasing of rats living in the rice fields leading to an epidemic of leptospirosis disease for the local areas [14, 15]. Significantly, normally the agriculturists, especially the rice farmers did not plow the rice stubble but considered the rice stubble burning process as the popular method because it could quickly eradicate rats that lived in the rice fields, so the farmers could do agriculture in the next round on the rice fields. Therefore, burning was an agricultural method that was extensively used in Thailand; unfortunately, in the year 2020, the country was faced with air pollution problems in the form of dust particles or PM 2.5 until it affected the health of the people in the North, Northeast, and Central areas. Hence, this was a health problem that created a health disparity among the general population until it became a national problem that all people had to work together to reduce the impact. Furthermore, nowadays, most farmers have not burned the rice stubble since the beginning of the year 2020 until now. This presented the causes and effects combined with the global trend for conserving the environment. Then, it could initiate the ideas of the development process for making use of rice stubble that required a simple process without any complications, and this enabled needy farmers in Thailand to apply these newly developed processes to meet their own need, including adding value to the economy as a form of income generated from rice. As a result, this could use products made from rice stubble in the form of wood substitute

| | Sum of the squares | df | Mean square | F | Sig. |
|---|---|---|---|---|---|
| Between Groups | 133.650 | 3 | 44.550 | 71.849 | 0.000 |
| Within Groups | 245.540 | 396 | 0.620 | | |
| Total | 379.190 | 399 | | | |

*Critical value (0.05. = F-table = 2.627).*

**Table 4.**
*Analysis of the differences for the four method variables.*

material and furniture made from rice stubble to create opportunities for increasing the income of farmers in the local countryside as another way, including helping bridge the social gap in Thailand.

### 5.2 Development of the transformation procedure of rice stubble as wood substitute material

This demonstrated that the creating of the transformation procedure for rice stubble resulted in suitable and environmentally friendly wood substitute material [16]. Additionally, this was based on the transformation to be a result of the wood substitute material with specific properties; such as. fineness and beauty, smoothness, strength and colors in the materials, and other related aspects [17]. Then, the researcher made the comparison of the boiling procedure to peel off the tissue by using the water immersing procedure for peeling tissue as two parts (**Figure 3** and **Table 5**).

**Table 5** shows the development of the transformation procedure for the rice stubble. The properties for the fiber were obtained by two methods. This were considered as suitable methods with the potential of the communities to bring the remaining rice stubble from the rice fields in the communities to apply to the transformation procedure. Later, it was transformed to be fiber with good properties; such as, fineness and suitability to be formed as wood substitute material for creating furniture products. After that, the fiber from the rice stubble utilized the pressing sheet procedure through the peeling of the tissue and bleaching methods.

This used the ratio of 7% of seven kilograms of rice stubble before mixing with isocyanate adhesive and spraying glue in a rotary grinder for 12 minutes. Later, the rice stubble tissue was mixed with the isocyanate adhesive again by being formed as wood substitute material as a wood block with the dimensions of 45 cm x 120 cm and a thickness of 15 ml. Then, it used heat pressing at 150°C under the pressing pressure of 35 kilograms/cm.$^2$ (**Figure 4**).

The test results of the wood substitute material from the newly developed remaining rice stubble had the standard of JIS A 5908–1994. Then, it was found that there was a specific gravity value of 0.75, and the properties of the density quantity was 8.85% that passed the standard, including the properties of the modulus of rupture (MOR) that had a level of 5.67 MPa, the properties of the modulus of elasticity (MOE) that had a level of 319.95 MPa, which was lower than the standard, the compression stress in the levels of 11.59 MPa and 5.97 MPa, and hardness in the level of 3,949.49 N by testing the decomposition of the wood substitute material from the rice stubble [18]. Therefore, this resulted from the decomposing phenomenon as the pattern of the brash tension and simple tension as the weakness of the wood substitute material made from the newly developed rice stubble at this time (**Figure 5**).

**Figure 3.**
*Transformation procedure of rice stubble fiber in the dry season.*

| Analysis factor | Boiling procedure | Immersing procedure |
|---|---|---|
| Line Separation Step of Rice Stubble Fiber  | 1. 1 cm of rice stubble was decomposed.<br>2. The rice stubble was boiled in water by mixing with 0.5% of sodium hypochlorite.<br>3. It was boiled under a temperature of 95–100°C for two hours.<br>4. 40 kg. of rice stubble was boiled in 30 liters of water. | 1. 1 cm of rice stubble was decomposed.<br>2. The rice stubble was immersed in water by mixing with 0.5% of sodium hypochlorite.<br>3. It was immersed and mixed in a 100-liter tank for 40 days.<br>4. 40 kg. of rice stubble with 30 liters of water. |
| Rice Stubble Fiber Characteristic  | 1. Fiber had the high level of fineness with a soft texture.<br>2. When the fiber was dried, it had a soft weight.<br>3. The fiber was spun and decomposed by screening through a 1.5 mm. sieve to gain 15.3 kg. | 1. Fiber had a moderate level of fineness.<br>2. Fiber had strength on the segments, bodies, and roots of the rice stubble.<br>3. When the fiber was dried, it had a soft weight.<br>4. The fiber was screened through a 1.5 mm. sieve to gain 16.8 kg. |
| Fiber Characteristic from the Transformation Procedure of the Rice Stubble | 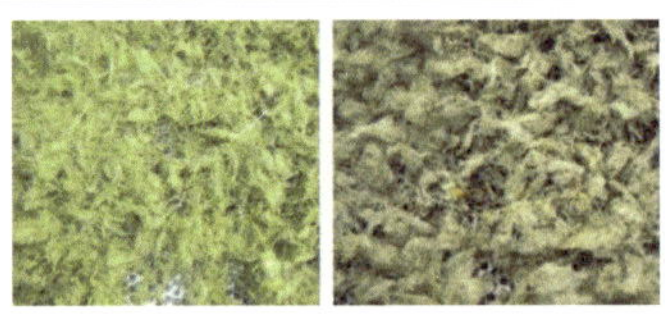 | 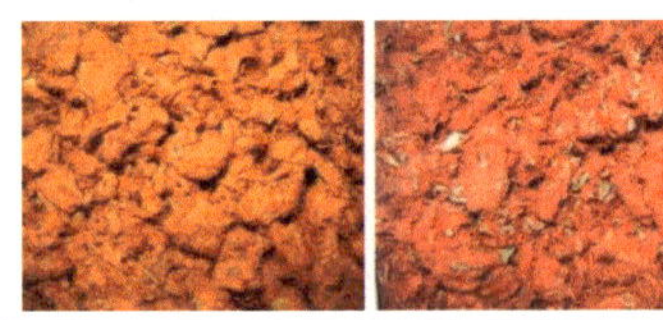 |
| Preparation procedure of the rice stubble fiber formed by using wood substitute material and isocyanate adhesive. 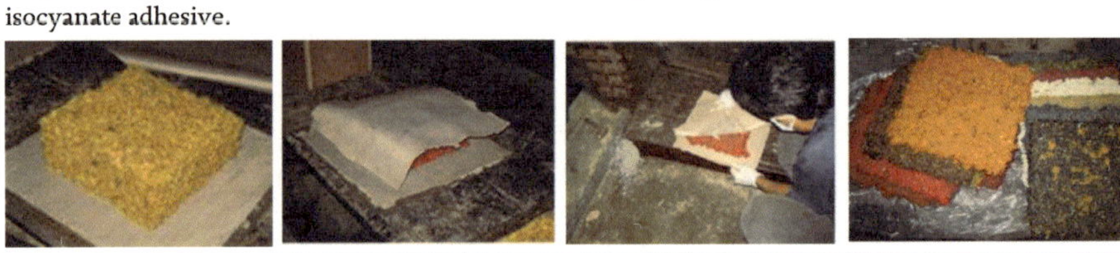 | | |

**Table 5.**
*Development of the transformation procedure of rice stubble.*

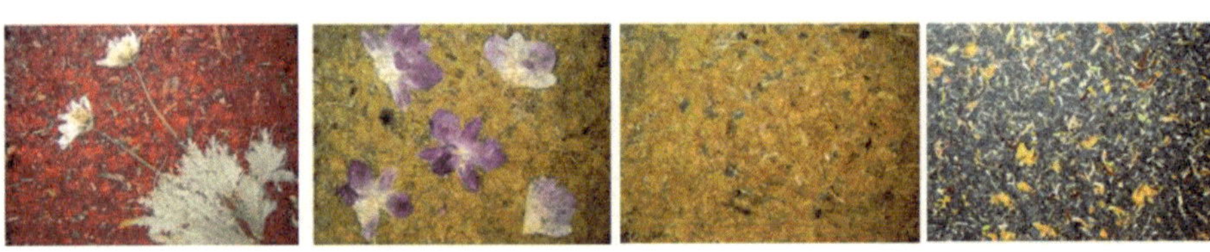

**Figure 4.**
*Wood substitute materials from heat pressing with rice stubble fiber.*

**Figure 5.**
*Decomposing trace of newly developed wood substitute material from rice stubble.*

From the using of wood substitute materials from rice stubble, this presented the resistance from pressing with the least level than the standard. Likewise, this was caused by the external peeling for the rice stubble surface fiber for its shortness and smallness appearance. Moreover, this resulted in the high level of flexibility for the wood substitute material through the heat pressing of the isocyanate adhesive, except for the less effectiveness from over pressing. Thus, this made the rice stubble fiber have softness and stickiness in a least level than normal with the effect on the wood substitute material to have a less bonding force and adhesion for the wood substitute material as the standard.

However, if the community group that wanted to apply rice stubble fiber in real life by producing products to sell for increasing the community's strength and force more than the past, they could so by using the following methods:

1. Increase the quantity of the isocyanate adhesive by 7% to have more mixing ratio by adding better properties of the wood substitute material from the rice stubble to gain the strength and active force with the higher level of the production costs.

2. Increase the length of the rice stubble fiber more than in the past to add the strength and effectiveness for the active force with the wood substitute material. The drying step to reduce the weight of the crushed rice stubble to 30 kilograms showed the remaining rice stubble to have a weight of seven to eight kilograms, which was suitable for being boiled or immersed to peel the tissue.

3. Creating the steps of the furniture products from rice stubble fiber

   This was based on the creative ideas in the designing step of creating the newly developed rice stubble products into two patterns. Then, it could respond to the furniture product requirements for the customer groups. Therefore, this depended on the presentation of the brainstorming procedure through the drafting of the furniture model products by using the wood substitute material from the rice stubble fiber (**Figure 6**).

   **Table 6** presented the test of the differences for the mean satisfaction to the furniture product patterns from the remaining rice stubble. The first pattern had the mean satisfaction at a moderate level of 3.38, and the furniture product patterns from the remaining rice stubble in the second pattern had the mean satisfaction at an excellent level of 4.18. The result of the t-test showed that the mean between the furniture product patterns of the

**Figure 6.**
*Draft of the furniture products made from wood substitute material from rice stubble.*

| New developed furniture | n | Mean | S.D. | Mean difference | t | df | Sig. 1 tailed |
|---|---|---|---|---|---|---|---|
| Design Development 1 | 100 | 3.38 | 0.81 | −0.80 | −7.139** | 198 | 0.000 |
| Design Development 2 | 100 | 4.18 | 0.77 | | | | |

***Statistical significance in the level of .01.*

**Table 6.**
*Selection result from the furniture product patterns of newly developed rice stubble.*

remaining rice stubble in the second pattern was at a higher level than the first pattern with a statistical significance level of .01.

Adopting the furniture product patterns in the second pattern could produce the rice stubble furniture products to be applied in real life. Thus, the researcher produced the models of the community groups in Khandong Subdistrict of Buri Ram province, where the community groups transformed the rice stubble to be wood substitute material for the handicraft industry. Finally, this enabled the researcher to make the summary of the economic values occurring from the rice stubble transformation (**Table 7**).

**Table 7** displays the value of the wood substitution material production from rice stubble per sheet with boiled and peeled tissue to gain the fiber. Then, it was brought to the pressing procedure from immersing the tissue to peel off the rice stubble fiber that amounted to 3 Baht per sheet. This was considered as gaining a lower level of production costs from the selling standard of the wood substitute material in the market. Then, the wood substitute material made from the rice stubble seemed to gain suitable properties for the handicraft industry in the communities, so that the agriculturists could bring the rice stubble to be purchased by businesses with a positive trend to increase more income for their communities.

Therefore, this could produce model furniture products from rice stubble by using the wood substitute materials with the dimension of 150 mm. As a result, it applied the abstract guideline and concrete guideline in real life prior

| Producing procedure | Boiling from peeling the tissue by using wood substitute material | Using wood substitute material from immersing to peel the tissue |
|---|---|---|
| 1. Preparing the material step (tissue material replacement of wood) | A. Fuel prices for boiling: 23 Baht<br>B. Coloring procedure: 32 Baht<br>C. Labor costs for 1 hour: 20 Baht<br>D. Material costs (rice stubble): 0 Baht | A. Tissue crushing procedure: 0 Baht<br>B. Coloring procedure: 32 Baht<br>C. Labor costs for 2 hours: 40 Baht<br>D. Material costs (rice stubble): 0 Baht |
| 2. Heat pressing step (isocyanate adhesive) | A. Glue: 35 Baht<br>B. Labor costs per sheet: 10 Baht | A. Glue: 35 Baht<br>B. Labor costs per sheet: 10 Baht |
| Total Costs | 120 THB (4 USD) | 117 THB (3.90 USD) |

**Table 7.**
*Comparison of the costs of the production with the sheet dimension of 120 X 60 cm. From wood substitute material of boiled rice stubble with peeling tissue and immersing the tissues to be peeled.*

**Figure 7.**
*Model furniture products from rice stubble fiber.*

to bringing the model to be assessed for the effectiveness and satisfaction scores (**Figure 7**).

4. Predicting the solution steps of the satisfaction values for rice stubble furniture customer groups

For the step of the influencing assessment factor of the rice stubble furniture production customer groups, the model furniture was brought to test with the targeted customer groups. For this study, it was brought to be tested in the furniture and home decoration booths in the Furniture and Electronic Fair 2018 in Khon Kaen province, Thailand. This also included making an assessment of the customers' feelings after testing the furniture product from the new rice stubble (**Table 8**).

As can be seen from **Table 8**, the assessment result of the four input factors demonstrated that the test group had the satisfaction from the most to the least levels; namely, factor X4 or the interestingness and the novelty had the most level of satisfaction (mean = 4.600; S.D. = 0.603), followed by factor X2 or worthiness to be transformed that had an excellent level of satisfaction (mean = 4.400; S. D. = 0.711), factor X1 or beauty also had an excellent level of satisfaction (mean = 3.970; S.D. = 0.784), and factor X3 or the convenience to apply had an excellent level of satisfaction (mean = 3.540; S.D. = 0.744). Therefore, when the researcher had classified the satisfaction assessment values with the rice stubble furniture product models, it presented the most level of satisfaction (mean = 4.530; S.D. = 0.627), including the result of the satisfaction assessment with the interestingness and novelty interests from the new procedure to the new model in similar levels. Moreover, the presentation of the Sig. value from the test of F = 0.00 demonstrated at least one pair of correlation gaining the differences of the level of significance with .05 by applying the multiple comparisons and LSD (**Table 9**).

From the conclusion of **Table 9**, it could be seen that the factors involved with the variables and satisfaction toward the furniture model from rice stubble for the customer groups consisted of worthiness to transformation (X2) and interestingness and novelty (X4). These two variables were considered as the correlation with the satisfaction to the customer groups while using the new rice stubble furniture model. There were also two noninfluencing factors; namely, beauty (X1) and convenience to apply (X3) that had a statistical significance of .05.

**Table 10** displayed the factor result with the co-efficient of the decision ($R^2$) by gaining the value of 0.4442, or the factor test with the rice stubble furniture designed at this time. Then, it could explain about the satisfaction change in the level of 44.42% by bringing the influencing factors to the satisfaction of the new

| Influencing factors of the satisfaction of furniture made from rice stubble | n = 100 People | | Satisfaction level |
|---|---|---|---|
| | Means | S.D. | |
| Satisfaction toward the furniture model from rice stubble. | 4.530 | 0.627 | Most |
| X1. Beauty. | 3.970 | 0.784 | Excellent |
| X2. Worthiness of the transformation. | 4.400 | 0.711 | Excellent |
| X3. Convenience to apply. | 3.540 | 0.744 | Excellent |
| X4. Interestingness and novelty. | 4.600 | 0.603 | Most |

**Table 8.**
*The mean, standard deviation (S.D,), and influencing factors toward the satisfaction of furniture of newly developed rice stubble from the customer groups.*

| Variable | X1 | X2 | X3 | X4 | Y |
|---|---|---|---|---|---|
| X1. Beauty. | 1.000 | | | | |
| X2. Worthiness of transformation. | 0.000* | 1.00 | | | |
| X3. Convenience to apply. | 0.000* | 0.000* | 1.00 | | |
| X4. Interestingness and novelty. | 0.000* | 0.033* | 0.000* | 1.00 | |
| Y. Satisfaction toward the furniture model from rice stubble. | 0.000* | 0.172 | 0.000* | 0.422 | 1.00 |

**Sig value: The test presented the significant level of .05; any pair that has less value than .05 showed the differences of significance to be .05.*

**Table 9.**
*Test summary of Fisher's least significant difference (LSD) between two-way ANOVA.*

| Model | n = 234 | | | |
|---|---|---|---|---|
| | R | R square | Adjusted R square | Std. error of the estimate |
| Test | 0.6665 | 0.4442 | 0.4208 | 0.477 |

**Table 10.**
*Co-efficient of the decision ($R^2$) for the influencing factors to the satisfaction of the newly designed rice stubble furniture model.*

design for determining the regression solution that was $\acute{y} = 1.473 + (0.271\ X1) + (0.312\ X2) + (-0.040\ X3) + (0.163\ X4)$.

**Table 11** was based on the predictors consisting of X1. beauty, X2. worthiness of transformation, X3. convenience to apply, and X4. interestingness and novelty with the dependent variable; such as, satisfaction to the newly designed rice stubble furniture model. According to the analysis result of the F-test = 18.978 > F-table = 2.490, it was found that the independent variable (X) had at least one variable that correlated with the dependent variable (Y). After that, the researcher analyzed the dependent variable before making a regression coefficient by using the statistical value of the t-test for testing (**Table 12**).

| Test | SS | df | MS | F | Sig. |
|---|---|---|---|---|---|
| Regression solution. | 17.282 | 4 | 4.321 | 18.978 | 0.000 |
| Discrepancy. | 21.628 | 95 | 0.228 | | |
| Total | 38.910 | 99 | | | |

**Table 11.**
*Analysis of the correlation between the designing factors with satisfaction.*

| Predicting variable | *b* | *S.E.b* | *B* | *t* | *P* |
|---|---|---|---|---|---|
| Constant | 1.473 | 0.405 | | 3.638 | 0.000 |
| X1. Beauty. | 0.271 | 0.080 | 0.340 | 3.392 | 0.001 |
| X2. Worthiness of transformation. | 0.312 | 0.086 | 0.353 | 3.607 | 0.000 |
| X3. Convenience to apply. | −0.040 | 0.079 | −0.048 | −0.505 | 0.615 |
| X4. Interestingness and novelty. | 0.163 | 0.092 | 0.157 | 1.768 | 0.080 |

**Table 12.**
*The multiple linear regression for predicting the satisfaction of new rice stubble furniture to the co-design factors.*

The multiple linear regression for predicting the variables with the customer groups' satisfaction to the new rice stubble furniture is presented in **Table 12**. This also involved the beauty (X1) to have a regression coefficient equal to 0.271 that regarded the importance of the beauty of rice stubble furniture by increasing by one unit.

Then, the customers' satisfaction to the newly designed rice stubble furniture had more chances to increase to 0.271 unit. Thus, according to the t-test where [X1] = 3.392 > t-table = 1.985, it was found that the beauty factor was correlated to the satisfaction factor of rice stubble furniture model.

The worthiness of transformation (X2) had multiple linear regression equal to 0.312. Then, whether there was importance on the transformation with the increasing of one unit, this would still indicate the customers' satisfaction to the new designed rice stubble furniture with more chance of 0.312 units. Additionally, according to the t-test where [X2] = 3.607 > t-table = 1.985, it was found that the worthiness of transformation was correlated with the satisfaction factors to the rice stubble furniture models.

The convenience to apply (X3) had multiple linear regression equal to −0.040. Then, whether there was the importance on the transformation by increasing one unit, this would present customer satisfaction to the newly designed rice stubble furniture with more chance of 0.312 units. Furthermore, according to the t-test where [X3] = −0.505 > t-table = 1.985, it was found that the convenience to apply was not correlated to the satisfaction factors to the rice stubble furniture models.

The interestingness and the novelty to apply (X4) had multiple linear regression equal to 0.163. Then, whether there was the importance on interestingness by increasing one unit would present the customer satisfaction to the newly designed rice stubble furniture with a greater chance of 0.163 units. Moreover, according to the t-test where [X4] = 1.768 ≤ t-table = 1.985, it was found that the interestingness and the novelty was not correlated to the satisfaction factors to the rice stubble furniture models. Therefore, this could make the following summary of the predicting solution of the rice stubble furniture design:

A. This could create the predicted solution from the raw scores as the coefficient of the decision that is $R^2$ by having the value of 0.4442 with the explanation to show the solution as follows:

$$\acute{y} = 1.473 + [0.271\ (\text{beauty})] + [0.312\ (\text{worthiness of transformation})] + [-0.040\ (\text{convenience to apply})] + [0.163\ (\text{interestingness and novelty})]$$

$$\acute{y} = 1.473 + (0.271\ X1) + (0.312\ X2) + (-0.040\ X3) + (0.163\ X4)$$

B. This could create the predicted solution as the standard scores from the coefficient of the decision to show the solution as follows:

$$Z = (0.340\ X1) + (0.353\ X2) + (-0.048\ X3) + (0.157\ X4)$$

$$Z = [0.340\ (\text{beauty})] + [0.353\ (\text{worthiness of transformation})] + [-0.048\ (\text{convenience to apply})] + [0.157\ (\text{interestingness and novelty})]$$

After that, the results of the assessment were used by classifying the details for the creative design. Then, this could be explained with the phenomenon of customer satisfaction to the transformation procedure of the rice stubble into product creation with patterns. In addition, it was based on applying the integration procedure by using the creative idea with multiplying the product patterns. Thus, a comparison between the rice stubble products for furniture and lamps could be made (**Figure 8** and **Tables 8** and **13**).

Home vase:

This soft weight and high force resistance model product was made from the fiber pressed with the rice stubble transformation of polystyrene (PS) using the forming technique to apply in multiple patterns.

**Figure 8.**
*Vase products from rice stubble with polystyrene.*

| Assessment list |  Furniture from rice stubble | | |  Vase from rice stubble | | | Comparison | |
|---|---|---|---|---|---|---|---|---|
| | Mean | S.D. | Satisfaction | Mean | S.D. | Satisfaction | t | Sig. |
| 1. Novelty | 4.18 | 0.69 | Excellent | 4.25 | 0.73 | Excellent | −0.698 | 0.243 |
| 2. Beauty | 4.04 | 0.74 | Excellent | 4.55 | 0.58 | Most | −5.453 | 0.000* |
| 3. Worthiness of the procedure | 4.03 | 0.76 | Excellent | 4.52 | 0.54 | Most | −5.324 | 0.000* |
| 4. Increasing values | 4.72 | 0.47 | Most | 4.55 | 0.52 | Most | 2.419 | 0.008* |
| 5. Customer requirements | 4.53 | 0.69 | Moderate | 4.36 | 0.67 | Excellent | 1.764 | 0.040 |
| 6. Environmentally friendly | 4.47 | 0.50 | Excellent | 4.32 | 0.71 | Excellent | 1.727 | 0.043 |
| 7. Saving the world | 4.06 | 0.65 | Excellent | 4.03 | 0.69 | Excellent | 0.317 | 0.376 |
| Overall | 4.29 | 0.70 | Excellent | 4.37 | 0.66 | Excellent | −2.167 | 0.015 |

**Statistical significance in the level of .01.*

**Table 13.**
*Analysis of the mean and standard deviation (S.D.) from customer satisfaction to remaining rice stubble materials in Thailand, (n = 100).*

As shown in **Table 13**, the overall satisfaction for the vase from rice stubble had the most level of satisfaction (mean = 4.37; S.D. = 0.66). As such, it had a higher level than the furniture from rice stubble with an excellent level of satisfaction (mean = 4.29; S.D. = 0.70) with no significance of the level of .01. This could be classified as follows:

1. The increasing value found that the rice stubble furniture had a higher level of customer satisfaction than the vases with significance in the level of .01.

2. The novelty value showed that the vase had a higher level of customer satisfaction value than the rice stubble furniture with no significance in the level of .01.

3. The beauty value displayed that the vase from the rice stubble had a higher level of customer satisfaction value than the rice stubble furniture with significance in the level of .01.

4. The worthiness procedure value demonstrated that the vase from the rice stubble had a higher level of satisfaction value than the rice stubble furniture with significance in the level of .01.

5. The customer requirement value found that the rice stubble furniture had a higher level of customer satisfaction value than the rice stubble furniture with significance in the level of .01.

| Basic knowledge | Guideline into new research |
|---|---|
| The wood replacement material often used heat compression that relies on digestion to contain small fragments of waste [19]. | This aimed to present a way to digest rice stubble into smaller fiber by using the water immersion method mixed with sodium hyper chloride to get fiber with wool: a smaller size of 0.1 mm; the fiber was white as a feature available for the dye onto the fiber, and the rice stubble was tinted. Then, the obtained fiber was spun to make it look fluffy before being bonded to form as a special sheet material to be different from the normal particle board material [20], and the entire new sheet from rice stubble had vibrant colors with the same colors on the unrough texture for both inside and outside of the panels. |
| Today's furniture molding is preferable for using three-layered particle board that differed in physical appearance. Moreover, the outermost layer had a resolution in the middle part to be rough, and the outer surface was covered with a laminated board, which had the characteristics of overlapping sandwiches. Then, this resulted in higher cost wastage for the furniture manufacturing industry from the process and the precoat of the board was preferable to coat the front surface of the board as a special one [21, 22]. | The sheet material was in the form of a particle board obtained from the rice stubble that had a unique feature: the boards were all the same color and texture. Therefore, it could reduce the cost of the precoat board by more than 3% of the normal production cost. Then, this helped to reduce the cost of expert technicians in painting, surface and application of baking furniture since the material made from rice stubble could be assembled as desired. Furthermore, this required only a smooth and varnished surface finish without having to paint on the furniture's surface because the material produced from the rice stubble was colorful, and the material was slick to be already in the permanent board [23]. |
| This created the higher pollution impacts than a normal wood substitu- tion process [24]. | These were the concepts of the transformation of waste from agricultural areas that were likely to cause impacts on the surrounding environment; such as, rice stubble. Then, if not converted to or encouraged to adopt, the farmers would have no choice but to destroy the stubble in their arable land [25, 26], and there was a tendency for farmers to burn the stubble in their own cultivated areas; this would affect the environment widely and severely, including affecting the environment, especially for the capital of Bangkok in the field from air pollution. |
| The coordination of wood substitutes in the production of partition boards would be divided into three layers of material with different textures, and the patch was then bonded with urea resin glue; when looking at the top surface, middle and bottom layers, it showed the difference of the texture of the resulting partitions, including the colors of the panels that were clearly different in color for each layer [27]. | The offered products that were made to serve low-income or low-growing farmers would help create sustainable income opportunities in small localities in rural areas [28, 29], and the transformation process used was an easy process with the rest of the materials being found in the community that could be applied; this was a sustainable solution to the inequality problem that appeared in the country in order to do it another way [30]. |
| | The physical properties of the newly developed sheet material was consistent in color across the sheet in both cross-sections, the front side of the sheet having the same color as the outer surface [31]. |

**Table 14.**
*The issue of extending the current concepts into further research.*

6. The friendliness value showed that the rice stubble furniture had a higher level of customer satisfaction value than the rice stubble furniture with significance in the level of .01.

7. The increasing value found that the saving of the world conformation had a higher level of customer satisfaction value than the rice stubble vases with no significance in the level of .01 (**Table 14**).

## 6. Discussion

In predicting the result for the years 2020–2021, Thailand would have a good trend of the expansion of the rice field area in the dry season because most people were eager to work. Furthermore, with the present situation, many jobs have ceased in large centers like Bangkok that has affected the movement of laborers back to their home town. As a consequence, the business sector in industrial locations has faced the retardation of the global economy after the COVID-19 crisis. This has resulted in many people moving back to work in agriculture, especially in cultivating the rice fields like their ancestors [32]. Therefore, the trend of remaining rice stubble would increase more than the previous cultivation season that would contribute to the preparation of the research and material development sector to gain more remaining rice stubble.

As a result, this could be applied to increase the economic value by reducing the ratio of the burning of rice stubble for preparing the cultivation area after harvesting, as well as reduce the chances of initiating the PM 2.5 problem in Thailand.

Therefore, the agriculturist groups cultivating rice in the North-eastern region of Thailand in the dry season had the trend of using the remaining rice stubble in cultivation, including transforming it to be handicraft products. Originally, they selected the tilling method that caused the problem of increasing rat infestation that resulted in the Leptospirosis infection. Then, the agriculturists selected to burn the rice stubble instead. However, provided that there was another method to remove the rice stubble from the cultivation, this would increase the income to return to the agriculturists in a suitable way. As the result, this was developed into creative rice stubble products in North-eastern Thailand.

The steps for the development of the procedure used this method to transform the remaining rice stubble to be wood substitute material by boiling. Then, the tissue was peeled for two hours, except for the immersing step that was conducted for 960 hours. These two procedures presented a similar result with small and soft fiber, especially for the fine spinning procedure into the pressing procedure [33]. Therefore, from the sheet pressing result, [34] the result of the analysis found the standard level value of JIS A 5908–1994. Similarly, the test result of the wood substitute material from rice stubble showed the specific gravity and the density quantity property through the standard that conformed with the test standard of industry products for particle board in Japan [35]. Additionally, the standard consisted of seven fields: 1) density, 2) quantity of the sheet density, 3) water absorption, 4) inflation during immersing with the mechanical property, 5) resistance of the bending strength, 6) internal bonding force, and 7) bonding force of the screw [35]. As a result, this showed the result from pressing when applying the wood substitute material procedure to use in the furniture industry [36].

When the remaining rice stubble was produced as the handicraft patterns, the customer groups showed satisfaction toward the furniture model from the rice stubble at the most level. Additionally, the customers displayed interest from the

*DOI: http://dx.doi.org/10.5772/intechopen.98739*

aspects of the products with the high level of interestingness and novelty by giving importance from the most to the least levels: namely, interestingness and novelty, worthiness of the transformation, beauty, convenience to apply, and others. Then, this was conformed to the concept of handicraft products by focusing on the wood substitute material procedure by increasing the values to promote the communities and local area in suitable ways [37].

The two variables of the worthiness of the transformation and interestingness and the novelty presented the correlation with the satisfaction for new rice stubble products to the customer groups. However, beauty and the convenience to apply displayed no relationship with the customer groups' satisfaction for creative ideas as the direction of designing management for economic designs [38]. Furthermore, this included a new product development method based on the potential of the community and society to learn and develop together, which was the basic alteration for the various groups to gain sustainable development opportunities in the future for their societies and local areas [39]. Thus, it would be essential to have congruence with the environmental and economic conditions in those areas for producing the wood substitute material, and to base this on the analysis of the product with the wood substitute material design as an economic direction [40].

However, the predicted solution with satisfaction of the newly designed rice stubble furniture model could explain the alteration of the customer groups' satisfaction for the rice stubble furniture products with 44.2% of the solution that was $\acute{y} = 1.473 + (0.271\ X1) + (0.312\ X2) + (-0.040\ X3) + (0.163\ X4)$. Thus, this could be considered as the predicting procedure of product images that could be considered for the future trend of newly developed products [41, 42].

The test for applying the peeling tissue procedure with rice stubble enabled the creation of different products, and a presentation of the decorated vase made of rice stubble was made [43, 44]. Then, this was used as the assessment of the comparison for the customer groups' satisfaction, and it was found that the customer groups had satisfaction to the rice stubble furniture and the decorated vases at an excellent level. As a result, there would be a better trend if the customers had an increased satisfaction level of 1.6% to the rice stubble products by focusing on some creative ideas compared to the first model products.

## Acknowledgements

This research received a grant from the budget of the Faculty of Industrial Education and Technology, King Mongkut's Institute of Technology Ladkrabang, Bangkok, Thailand.

**Author details**

Somchai Seviset* and Songwut Egwutvongsa
Faculty of Industrial Education and Technology, Department of Architecture Education and Design, King Mongkut's Institute of Technology Ladkrabang, Bangkok, Thailand

*Address all correspondence to: kssomcha@yahoo.com

## References

[1] Pongsrihadulchai, Apichart. Thailand Agricultural Policies and Development Strategies [Internet]. 2019. Available from https://ap.fftc.org.tw/article/1393

[2] Siwatt Pongpiachan, Mattanawadee Hattayanone, Junji Caoc: Effect of agricultural waste burning season on PM2.5-bound polycyclic aromatic hydrocarbon (PAH) levels in Northern Thailand. Atmospheric Pollution Research. 2017; 8(6): 1069-1080. https://doi.org/10.1016/j.apr.2017.04.009

[3] Kittiphong Phontip: The 12th National Economic and Social Development Plan and Nurse Role in Health Development. The Journal of Baromarajonani College of Nusing. 2018; 24(1): 144-150. https://he02.tci-thaijo.org/index.php/Jolbcnm/article/view/131738

[4] Roongrattana Jaroenjitt: Legal implications for agricultural land laws of the National Economic and Social Development Plan No. 1-12. Rajapark Journal. 2019; 13(28): 96-113. https://so05.tci-thaijo.org/index.php/RJPJ/article/view/187345

[5] Sharma R.:Sustainable Development: The Way for Future, Where are we?. Indian journal of community medicine : official publication of Indian Association of Preventive & Social Medicine. 2009; 34(4): 276–278. https://doi.org/10.4103/0970-0218.58381

[6] Kari Alanne and ArtoSaari: Distributed energy generation and sustainable development. Renewable and Sustainable Energy Reviews. 2006; 10(6): 539-558. https://doi.org/10.1016/j.rser.2004.11.004.

[7] IT Technology Center and Communication. Agriculturist Registration Database of Thailand. Bangkok: Department of Agricultural Extension; 2019. 112p.

[8] Yamane, Taro. Statistics, An Introductory Analysis, 2nd Ed. New York : Harper and Row; 1967. 95p.

[9] Keativipak, K., Seviset, S., Eakwutvongsa, S: Principles of thought to adopt cultural dimension into creating design in Thailand. Revista espacios. 2020; 41(31): 66-80. https://www.revistaespacios.com/a20v41n31/a20v41n31p 06.pdf

[10] Egwutvongsa, S., Seviset, S., Charoensettasilp S.: Results of artificial soil from wet garbage produced from thermal drum reprocess machine on the growth of Green Oak. AIP Conference Proceedings. 2018; 1986(1). https://doi.org/10.1063/1.5047124

[11] Jewpairojkit, P., Rattanolarn, T., Egwutvongsa, S.: Factor Analysis of Interior Design Skills in Higher Education According to the 20 Year National Strategy. International Journal of the Computer, the Internet and Management. 2019; 28(2): 11-19. https://doi.org/10.2478/mjss-2019-0048

[12] Nourbakhsh, A., Ashori, A.: Wood plastic composites from agro-waste materials: Analysis of mechanical properties. Bioresource Technology. 2010; 101(7): 2025-2028. https://doi.org/10.1016/j.biortech.2009.11.040.

[13] Lionetto, F., Del Sole, R., Cannoletta, D., Vasapollo, G., Maffezzoli, A.: Monitoring Wood Degradation during Weathering by Cellulose Crystallinity. Materials. 2012; 5(10): 1910-1922. https://doi.org/10.3390/ma5101910.

[14] W. Tangkanakul, HL. Smits, S. Jatanasen, DA. Ashford: Leptospirosis: an Emerging Health Problem in Thailand. Southeast Asian J Trop Med Public Health. 2005; 36(2): 281-288. PMID: 15916031.

[15] Soawapak Hinjoy, Somkid Kongyu, Pawinee Doung-Ngern, Galayanee Doungchawee, Soledad D. Colombe, Royce Tsukayama, Duangjai Suwancharoen: Environmental and Behavioral Risk Factors for Severe Leptospirosis in Thailand. Tropical Medicine and Infectious Disease. 2019; 4(2): 79. doi: 10.3390/tropicalmed4020079

[16] Chaowana, Pannipa: Bamboo: An Alternative Raw Material for Wood and Wood-Based Composites. Journal of Materials Science Research. 2013; 2(2): 90-102. DOI: 10.5539 / jmsr.v2n2p90.

[17] Song, J., Chen, C. and Zhu, S. et al,: Processing bulk natural wood into a high-performance structural material. Nature. 2018; 554(1): 224-228. https://doi.org/10.1038/nature25476

[18] Sumardi, I., Ono, K., Suzuki, S.: Effect of board density and layer structure on the mechanical properties of bamboo oriented strandboard. Journal of Wood Science. 2007; 53(1): 510-515. http://dx.doi.org/10.1007/s10086-007-0893-9.

[19] Apri Heri, I., Febrianto, F., Hadi, Y., Ruhendi, S., Hermawan, D.: The Effect of Pressing Temperature and Time on the Quality of Particle Board Made from Jatropha Fruit Hulls Treated in Acidic Condition. MAKARA. 2013; 17(1): 145-151. https://doi.org/10.7454/mst.v17i3.2930

[20] Kim, Taeho: Production Planning to Reduce Production Cost and Formaldehyde Emission in Furniture Production Process Using Medium-Density Fiberboard. Processes. 2019; 7(8): 529. https://doi.org/10.3390/pr7080529

[21] Pirayesh Hamidreza, Moradpour Payam, Sepahvand Sima: Particleboard from wood particles and sycamore leaves: Physico-mechanical properties. Engineering in Agriculture, Environment and Food. 2014; 8(1). http://dx.doi.org/10.1016/j.eaef.2014.07.003

[22] Laskowska, Agnieszka, Maminski, Mariusz: Properties of particleboard produced from post-industrial UF- and PF-bonded plywood. European Journal of Wood and Wood Products. 2018; 76(2): 427-435. https://doi.org/10.1007/s00107-017-1266-8

[23] Rofii, M.N., Yumigeta, S., Kojima, Y., et al,: Effect of furnish type and high-density raw material from mill residues on properties of particleboard panels. J Wood Sci. 2013; 59(1): 402-409. https://doi.org/10.1007/s10086-013-1353-3

[24] Petersen, Ann Kristin, Solberg, Birger: Environmental and economic impacts of substitution between wood products and alternative materials: a review of micro-level analyses from Norway and Sweden. Forest Policy and Economics. 2005; 7(3): 249-259. https://doi.org/10.1016/S1389-9341(03)00063-7

[25] Agapol Junpen, Jirataya Pansuk, Orachorn Kamnoet, Penwadee Cheewaphongphan, Savitri Garivait: Emission of Air Pollutants from Rice Residue Open Burning in Thailand, 2018. Atmosphere. 2018; 9(449): 1-23. doi:10.3390/atmos9110449

[26] Danutawat Tipayarom, Nguyen Thi, Nguyen Thi Kim Oanh: Effects from Open Rice Straw Burning Emission on Air Quality in the Bangkok Metropolitan Region. Science Asia. 2007; 33: 339-345. doi:10.2306 / scienceasia1513-1874.2007.33.339

[27] Czarnecki, Rafał., Dziurka, Dorota, Lecka, Janina: The use of recycled boards as the substitute for particles in the centre layer of particleboards. Electronic Journal of Polish Agricultural Universities. 2003; 6. https://www.researchgate.net/publication/265195400

[28] MATHAI, P M., Franck, Robert E.: Coir. Bast and Other Plant Fibres. Wood head Publishing Series in Textiles. 2005: 274-314p. https://doi.org/10.1533/9781845690618.274

[29] MÜSSIG, J., KARUS, M., FRANCK, R R., Franck, Robert E. CHAP. Bast and Other Plant Fibres. Wood head Publishing Series in Textiles. 2005: 345-376p. https://doi.org/10.1533/9781845690618.345

[30] Anil B. Deolalikar. ERD Working Paper No. 8: Poverty, Growth and Inequality in Thailand. Manila Philippines: Asian Development Bank; 2002. 57p.

[31] Pannarunothai, Supasit, Mills, Anne: The poor pay more: Health-related inequality in Thailand. Social Science & Medicine. 1997; 44(12): 1781-1790. https://doi.org/10.1016/S0277-9536(96)00287-0

[32] Department of Agricultural Extension. Agriculture Techniques. Bangkok: Chutimakarnpim; 1991. 35p.

[33] H. Bouafif, A. Koubaa, P. Perre, A. Cloutier: Effects of Fiber Characteristics on the Physical and Mechanical Properties of Wood Plastic Composites. Composites Part A: Applied Science and Production. 2009; 40(12): 1975-1981. https://doi.org/10.1016/j.compositesa.2009.06.003

[34] Preecha Kertkrachai. Glue and Wood Bonding. Bangkok: Kasertsart University Printing; 1988. 97p.

[35] Japan Standards Association. JIS–A 5905: Particleboards. Tokyo, Japan: 2003. https://archive.org/details/jis.a.5905.e.2003

[36] Woratham Aunchittichai. Composite Polymer from Phayasatbun Forests in Research Report of A.D. 2006. Bangkok: Forest Department Printing; 2006. 129p.

[37] Udomsak Saribud. Industrial Product Technology. Bangkok: Odean Store Printing; 2006. 116p.

[38] A Design Business Association. Professional Practice in Design Consultancy. London: The Design Council; 1992. 38p.

[39] Braxter Mike. A Practical Guide to Systematic Methods of New Product Development. London: Chapman & Hall; 1995. 105p.

[40] Ann Kristin, Birger Solberg: Environmental and economic impacts of substitution between wood products and alternative materials: a review of micro-level analyses from Norway and Sweden. Forest Policy and Economics. 2005; 7(3): 249-259. https://doi.org/10.1016/S1389-9341(03.00063-7.

[41] Cross, Nigel. Engineering Design Method: Strategies for Product Design. New York: John Wiley & Sons; 2000. 37p.

[42] Green, Peter. Design Education. London: The anchor Press. Ltd; 1979. 118p.

[43] Thai Industrial Standards Institute. Particle Board Industry Product Standard JIS A 5908-2549. Bangkok: Industrial Ministry Printing; 2007. 17p.

[44] Sparke, Penny. An introduction to DESLGN AND CULTURE. London: Twentieth Century; 1987. 59p.

Chapter 14

# Sustainable Biowaste Management in Cereal Systems: A Review

*Sanat Kumar Dwibedi, Mahendra Kumar Mohanty, Vimal Chandra Pandey and Donakonda Divyasree*

## Abstract

Among the field crops, cereals being the staple food for humans and feed for cattle, occupy 50.8 per cent of the cultivated land and contribute 52.5 per cent of the body calories. Cereals are the good source of carbohydrate, minerals, and dietary fibre for humans and animals. With the ever growing human population the agricultural production and agri-wastes are increasing across the globe. In Asia, Africa and Latin America, near about 66, 21 and 13 per cent of total estimated 2,060 Tg of biomass are generated every year. Burning has been the cheapest, simplest, easiest and quickest way of eliminating bulky unwanted biomass *in-situ* before raising of the succeeding crop(s). Rice, wheat, sugarcane and maize constitute 24, 23, 5 and 48 per cent of the global burnt residues. Although killing of problematic weeds, insects, and pathogens, and addition of valuable plant nutrients are the very basic objectives of this anthropogenic post-harvest residue management strategy but it releases noxious gases into the atmosphere polluting air and contributing to the global warming. Shorter sowing windows very often compel the farmers to remove crop residues through burning, especially in absence of alternative options for its productive and profitable disposal. Rising labour cost and their seasonal scarcity sometimes also insist the farmers to burn crop residues. However, stringent punitive actions have yet failed to curb such open burning in many countries in absence of the farmers' friendly and financially viable options of crop residue management. In this chapter, attempts have been made to elucidate various sustainable crop residue management strategies in cereal systems.

**Keywords:** biofuel, biomass, bio-waste, cereal system, residue management

## 1. Introduction

In this 21st century, feeding the teeming millions is the greatest challenge before us. The green revolution of 1960s although could alleviate the growing demands for food to a great extent but at the cost of food quality and environmental health. Long term applications of chemical fertilizers in crop production systems have been resulting in unpredictable reduction in yields and increase in the cost of cultivation. However, with the shrinking land, the burgeoning competition for water, land and other resources from non-agriculture sector might aggravate agricultural production in the near future. Hence, intensive farming through enhanced cropping intensity has been the way out for mitigating the population driven demands for food, feed, fodder and fibre without much scope for recycling of the agricultural



IntechOpen

wastes in majority of cases. For raising of multiple crops in a year the farmers burn farm residues *in-situ* in absence of appropriate sustainable recycling technologies that ultimately pollute the environment and increase carbon foot print. Globally, annual plant waste production is estimated at around 5.5 billion tons in 2013 that accounts for 13 per cent of the greenhouse gas emission from agriculture sector [1]. Some of these wastes are although used as cattle feed and organic manure but a plenty are still available for alternative uses. Hence, it is high time to adopt the best possible technologies for recycling of bio-wastes from crop fields for harnessing the nutrients and green energy as well.

Near about half of the habitable land on this planet is under agriculture [2]. Of the 1,600 million hectares of cultivated land [3], 50.8 per cent is occupied by cereals [4]. About 52.5 per cent calories for humans are available from cereals at global scale [4] with major contributions from corn (1,116.34 million tons), wheat (764.49 Mt), rice (495.78 Mt), barley (156.41 Mt), sorghum (57.97 Mt), oat (22.83 Mt) and rye (12.17 Mt) [5]. Cereals are special not because of their uses as staple food but due to production of ethanol and cattle feed in many advanced nations. However, in many underdeveloped and developing countries this precious wealth has not yet been fully utilized [6]. It is high time to use this precious waste from crop field in judicious manner not only to recycle the carbon and sequester it back into the soil but also to harness clean and green energy out of it through appropriate measures.

## 2. Types of agricultural biowastes

Farm residues can be broadly divided into crop residues, and wastes from livestock and aquaculture depending on the activities carried out. Field crop residues are plant parts left over in the field without much attention unless otherwise is immediately followed by a succeeding crop. Crop residues can be put under

| Agro-industrial wastes | Chemical composition (%) | | | | | | References |
|---|---|---|---|---|---|---|---|
| | Cellulose | Hemi-cellulose | Lignin | Ash (%) | Total solids (%) | Moisture (%) | |
| Sugarcane bagasse | 30.2 | 56.7 | 13.4 | 1.9 | 91.66 | 4.8 | [8, 9] |
| Rice straw | 39.2 | 23.5 | 36.1 | 12.4 | 98.62 | 6.58 | [8] |
| Corn stalks | 61.2 | 19.3 | 6.9 | 10.8 | 97.78 | 6.40 | [8] |
| Sawdust | 45.1 | 28.1 | 24.2 | 1.2 | 98.54 | 1.12 | [8, 10] |
| Sugar beet waste | 26.3 | 18.5 | 2.5 | 4.8 | 87.5 | 12.4 | [8] |
| Barley straw | 33.8 | 21.9 | 13.8 | 11 | _ | _ | [9] |
| Cotton stalks | 58.5 | 14.4 | 21.5 | 9.98 | _ | 7.45 | [9] |
| Oat straw | 39.4 | 27.1 | 17.5 | 8 | _ | _ | [10] |
| Soya stalks | 34.5 | 24.8 | 19.8 | 10.39 | _ | 11.84 | [11] |
| Sunflower stalks | 42.1 | 29.7 | 13.4 | 11.17 | _ | _ | [11] |
| Wheat straw | 32.9 | 24.0 | 8.9 | 6.7 | 95.6 | 7.0 | [9, 10] |

**Table 1.**
*Chemical composition of agri-industrial wastes [7].*

agricultural and agri-industrial categories. Agricultural residues remaining after threshing and separation of the economic plant part(s) can be of (a) processed residues such as husks and hay and (b) field crop residues such as stalks and stubbles. Husk and hay are often left over in the crop field due to engagement of crop combined harvesters and axial flow threshers. Sometimes, the distance between crop field and farm house plays a decisive role in stacking of hay and husk in the crop field after threshing. That apart, farm mechanization in many developed countries has also shifted from animal driven to fossil fuel based farm power and thus excluding the need for gathering feedstock in the haystack. Furthermore, the risk of fire in haystack due to storage of dried crop residues can also not be eliminated completely. Hence, many farmers are not interested in transporting such bulky by-products from crop field to farmhouse. In mono cropped areas, natural weathering and decomposition by soil organisms usually degrade the field crop residues but the residue management challenge is mostly under sequential cropping.

Agri-industrial residues in the other hand are derived from industries such as peels of potato, orange, and cassava; bagasse and molasses of sugarcane; oilcakes of groundnut, mustard, sunflower, sesame, soybean and coconut; and husks and bran of rice [7]. Huge quantities of organic wastes are produced by food and vegetable oil processing industries every year- but if left untreated and unutilized, may cause environmental pollution as well as human and animal health issues. A representative chemical analysis report of few agri biomass are depicted under **Table 1** for comparative studies.

## 3. What is agricultural burning?

Agricultural burning is the intentional setting of fire in the open field for preparation of the land for the next crop or killing the weeds and insect pests. Natural causes such as lightening and planned anthropogenic fire account for only 10–20 per cent of the total open burning across the globe [12]. Burning of agricultural residues is different from fire in forests, grasslands or any vegetation.

## 4. Drivers of open burning of crop residues

Slash and burn cultivation has been a traditional system in agriculture to clean up vegetation on virgin land and cultivate crops for a few years before shifting to a new area. Tradition, timing, ease, weather and location factors encourage the farmers to burn residues in many regions. Burning is the cheapest and quickest way of eliminating unwanted thrash from the crop fields. Addition of plant nutrients and killing of pathogens, insects and weed species also influence decision to burn residues *in-situ* [12]. Moreover, shorter sowing window for one or two weeks compels the farmers to remove crop residues through burning, especially in absence of alternative options for its productive and profitable disposal. So also, rising labour cost and their seasonal scarcity insist the farmers to burn residues *in situ*. However, absence of stringent punitive action very often fails to curb such open burning of crop residues.

## 5. Status of agricultural burning in the world and India

Burning of crop residues *in situ* has been a traditional practice in many countries as it is the cheapest, easiest and quickest way of getting rid of such bulky materials immediately before raising the succeeding crop. It also checks weed and pest infestation in the succeeding crop. More so over, it does not require much technical skill and expertise in doing so. This practice was widespread and popular across the globe until 1990s when many governments restricted open burning of crop residues. In China and England, stubble burning is banned but in Australia, it is restricted to only need-based burning. In America and Canada, residue burning is still allowed in some counties and provinces. China, India and America are in the forefront of burning of crop residues followed by Brazil, Indonesia and Russian federation. Some African countries have top rated in the global ranking of intensive burning per hectare. Mexico and Tanzania are at the top position in intensive burning followed by Brazil, United States and Nigeria [12]. Globally, 24 per cent of the burnt residues are from rice, 23 per cent from wheat, 5 per cent from sugarcane and 48 per cent from maize [12]. An estimate in 1985 indicated burning of 2060 Tg of biomass in the developing world; of this, Asia, Africa and Larin America contributed 66, 21 and 13 per cent, respectively [13].

Large scale burning of paddy stubbles in Punjab, Haryana and western Uttar Pradesh in India in the month of late October and November every year is estimated to be 35 Mt. This practice is spreading to other parts of the country like wildfire due to the advent of precision farm-equipments that allow resowing with the minimum soil disturbance. The crop field is made ready for the succeeding zero till wheat crop by burning of straw and stubbles leftover in the field from the crop harvested by combined harvester. India generates around 500 Mt. of residues from rice, wheat, sugarcane, maize, millet and other crops every year [14] of which 142 Mt. are

| Country | Agricultural waste generated (million tons/year) |
|---|---|
| Myanmar | 19 |
| Indonesia | 55 |
| Bangladesh | 72 |
| India | 500 |

**Table 2.**
*Agricultural waste generation in India and adjacent countries [15–17].*

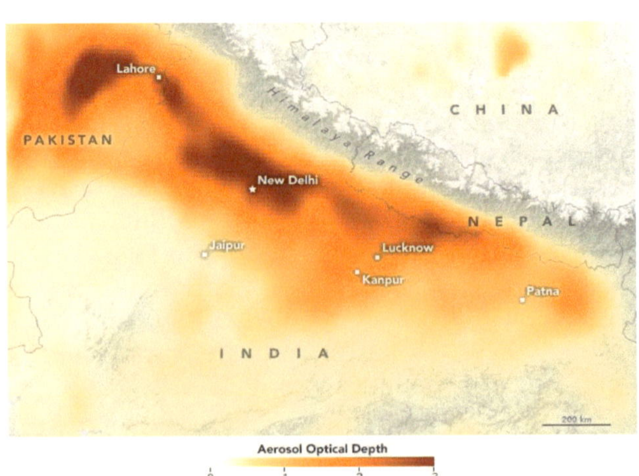


**Figure 1.**
*NASA Earth Observatory image of the aerosol pollution in India, Pakistan and Nepal on 7 November 2017 [21].*

leftover after fuel, fodder and industrial uses [15] and 92 Mt. are burnt every year across the country. **Table 2** compares the agricultural wastes generated in India and its adjacent countries which reveal that the volume of waste is far more than the total waste generated by other countries.

Near about 70 per cent of crop residues in India are cereals of which 34 per cent come from rice, and 22 per cent from wheat crops [14]. Estimation indicated burning of about 80 per cent of the total 20 Mt. of rice stubble in Punjab alone [14]. Whereas another estimate indicated 9.8 and 1.23 Mt. of rice residue-burning in Punjab and Haryana, respectively [18]. Burning of rice is more compared to wheat in the North West India as rice contains more silica (12–16 per cent vs. 3–5 per cent) which is not easily digestible. About 75 per cent of wheat straw is collected and stored as fodder. Rice stem contains lower silica than leaves and hence rice is to be cut as close to ground if used for feeding animals [19]. Management of rice straw is difficult compared to wheat due to shorter window for sowing of wheat and low temperature which compels the farmers to resort to burning during October–November every year [20]. Several major cities of North India—including New Delhi, Lucknow, and Kanpur—faced elevated levels of aerosol pollution [21]. The extent of aerosol pollution in India, Pakistan and Nepal region, mostly from crop residue burning, can be observed from the captured image of the NASA's Moderate Resolution Imaging Spectroradiometer (MODIS) on Aqua satellite on 7 November 2017 (**Figure 1**).

## 6. Legal implications

Agriculture comes under the state list of the Seventh Schedule of the Constitution of India and hence, the State Governments have to take austerity measures against residue burning. Burning of crop residues is a crime in India according to the Air Pollution Act, 1981 and Section 188 of the Indian Penal Code [14]. Courts in India have banned open burning of crop residues and made provisions of penal actions by collecting fines from the errant farmers. In 2018, the national green tribunal (NGT) of India imposed penalty of Rs.2,00,000 on the Delhi government for not filing an action plan for incentives and infrastructural assistance against stubble burning [22]. Subsequently, the NGT asked the Delhi government to deposit 250 million rupees (INR) with the Central Pollution Control Board (CPCB) as performance guarantee [23]. Consequent up on Public Interest Litigation in M. C. Mehta vs. Union of India (order IA No.158129 and 158129 of 2019 in writ petition (C) No.13029 of 1985) [24] an ordinance dissolving the Environment Pollution (Control and Prevention) Authority has been passed by the Indian government to set up a new Commission with over 20 members to regulate pollution in Delhi-NCR region [25]. In this ordinance, the Ministry of Law and Justice has made provisions for imprisonment up to five years or with fine up to rupees one crore or both for abrogation of the rule/provisions or order/directions of the Commission [25]. The Hon'ble Supreme Court of India has also realized the need for incentives to small and marginal farmers those abiding to the rules by paying a sum of Rs.100 per quintal of crop residues [24].

In the United States, agricultural burning policy has been formulated to monitor open burning of agricultural wastes and weeds for fire, weed and pest control adjacent to the crop field so as to allow regulated burning in small scale to maintain agricultural production but without impairing public health and air quality parameters. The agricultural burning managers are authorized to monitor burning at state, local and tribal level and no burning should be carried out without approval from the competent authority [26].

**Figure 2.**
*Open burning of rice straw before land preparation for second rice crop in Bargarh district of Odisha.*

## 7. Environmental impacts of open crop residue burning

Cereals generate huge agriculture as well as agri-industrial wastes across the globe. If not managed judiciously then in long run, that may lead to the environmental pollution and global warming. Open burning of agricultural wastes is detrimental to both environment and human health. Poisonous gases like carbon dioxide ($CO_2$), carbon monoxide (CO), sulfur oxides ($SO_x$), nitrogen oxides ($NO_x$), methane ($CH_4$) and particulate matters ($PM_{2.5}$ and $PM_{10}$) are released into the atmosphere (**Figure 2**). An estimate reveals burning of crop residues release 149.24 Mt. of $CO_2$, 9 Mt. of CO, 0.25 Mt. of SOx, 1.28 Mt. of PM and 0.07 Mt. of black carbon [14]. The situation is austere in India due to intensive rice-wheat cropping system [27]. One ton of stubble burning leads to the loss of 5.5 kg nitrogen, 2.3 kg phosphorous, 25 kg potassium and 1 kg sulfur besides organic carbon [14]. As per an estimation, stubble burning releases substantial quantity of heat that elevates the surface temperature from 33.8 to 42.2°C killing soil fertility maintaining biota [14]. The population of microorganisms, earthworms and beetles get reduced drastically in the upper layer of soil affecting the rate of soil formation. The population of beneficial insects reduces drastically and the enemy inset population increases to a great extent.

Stubble burning increases the particulate matters in the air creating pulmonary diseases (COPD), bronchitis, lung capacity loss, emphysema, cancer, etc. [27] in humans and animals besides irritation in eyes, nose and throat [14]. The Ministry of Earth Sciences' monitoring agency SAFAR in Delhi has estimated the share of stubble burning in $PM_{2.5}$ pollution as high as 36 per cent [28]. In Punjab (India) alone an estimated 760 million rupees (INR) is spent annually to alleviate stubble burning related diseases [27]. The Energy and Resources Institute (2019) reported 5 million deaths in South Asia in 2012 due to air pollution which was around 22 per cent of the total deaths in the region [27].

## 8. Agri-waste management options

The crop stubbles and agricultural wastes, if managed properly, could generate profits to the farmers and protect the environment from the severe pollution as well. Some of the available alternative management practices include soil incorporation, compost and biochar making, thermal power generation, pulp and

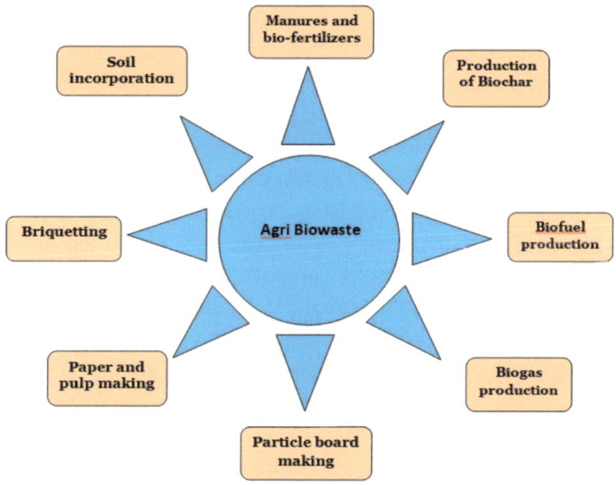


**Figure 3.**
*The management of agricultural wastes.*

paper manufacturing, cement brick making, mushroom production, or biofuel production (**Figure 3**) [27]. However, most of the farmers in North India are not yet fully aware of many such alternatives that lead to *in-situ* open burning of crop residues.

### 8.1 Residue incorporation

Since long back, *in-situ* incorporation of crop residues has been the simplest, easiest, quickest and cheapest technique of agri-waste management next to open burning. Rotavators, soil turning or mould board ploughs, and puddlers are most effectively and widely used these days to shred down larger plant parts and to incorporate into the soil. Rotavator, a low cost precision implement of around 0.1 to 0.12 million rupees (INR) with the working efficiency of 5–6 hectares per day is suitable for both *kharif* and *rabi* crops. Residue incorporation improves soil bio-physicochemical properties and increases crop productivity as well. Rotavator readily and economically incorporates biomass of weeds and green manuring crops as manual removing or chopping and mixing would cost higher [29].

Incorporation of maize crop residues in clayey Andosol in Ethiopia at 6 Mg per hectare for consecutive three years indicated 22–52 per cent reduction in penetration resistance in top 5–10 cm soil, 39–57 per cent lower evaporative flux and elevated (22 per cent) macro and meso porosity [30]. After 17–18 cycles of residue incorporation in rice-wheat system, the mean weight diameter (MWD) of water stable aggregates, bulk density (BD), and water holding capacity (WHC) of soil increased [31].

Crop residues on the soil surface protect the soil from erosion, act as mulch that keep the soil cool and improves soil tilth [32]. In the USA, near about 40 per cent cropland are under no till farming with minimal investment and more than 10 M ha has been sown under cover crops with basic objectives to incorporate residues *in situ* and regenerate crop without tillage [32]. In Indo-Gangetic plain of India, mulching with the preceding rice residues has been a good agronomic practice in absence of tillage that increases soil organic carbon in long run, WHC, water use efficiency and profitability in wheat [33]. Surplus residues of the wheat crop can also be incorporated into the next rice crop with improvement in physical, chemical and biological properties of the soil [33].

Rapid reduction in the soil organic carbon (SOC) across the globe due to intensive monocropping without biomass incorporation has been the greatest challenge before us in this 21st century. With the changing climate and advent of chemical farming, the role of soil in maintaining the ecosystem services has brought forth so many issues and if left unattended may end in peril. About 29 and 60 per cent increase in carbon stocks in silt-loam and clayey soil in top 20 cm soil whereas the effect was seen in the upper horizon only in sandy soil has been reported [34].

Residue incorporation needs energy and time. Extra N at the time of incorporation is needed for preventing temporary immobilization of nutrients (mostly N) and correcting high C:N ratio of substrates [35]. The rate of immobilization lasts for four to six weeks under favorable soil type, moisture and temperature conditions and management factors. Starter N dose of 15 to 20 kg $ha^{-1}$ could very well increase the yield of succeeding wheat or rice crop without any adverse effect on the next crop. Wheat yield depression of 0.54 to 0.08 t $ha^{-1}$ has been reported with soil application of 60 and 180 kg N $ha^{-1}$, respectively [36]. However, release of greenhouse gases such as $CO_2$ and $CH_4$ that leads to global warming can also not be set aside. Incorporation of cereal straw (having wide C:N ratio) with green manure (having narrow C:N ratio) facilitates decomposition before rice transplanting. Wheat yield reduction in initial 2 to 3 years of rice straw incorporation a month before wheat planting were although reported but in subsequent years, straw incorporation had no significant adverse effect on wheat yield. Rather, wheat yield increased by 0.6 t $ha^{-1}$ over 2.91 t $ha^{-1}$ with straw removal [36]. In contrast, yield advantage in wheat sown after 3 weeks of rice straw incorporation was reported in clay loam soil but not in sandy loam soil. After incorporation of rice straw, about 10–20 per cent of it is assimilated by the rice crop itself, 10–20 per cent is lost to the atmosphere through various pathways and 60 to 80 per cent is immobilized in soil [36]. Nutrient up to 40 kg $ha^{-1}$ could be harnessed through incorporation of 10 t of rice straw 4 to 5 weeks before transplanting of rice in the main field [36]. Residue incorporation increases soil N and available P and K [36]. Long-term comparative studies on wheat crop residue incorporation versus inorganic fertilizer application in India showed significantly higher yield in rice and wheat through inorganic nutrition but in subsequent years, the yield under residue incorporation plus inorganic fertilizer was at par with sole inorganic one. In the fourth year, the combined mode of nutrition out-yielded the inorganic one [36].

### 8.2 Composting

Composting is the method of aerobic or anaerobic decomposition of organic solid wastes. It is not new; rather, it has been the oldest practice of recycling the plant nutrients in the soil. Small scale backyard composting is a usual practice in many developing and underdeveloped countries. Up till now, composting had not gained the status of agriculture industry. But with the gaining popularity of organic farming or eco-farming, its demand has increased these days. Its bulkiness, low nutrient content and high labour requirement are the major challenges in undertaking such organic waste composting projects. However, on-site composting without transportation of crop residues could be the befitting answer for maintaining soil fertility and sustaining crop production in long run. Compost improves bio-physiochemical properties of the soil while the need for synthetic fertilizers and plant protection measures could be eliminated completely. Its application improves nutrient uptake and cycling, soil microbial activity and biodiversity, and deficit moisture stress conditions as it regulates soil pH, improves soil texture, structure and aggregates, increases water holding capacity, cation ion exchange capacity and soil biodiversity [37]. It reduces soil erosion, protects crop against soil borne diseases, increase carbon sequestration and reduce compaction [37]. Composting releases heat during thermophyllic stage that kills most of the pathogens, insect larvae and eggs, and weed seeds [37].

On decomposition, biomass turns into a humus like substance called compost. The rate of compositing depends on the type of substrate and microbes, ambient air temperature, moisture level, aeration, presence or absence of toxic chemicals and heavy metals and surface area of the residue. Aerobic decomposition releases $CO_2$ and $H_2O$ while anaerobic composting releases $CH_4$.

$$C_6H_{12}O_6 + 6O_2 \rightarrow CO_2 + H_2O + \Delta E\left(3{,}880\,\text{kJ mol}^{-1}\right) \quad \text{Aerobic decomposition}$$

$$C_6H_{12}O_6 + 2H_2O \rightarrow CO_2 + H_2O + \Delta E\left(405\,\text{kJ mol}^{-1}\right) \quad \text{Anaerobic decomposition}$$

The total carbon and nitrogen (C:N) ratio of the substrate is important for deciding the rate of decomposition of organic matter. Higher the ratio then longer is the duration for degradation. The desired C:N ratio for decomposition is 24:1 [38]. This 24 part of carbon is divided into 16 parts for energy and 8 parts for microbial body as most microbes have a body with C:N of 8:1 [38]. When C:N ratio exceeds 24 then microbes explore other available sources with moderate ratio. Immediately after addition of biomass, the microbial population increases resulting in immobilization i.e. transformation of N from available form to non available form. When these microbes die and decompose, the N mineralizes and becomes available for crop removal. Cereals have higher C:N ratio than legumes and hence, legumes decompose faster [38]. The **Table 3** depicts C:N ratio of different agricultural crops.

The C:N ratio changes with stage of the crop. It also differs in different plant parts and with the progression of decomposition [38]. Cereals take longer period for composting that can be reduced by mixing with legumes or supplementing nitrogenous fertilizers. In compost pits cereal substrates are put in alteration with the vegetables or pulse residues. For example, rice straw and grass put together resulted in the highest rate of vermicompost production at the end of 120 days cycle compared to either of these substrate composted separately [39]. Similarly, [40] suggested addition of food stuff with rice bran for getting superior vermicompost with average C:N ratio of 20.85, 183.3, 16.86 and 15.16 from 1:1, 1:2, 1:3 and 1:5 ratio of rice bran: food stuff, respectively.

Crop residues are used for vermicomposting, enriched composting, farm yard manure, etc. Vermicomposting is the biological degradation of substrates by combined action of earthworms and microorganisms. Windrows method of vermicomposting is popular and widely practiced by adding rice straw, animal manure, and

| Name of crops | C:N ratio |
|---|---|
| Wheat straw | 80:1 |
| Rye straw | 82:1 |
| Oat straw | 70:1 |
| Rice straw | 67:1 |
| Corn stover | 57:1 |
| Legume hay | 17:1 |

**Table 3.**
*C:N ratio of different agricultural crops at harvest [38].*

shredded banana trunks and maintaining the moisture at 60 per cent [41]. Tank, pit or heap method of vermicomposting can be followed as per convenience and quantity of available residues to be managed. Spent straw from mushroom farm containing C and N of 14.3 and 0.7 per cent can also be recycled through composting [41]. *Eisenia fetida* is the most widely used species of earthworm for vermicomposting in many parts of the world. However, *Lampito mauritii*, *Lumbricus rubellus*, *Eudrilus eugeniae* and *Perionyx excavates* are also inoculated depending on the purpose of composting, availability of culture and ecological conditions.

Unlike open burning, composting preserves essential plant nutrients and almost all nutrients remain inside the compost. Only the loss of N occurs in form of ammonia and nitrous oxide due to volatilization [42]. As much as 75 per cent of total N in manure is lost in form of $NH_3$ and 1.5 to 7.3 per cent in form of $N_2O$ [43, 44]. Most composts do not contain more than 2 per cent N and its release depends on the C:N ratio, soil temperature, moisture and microbial activity [44]. Composts are better supplements for crop plants unlike most chemical fertilizers that are devoid of trace or micronutrients. The CHNS analyses of rice straw and its compost revealed increase in oxygen, sulfur and moisture but reduced total organic carbon, hydrogen and nitrogen [45]. Application of effective microorganisms (EM) to composting rice is reported to have increased macro and micronutrient content. The N, P and K content of the rice-compost is higher with EM and the Fe content was significantly higher without significant increase in Zn and Cu [46].

### 8.3 Production of biochar

Production of biochar or pyrogenic carbon was the age-old practice in the Amazonian river bank which was evident from the *Terra preta* culture to enrich soil and cultivate crops sustainably. Modern researchers are constantly looking for imbibing such technology to smother greenhouse gas emissions and increase carbon sequestration as well. The process of carbon sequestration needs higher residence time and resistance to chemical oxidation of carbon to carbon dioxide or methane [47]. Use of agricultural wastes for biochar-making could be a viable option in the era of massive deforestation and loss of habitats. Biochar is a porous fine-grained carbonaceous material released from thermo-chemical conversion of biomass called pyrolysis at relatively moderate temperature [48]. It contains carbon, hydrogen, oxygen, nitrogen, sulfur and ash. On its addition, the bio-physicochemical properties of the soil improve and crop yield enhances. Apart from agriculture-use, biochar is used in water treatment plant, food and cosmetic industry, metallurgy, construction industry and many more purposes.

Researchers have observed that the pyrolytic temperature of 400°C brings in high alkalinity, cation exchange capacity, high level of available P and exchangeable cation in rice straw biochar which is suitable for soil amendment and used as fertilizer [49]. At this temperature, rice straw biochar shows the largest Cu (II) absorption capacity (0.37 mol $kg^{-1}$) that is mostly of non electrostatic absorption [50]. Corn stalk biochar can also be used as efficient absorber of $Pb^{+2}$ [51] and $Cd^{+2}$ [52]. Continuous application of rice straw biochar and rice straw has positive influence on soil physicochemical properties with 26.9 and 70.2 per cent increase in total porosity and air permeability [53]. Its application increases soil microbial biomass carbon and nitrogen [53] and increases wheat productivity and accumulate P in grain [54]. Corn cob biochar is reported to have increased the pH, organic matter, soluble and available K in calcareous sandy soil [55]. Maize straw biochar application to soil reduced harmful bacteria diversity but selectively promoted community of functional bacteria population [56]. The C sequestration capacity of corn stalk (0.26) was increased to 0.64 to 1.0 on charring as resistance of char to decomposition prohibits C losses during charring [57].

### 8.4 Biofuel production

Plant residues contain cellulose, hemicelluloses and lignin with small fractions of sugars, pectin, protein, nitrogenous, lipids, tanins and inorganic materials [58]. Lgnin mostly provides the structural support and is almost resistant to chemical reactions and biological degradation compared to cellulose and hemicelluloses and thus resists fermentation [59–61]. In crop plants, the nonfood portion such as stalk, husk, straw, stover and bare corn cob contain lignocellulosic biomass. As in agriculture, cereals occupy the maximum area and production so also the largest quantity of such lignocellulosic materials. The residue management in cereal-cereal system such as in rice-rice and rice-maize/wheat is the biggest challenge before researchers. Very often the farmers opt for onsite open burning of the crop residues to get rid of huge biomass with higher lignocellulosic materials in it [62]. But with the advent of innovative green energy technologies, such so called wastes are now converted into precious biofuels to mitigate the growing demands.

Biofuels are produced through pretreatment of lignocellulosic materials by fungi, bacteria and enzymes that break down the lignin, a complex polymer and degrade cellulose and hemicelluloses to corresponding monomers and sugars for effective fermentation and fuel conversion [63]. The pretreatment is mostly chemical or biological but it could be mechanical and physicochemical too that result in increased surface area and porosity, and decrease in crystalinity. Biomass degradation results into ethanol, biodiesel, biobutanol, syngas, and woodtar/oil. The ethanol produced from crop residues is known as 2G bioetahol. Depending on the feedstock and process design, several by-products such as stillage, evaporator condensate and solubles, spent cake and/or distiller's grains are produced which can be used in agricultural amendment, civil construction or sanitary landfills. Stillage is a nutrient rich biodegradable material rich in both total suspended solids (TSS) and Chemical Oxygen Demand (COD) that requires significant processing for remediation. Lignin, a waste product from bioethaol plant is used for generating heat energy required for other processes and thus the final produce is in a form of ash. Ash is alkaline in reaction with significant quantities of Si, P, K, Ca, AL, Fe, and Mg in it which can very well be used in agriculture. In **Figures 4** and **5**, the harvested paddy straw is gathered by square baler and stacked in the collection centre at Thuapali village of Bargarh distract in Odisha, India as a pilot study programme under the direct supervision of the BPCL, India.

**Figure 4.**
*Post harvest operation of square baler in farmer's field in Bargarh district of Odisha, India.*

**Figure 5.**
*Stacking of square bales of rice straw in stockyard in Bargarh district of Odisha.*

## 8.5 Biogas production

Anaerobic digestion of biomass produces biogas, a renewable energy containing methane as primary constituent and a final solid nutrient rich residue. Stages of anaerobic digestion include hydrolysis, acidogenesis, acetogenesis and methanogenesis. In hydrolysis, the water splits into $H^+$ and $OH^-$. Larger polymers such as proteins, fats and carbohydrates breakdown to smaller monomers such as amino acids, simple sugars, and fatty acids in presence of an acid catalyst. In acidogenesis, acidogenic bacteria further break down organic matter still too large for methane production. Acetogenesis is the formation of acetate by acetogens for further breaking down of the biomass to a point from where methanogens can further act and degrade the remaining material to generate methane as biofuel [64]. Dried cereal crop residues should not be directly injected into the biogas unit rather mixing of animal dung in partial combination is preferable to increase the biogas efficiency. However, maize silage can be directly used for biogas production [65]. Biogas generation technology is older than biofuel production technology. The methane production potential of wheat straw is of 0.145 to 0.39 $m^3\ kg^{-1}$ and rice straw of 0.241 to 0.367 $m^3\ kg^{-1}$ [15]. By 2030, grasses and cereals could be the primary source of biomass for the biogas plants across the globe [66]. **Table 4** enlists the major composition of bio-wastes from major crops.

## 8.6 Particle/composite board making

Rice husk and cereal straw are used for making of particle boards. Rice husk is cleaned and cereal straw thus defibred into particles is mixed with rice husk

| Source | Composition |
|---|---|
| Rice | Husk, bran and straw |
| Wheat | Bran and straw |
| Maize | Stover, husk, and skins |
| Millet | Stover |

**Table 4.**
*Residues produced from major crops [67].*

at desired proportion and then blended with cashew nut shell liquid or cardanol phenol formaldehyde resin [68]. The mixture is spread into a mat or layer of uniform desired thickness and hot pressed like conventional method of particle board making [68]. Rice husk is 20 per cent of total rice produced which can be used as cheaper, lighter, denser, stronger, durable and more uniform substitute for conventional wooden and ply boards thereby protect against deforestation and environmental degradation. Because of high Si content rice husk is difficult to burn. Apart from rice husk, rice and wheat straw can also be used for making straw-wood particle composite boards and insulation boards. However, use of rice husk in comparison to bamboo for particle board making resulted in poor quality due to higher Si content in rice husk and non-availability of suitable blender for effectively binding rice husk [69]. Advance researches are still continuing to develop an efficient and effective adhesive for rice husk boards.

### 8.7 Paper making and packaging

Rice straw can be used as raw material for making quality paper. It contains lesser lignin compared to conventional wood and thus requires milder chemical pre treatment. Cheaper soda and soda-AQ methods are used for making paper in many developing countries but blending pollutes water by releasing more than 500 chlorinated compounds that are highly toxic, bioaccumulative and carcinogenic [70]. The graduates of IIT, Delhi have developed a pulp making process in a start up called Kriya Labs [71] that can be used in making paper, plates and cups [72].

Bio-Lutions India in Bengaluru purchases crop wastes from farmers and transforms them into biodegradable packaging materials for fruits and vegetables which can be degraded completely within three months [72]. Bio-plastics, derived from rice straw by mixing with starch, cellulose, glycerol and protein are ready to substitute the conventional plastic very shortly as it is readily biodegradable within 180 days of use compared to 500 years required for plastics to degrade [72].

### 8.8 Briquetting

The straw functions very well as bedding for animals such as horses. When briquetted, straw absorbs 5 times more fluid than normal straw for bedding. This minimizes the cleaning work in the stable, and creates a better environment for the animals. Furthermore, briquetted straw is useful for burning, and it is an excellent source of energy through generating heat, steam and electricity in conventional boilers or gasification plants.

## 9. Conclusion and future outlook

With the advent of modern scientific agricultural practices, the agriculture and agri-waste production have increased at exponential rates across the globe. Cereals, being the staple food for humans as well as feed for cattle, contribute the most to the pool of such agri-wastes. Sustainable management of crop residues, especially in cereal systems, has been the greatest challenge before us in this world with the ever burgeoning population, agricultural production and economic growth. Rice and wheat contribute the most to the agri bio-waste pool due to wider cultivation and large scale production. However, many countries in Asia, Africa and America, at present, have failed to cope up with the large volume of crop residues although a majority of these are used as fodder and fuel. In India, northern states such as Punjab, Haryana, and western Uttar Pradesh burn crop residues in the month of

October and November every year thereby releasing toxic fumes into the atmosphere that are very often drifted to the adjacent cities and states. Most of these residues are byproducts of wheat and rice. Small farmers usually resort to burning of crop residues as it is the inexpensive alternative in absence of technical know-how on any other better profitable and sustainable residue management or disposal opportunities.

Large scale burnings of crop residues shockingly increase air pollution and serious health issues. In the past few decades, the authorities have relentlessly tried to explore multiple waste management options to cater such unequivocal but perilous agri-wastes from the cereal systems. The possibilities of waste incorporation and decomposition through soil addition and composting are few preferred acceptable alternatives. Penal actions have also been provisioned against the errant promoters of open burning. In this line, in India, the National Remote Sensing Agency (NRSA) and Central Pollution Control Board (CPCB) have come together to monitor open burning of crop residues through aerial surveillance and to penalize farmers for doing so. However, continued air pollution in the month of November and December in spite of much touted successful, sustainable and effective actions against open burning has raised many eyebrows. Hence, efforts are being made to explore farmers' friendly and financially viable options of residue management such as composting, biochar making, biofuel and biogas production, particle and composite board making, paper manufacturing, etc. In many developed countries, 1G and 2G ethanol production have now gained momentum that use waste biomass judiciously for generation of liquid and gaseous fuels. Corporate social responsibility (CSR) funds are being allocated in many countries for conducting research and development on large scale profitable biofuel production. It is high time to develop national gas-grid line with the support of remote sensing and GIS tools to monitor and regulate biomass production and utilization. Community biomass collection centres could facilitate easy and speedy collection and back up storage of biomass for further residue management strategies. And importantly, the residue management options should involve environment, education, social, and economic sectors holistically in addition to agriculture and energy sectors beyond the disciplinary boundaries.

## Conflict of interest

The authors declare no conflict of interest.

## ORCID of the authors

https://orcid.org/0000-0003-2093-7397 (SK Dwibedi).
https://orcid.org/0000-0003-2250-6726 (VC Pandey).

## Author details

Sanat Kumar Dwibedi[1*], Mahendra Kumar Mohanty[2], Vimal Chandra Pandey[3] and Donakonda Divyasree[1]

1 Department of Agronomy, College of Agriculture, Odisha University of Agriculture and Technology, Bhubaneswar, India

2 Department of Farm Power and Machinery, College of Agricultural Engineering, Odisha University of Agriculture and Technology, Bhubaneswar, India

3 Department of Environmental Science, Babasaheb Bhimrao Ambdekar University, Lucknow, Uttar Pradesh, India

*Address all correspondence to: sanatdwibedi@rediffmail.com

## References


[1] Wight A. Scientists around the world are turning agricultural waste into food, packaging and skincare products. 2017, May 17. URL: https://www.greenbiz.com/article/scientists-around-world-are-turning-agricultural-waste-food-packaging-and-skincare-products

[2] Ritchie H. Half of the world's habitable land is used for agriculture. 2019, November 11. URL: https://ourworldindata.org/global-land-for-agriculture

[3] FAO. The state of the world's land and water resources for food and agriculture (SOLAW) - Managing systems at risk. Food and Agriculture Organization of the United Nations, Rome and Earthscan, London. 2011; **1**:21-60. URL:http://www.fao.org/3/i1688e/i1688e03.pdf

[4] FAO. World cropped area, yield and production of cereals. 2021. URL: http://www.fao.org/economic/the-statistics-division-ess/chartroom-and-factoids/chartroom/35-world-cropped-area-yield-and-production-of-cereals/en/

[5] Shahbandeh M. Grain production worldwide 2019/20, by type. 2020, November 10. URL:https://www.statista.com/statistics/263977/world-grain-production-by-type/

[6] Shahane A and Shivay YS. Cereal Residues - Not a waste until we waste it: A review. International Journal of Bio-resource and Stress Management. 2016; 7(1):162-173 DOI: https://doi.org/10.23910/IJBSM/2016.7.1.1401b

[7] Sadh PK, Duhan S and Duhan JS. Agro-industrial wastes and their utilization using solid state fermentation: a review. Bioresour. Bioprocess. 2018;**5**: 1. DOI:https://doi.org/10.1186/s40643-017-0187-z

[8] El-Tayeb TS, Abdelhafez AA, Ali SH, Ramadan EM. Effect of acid hydrolysis and fungal biotreatment on agro-industrial wastes for obtainment of free sugars for bioethanol production. Brazilian Journal of Microbiology. 2012; **43**(4):1523-1535. DOI:https://doi.org/10.1590/S1517-83822012000400037

[9] Nigam PS, Gupta N, Anthwal A.) Pre-treatment of agro-industrial residues. In: Nigam PS, Pandey A (eds) Biotechnology for agro-industrial residues utilization. Springer, Heidelberg. 2009; 13-33. DOI: https://doi.org/10.1007/978-1-4020-9942-7_2

[10] Martin JGP, Porto E, Correa CB, Alencar SM, Gloria EM, Cabral ISR, Aquino LM. Antimicrobial potential and chemical composition of agro-industrial wastes. Journal of Natural Products. 2012; **5**:27-36. URL:https://www.semanticscholar.org/paper/Antimicrobial-potential-and-chemical-composition-of-Guilherme-Martin/9c9ee72b0628030c160f9293333c4d8ea551d129

[11] Motte JC, Trably E, Escudié R, Hamelin J, Steyer JP, Bernet N, Delgenes JP, Dumas C. Total solids content: a key parameter of metabolic pathways in dry anaerobic digestion. Biotechnology for Biofuels. 2013;**6**:164. DOI:https://doi.org/10.1186/1754-6834-6-164

[12] World Bank Group. Agricultural pollution field burning. URL:http://documents1.worldbank.org/curated/en/989351521207797690/pdf/124342-repl-WB-Knowledge-Burning.pdf

[13] Yevich R and Logan J. An assessment of biofuel use and burning of agricultural waste in the developing world. Global Biogeochemical Cycles. 2003; **17**(4):1095. 2003, October 10. DOI:https://doi.org/10.1029/2002GB001952

[14] Yadav RS. Stubble burning: A problem for the environment,

agriculture and humans. DownToEarth. 2019 June 4. URL:https://www.downtoearth.org.in/blog/agriculture/stubble-burning-a-problem-for-the-environment-agriculture-and-humans-64912

[15] Bhuvaneshwari S, Hettiarachchi H Meegoda JN. Crop residue burning in India: Policy challenges and potential solutions. International Journal on Public Research and Public Health. 2019; *16*(5):832. DOI:https://doi.org/10.3390/ijerph16050832

[16] NPMCR. In: National Policy for Management of crop residues (NPMCR), Government of India, Ministry of Agriculture.2014; URL:http://agricoop.nic.in/sites/default/files/NPMCR_1.pdf

[17] Jeff S, Prasad M, Agamuthu P. Asian Waste Management Outlook; United Nations Environment Programme: Nairobi, Kenya, 2017. URL:http://hdl.handle.net/20.500.11822/27289

[18] Prakash S. Stubble-burning: President promulgates ordinance to deal with air pollution in Delhi-NCR In: The Tribune, New Delhi. 2020, October 29. URL: https://www.tribuneindia.com/news/nation/stubble-burning-president-promulgates-ordinance-to-deal-with-air-pollution-in-delhi-ncr-162819

[19] Singh Y, Sidhu HS. Management of Cereal Crop Residues for Sustainable Rice-wheat Production System in the Indo-gangetic Plains of India. Proceedings of the Indian National Science Academy. 2014; **80**(1) 95-114. DOI: 10.16943/ptinsa/2014/v80i1/55089

[20] Singh Y, Singh M, Sidhu HS, Khanna PK, Kapoor S, Jain AK, Singh AK, Sidhu GK, Singh A, Chaudhary DP, Minhas PS. Options for effective utilization of crop residues. In: Research Bulletin. Directorate of Research, Punjab Agricultural University, Ludhiana, India. 2010

[21] Stevens, J., & Voiland, A. (2017). Haze blankets Northern India. NASA Earth Observatory image. URL: https://earthobservatory.nasa.gov/images/91240/haze-blankets-northern-india (accessed on 14 March 2021)

[22] Scroll. NGT fines Delhi government Rs 2 lakh for not filing action plan on curbing pollution. 2018, March 21. URL:https://scroll.in/latest/872798/ngt-fines-delhi-government-rs-2-lakh-for-not-filing-action-plan-on-curbing-pollution

[23] The Economics Times. EnergyWorld. Air pollution: NGT slaps 25 crore fine on Delhi government. URL:https://energy.economictimes.indiatimes.com/news/coal/air-pollution-ngt-slaps-25-crore-fine-on-delhi-government/66920962

[24] Advani P. India needs more than laws to stamp out farm fires. In:The Businessworld. 2020, December 21. URL:http://www.businessworld.in/article/India-Needs-More-Than-Laws-To-Stamp-Out-Farm-Fires/21-12-2020-356271/

[25] Taskin B. All about stubble burning, its alternatives and steps taken by Centre and state govts. In: The Print. 2020, November 8. URL:https://theprint.in/theprint-essential/all-about-stubble-burning-its-alternatives-and-steps-taken-by-centre-and-state-govts/538466/#:~:text=According%20to%20the%20ordinance%2C%20released,one%20crore%20or%20with%20both%E2%80%9D

[26] USDA-Natural Resources Conservation Policy, 1999. In: Agricultural burning policy-Recommendations from the agricultural air quality task force. URL:https://www.nrcs.usda.gov/wps/portal/nrcs/

detailfull/national/air/?&cid=nrcs143_008984

[27] Muhammad IA, Sukalpaa C and Gaurav S. Stubble burning: Effects on health & environment, regulations and management practices. Environmental Advances. 2016; **2**:100011. DOI: https://doi.org/10.1016/j.envadv.2020.100011

[28] Ray K. Regulations for stubble burning in the offing – Penalty may include 5 year jail term, Rs one crore fine. In:The Deccan Herald. 2020, October 29. URL:https://www.deccanherald.com/national/regulations-for-stubble-burning-in-the-offing-penalty-may-include-5-year-jail-term-rs-one-crore-fine-908838.html

[29] Prasad YG, Maheswari M, Dixit S, Srinivasarao C, Sikka AK, Venkateswarlu B, Sudhakar N, Kumar SP, Singh AK, Gogoi AK, Singh AK, Singh YV and Mishra A. In situ incorporation of biomass and crop residues for improving soil health. In: Smart practices & technologies for climate resilient agriculture. Vikaspedia. 2014;68-69. URL: https://vikaspedia.in/agriculture/crop-production/crop-management-for-aberrant-weather-conditions/smart-practices-technologies/in-situ-incorporation-of-biomass-and-crop-residues-for-improving-soil-health

[30] Zeleke TB, Grevers MCJ, Si BC, Mermut AR and Beyene S. Effect of residue incorporation on physical properties of the surface soil in the South Central Rift Valley of Ethiopia. Soil and Tillage Research. 2004; **77**(1): 35-46. DOI:https://doi.org/10.1016/j.still.2003.10.005

[31] Kumar K, Prasad J. Solanki IS, Chaudhary R. Long-term effect of crop residues incorporation on yield and soil physical properties under rice - wheat cropping system in calcareous soil. Journal of Soil Science Plant Nutrition. 2018; **18**(1):27-40. DOI: http://dx.doi.org/10.4067/S0718-95162018005000103

[32] Hajdu I. The best practices for using plant residues. AGRIVI. URL:https://blog.agrivi.com/post/the-best-practices-for-using-plant-residues

[33] Zhang H, Yu X, Jin Z, Zheng W, Zhai B, Li Z. Improving grain yield and water use efficiency of winter wheat through a combination of manure and chemical nitrogen fertilizer on the Loess plateau, China. Journal of Soil Science and Plant Nutrition. 2017; **17**(2). DOI:http://dx.doi.org/10.4067/S0718-95162017005000033

[34] Pituello C, Berti A, Morari F. Crop residue incorporation for increasing SOC stock. Is it worth it? EGU General Assembly 2015, held 12-17 April, 2015 in Vienna, Austria. Id.14946 URL:https://ui.adsabs.harvard.edu/abs/2015EGUGA..1714946P/abstract

[35] Singh Y, Singh B, Timsina J. Crop residue management for nutrient cycling and improving soil productivity in rice-based cropping systems in the tropics. Advances in Agronomy. 2005; **85**:269-407. DOI:https://doi.org/10.1016/S0065-2113(04)85006-5

[36] IRRI. Crop residue management in rice-wheat system. Knowledgebank. URL: http://www.knowledgebank.irri.org/images/docs/crop-residue-management-in-rice-wheat-system.pdf

[37] Compost. Solution Center for Nutrient Management, Division of Agriculture and Natural Resources, University of California. Accessed online on 10 March, 2021. URL:https://ucanr.edu/sites/Nutrient_Management_Solutions/stateofscience/Compost/

[38] James A. Understanding the Carbon: Nitrogen Ratio (C:N). Advance Cover Crops. URL:https://advancecovercrops.

com/resources-advanced-cover-crops/carbon-nitrogen-ratio/

[39] Ramnarain YI, Ansari AA, Ori L. Vermicomposting of different organic materials using the epigeic earthworm *Eisenia foetida*. The *International Journal* of *Recycling* of Organic *Waste* in *Agriculture*. 2019; **8**:23-36. DOI:https://doi.org/10.1007/s40093-018-0225-7

[40] Pourzamani H, Ghavi M. Effect of rice bran on the quality of vermicompost produced from food waste. International Journal of Environmental Health Engineering. 2016;5:13. DOI: https://doi.org/10.4103/2277-9183.190639

[41] Nghi NT, Ryan R, Romasanta Hieu NV, Vinh LQ, Du NX, Ngan NVC, Chivenge P, Hung NV. Rice Straw-Based Composting. In: Gummert M., Hung N., Chivenge P., Douthwaite B. (eds) Sustainable Rice Straw Management. Springer, Cham.2020; 33-41. DOI:https://doi.org/10.1007/978-3-030-32373-8_3

[42] Prochnow LI, Kiehl JC, Pismel FS, Corrente JE. Controlling ammonia losses during manure composting with the addition of phosphogypsum and simple superphosphate. Scientia Agricola (Piracicaba, Braz.).1995; **52**(2): DOI: https://doi.org/10.1590/S0103-90161995000200024

[43] Martins O, Dewes T. Loss of nitrogenous compounds during composting of animal wastes. Bioresource Technology. 1992; **42**(2): 103-111. DOI:https://doi.org/10.1016/0960-8524(92)90068-9

[44] Jiang T, Schuchardt F, Li G, Guo R, Zhao Y. Effect of C/N ratio, aeration rate and moisture content on ammonia and greenhouse gas emission during the composting. Journal of Environmental Sciences.2011; **23**(10): 1754-1760. DOI: https://doi.org/10.1016/S1001-0742(10)60591-8

[45] Paul N, Sinha AK, Developing and effective method for composting of rice straw. Course report, CSIR-Indian Institute of Petroleum, Dehradun. 2020. URL:https://www.iip.res.in/csir-harit/wp-content/uploads/sites/51/2020/10/PAL1602179347-HIDeH8cZZuZq.pdf

[46] Jusoh ML, Manaf LA, Latiff PA. Composting of rice straw with effective microorganisms (EM) and its influence on compost quality. Iranian Journal of Environmental Health Science and Engineering. 2013; **10**(1):17. DOI: https://doi.org/10.1186/1735-2746-10-17

[47] Srinivasarao CH, Venkateswarlu B. Lal R, Singh AK, Sumanta K. Sustainable management of soils of dryland ecosystems for enhancing agronomic productivity and sequestering carbon. Advances in Agronomy. 2013; **121**:253-329. DOI:https://doi.org/10.1016/B978-0-12-407685-3.00005-0

[48] Amonette J, Joseph S, Characteristics of biochar: Micro-chemical properties. In: Lehmann J., Joseph S., editors. Biochar for Environmental Management: Science and Technology. Earth Scan; London, UK: 2009. pp. 33-52. URL:https://www.osti.gov/biblio/985016

[49] Wu W, Yang M, Feng Q, McGrouther K, Wang H, Lu H, Chen Y, Chemical characterization of rice straw-derived biochar for soil amendment. Biomass and Bioenergy 2012; **47**:268-276 DOI:https://doi.org/10.1016/j.biombioe.2012.09.034

[50] Jiang J, Peng Y, Yuan M, Hong Z, Wang D, Xu R. Rice straw-derived biochar properties and functions as Cu(II) and cyromazine sorbents as influenced by pyrolysis temperature. Pedosphere. 2015; **25**(5):781-789. DOI:https://doi.org/10.1016/S1002-0160(15)30059-X

[51] Kwak JH, Islam MS, Wang S, Messele SA, Naeth MA, El-Din MG, Chang SX. Biochar properties and lead (II) adsorption capacity depend on feedstock type, pyrolysis temperature, and steam activation. Chemosphere. 2019; **231**:393-404. DOI:https://doi.org/10.1016/j.chemosphere.2019.05.128

[52] Ma F, Zhao B, Diao J. Adsorption of cadmium by biochar produced from pyrolysis of corn stalk in aqueous solution. Water Science & Technology. 2016; **74**(6):1335-1345. DOI: https://doi.org/10.2166/wst.2016.319

[53] Zheng Y, Han X, Li Y. Li N, An N. Effects of Biochar and Straw Application on the Physicochemical and Biological Properties of Paddy Soils in Northeast China. Scientific Reports. 2019; **9**:16531. DOI:https://doi.org/10.1038/s41598-019-52978-w

[54] Iqbal MT, Ortaş I, Ahmed IAM, Isik M, Islam MS. Rice straw biochar amended soil improves wheat productivity and accumulated phosphorus in grain, Journal of Plant Nutrition.2019;**42**(14):1605-1623, DOI:https://doi.org/10.1080/01904167.2019.1628986

[55] El-Eyuoon, Amin AZ. Impact of Corn Cob Biochar on Potassium Status and Wheat Growth in a Calcareous Sandy Soil, Communications in Soil Science and Plant Analysis, 2016; **47**(17):2026-2033, DOI:https://doi.org/10.1080/00103624.2016.1225081

[56] Liu J, Ding Y, Ji Y, Gao G, Wang Y. Effect of Maize Straw Biochar on Bacterial Communities in Agricultural Soil. Bulletin of Environmental Contamination and Toxicology. 2020; **104**:333-338. DOI:https://doi.org/10.1007/s00128-020-02793-1

[57] Calderón FJ, Benjamin J, Vigil MF. A Comparison of Corn (*Zea mays* L.) Residue and Its Biochar on Soil C and Plant Growth. PLoS ONE. 2015. **10**(4): e0121006. DOI: https://doi.org/10.1371/journal.pone.0121006

[58] de Leeuw JW, Largeau C. In: Engel MH, Macko SA (ed.). Organic geochemistry, New York-London: Plenum Press. 1993; pp.23.

[59] Taherzadeh MJ. Ethanol from Lignocellulose: Physiological Effects of Inhibitors and Fermentation Strategies. Ph.D. Thesis, Biotechnology, Chemical Reaction Engineering, Chalmers University of Technology, Gothenburg, Sweden, 1999. URL:https://publications.lib.chalmers.se/records/fulltext/194558/194558.pdf

[60] Palmqvist E, Hahn-Hägerdal B. Fermentation of lignocellulosic hydrolysates II: Inhibitors and mechanisms of inhibition. Journal of Bioresource Technology. 2000; **74**:25-33. DOI:https://doi.org/10.1016/S0960-8524(99)00161-3

[61] Perez J, Dorado JM, Rubia TD, Martinez J. Biodegradation and biological treatment of cellulose, hemicellulose and lignin: An overview. International Microbiology. 2002; **5**:53-56.DOI:https://doi.org/10.1007/s10123-002-0062-3

[62] Bhuvaneshwari S, Hettiarachchi H, Meegod JN. Crop Residue Burning in India: Policy Challenges and Potential Solutions. *International Journal of Environmental Research and Public Health*. 2019; **16**(5): 832. DOI: https://doi.org/10.3390/ijerph16050832

[63] Chandra R, Takeuchi H, Hasegawa T. Methane production from lignocellulosic agricultural crop wastes: A review in context to second generation of biofuel production. Renewable & Sustainable Energy Reviews. 2012; **16**:1462-1476. DOI:https://doi.org/10.1016/j.rser.2011.11.035

[64] Biarnes M. Biomass to Biogas—Anaerobic Digestion. E-instruments.

URL:https://www.e-inst.com/training/biomass-to-biogas/

[65] Hutnan Miroslav. Maize silage as substrate for biogas production. Published online in Intechopen. 2016. DOI:https://doi.org/10.5772/64378

[66] Bjornsson L, Prade T, Lantz M. Grass for biogas Arable land as a carbon sink. Energiforsk. 2016. URL:https://energiforskmedia.blob.core.windows.net/media/21293/grass-for-biogas-energiforskrapport-2016-280.pdf

[67] Phonbumrung T, Khemsawas C. Agricultural Crop Residue. In Proceedings of the Sixth Meeting of Regional Working Group on Grazing and Feed Resources for Southeast Asia, Legaspi, Philippines, 5-9 October 1998; pp. 183-187. URL: https://www.mdpi.com/1660-4601/16/5/832/htm

[68] George J. Novel rice husk and cereal straw and/or bagasse particle board and a method of making the same. Patent number-193404. URL:https://www.allindianpatents.com/patents/193404-novel-rice-husk-and-cereal-straw-and-or-bagasse-particle-board-and-a-method-of-making-the-same

[69] de Melo RR, Stangerlin DM, Santana RRC, Pedrosa TD. Physical and Mechanical Properties of Particleboard Manufactured from Wood, Bamboo and Rice Husk. Materials Research. 2014; **17**(3): 682-686. DOI:http://dx.doi.org/10.1590/S1516-14392014005000052

[70] Kaur D, Bhardwaj NK, Lohchab RK. Prospects of rice straw as a raw material for paper making. Waste Management.2017; **60**:127-139. DOI:https://doi.org/10.1016/j.wasman.2016.08.001

[71] Singh K, Bhaskar, S. Paddy straw to bio-degradable paper: IIT-Delhi graduates develop a technology to help farmers turn crop residue into useful products. BanegaSwasthIndia. 2018, April 23. URL:https://swachhindia.ndtv.com/paddy-straw-to-bio-degradable-paper-iit-delhi-graduates-develop-a-technology-to-help-farmers-turn-crop-residue-into-useful-products-19074/

[72] Sain M. Use of Rice Straw for Packaging Materials - The Best Option. FnBnews.com. 2016, December 10. URL:http://www.fnbnews.com/Top-News/use-of-rice-straw-for-packaging-materials--the-best-option-53480